ASTROBIOLOGY, DISCOVERY, AND SOCIETAL IMPACT

The search for life in the universe, once the stuff of science fiction, is now a robust worldwide research program with a well-defined roadmap probing both scientific and societal issues. This volume examines the humanistic aspects of astrobiology, systematically discussing the approaches to, and critical issues and implications of, discovering life beyond Earth. What do the concepts of life and intelligence, culture and civilization, technology and communication mean in a cosmic context? What are the theological and philosophical implications if we find life – and if we do not? Steven J. Dick argues that, given recent scientific findings, the discovery of life in some form beyond Earth is likely and so we need to study the possible impacts of such a discovery and formulate policies to deal with them. The remarkable and often surprising results are presented here in a form accessible to disciplines across the sciences, social sciences, and humanities.

STEVEN J. DICK is one of the best-known and most qualified writers on topics relating to humanity's thoughts on extraterrestrial life. He held the 2014 Baruch S. Blumberg NASA/Library of Congress Chair in Astrobiology at the John W. Kluge Center of the Library of Congress. In 2013, he testified before Congress on the subject of astrobiology. He served as the Charles A. Lindbergh Chair in Aerospace History at the National Air and Space Museum from 2011 to 2012, and as the NASA Chief Historian and Director of the NASA History Office from 2003 to 2009. He is the recipient of numerous awards, including the NASA Exceptional Service Medal and the Navy Meritorious Civilian Service Medal, and is author or editor of twenty books, including *The Biological Universe*. He was awarded the 2006 LeRoy E. Doggett Prize for Historical Astronomy by the American Astronomical Society. In 2009 the International Astronomical Union designated minor planet 6544 stevendick in his honor.

Cambridge Astrobiology

Series Editors

Bruce Jakosky, Alan Boss, Frances Westall, Daniel Prieur, and Charles Cockell

Books in the Series

1. *Planet Formation: Theory, Observations, and Experiments*
   Edited by Hubert Klahr and Wolfgang Brandner ISBN 978-0-521-18074-0

2. *Fitness of the Cosmos for Life: Biochemistry and Fine-Tuning*
   Edited by John D. Barrow, Simon Conway Morris, Stephen J. Freeland, and Charles L. Harper, Jr. ISBN 978-0-521-87102-0

3. *Planetary Systems and the Origin of Life*
   Edited by Ralph Pudritz, Paul Higgs, and Jonathan Stone ISBN 978-0-521-87548-6

4. *Exploring the Origin, Extent, and Future of Life: Philosophical, Ethical, and Theological Perspectives*
   Edited by Constance M. Bertka ISBN 978-0-521-86363-6

5. *Life in Antarctic Deserts and Other Cold Dry Environments*
   Edited by Peter T. Doran, W. Berry Lyons, and Diane M. McKnight
   ISBN 978-0521-88919-3

6. *Origins and Evolution of Life: An Astrobiological Perspective*
   Edited by Muriel Gargaud, Purificación Lopez-Garcia, and Hervé Martin
   ISBN 978 0521-76131-4

7. *The Astrobiogical Landscape: Philosophical Foundations of the Study of Cosmic Life*
   Milan M. Ćirković ISBN 978 0521-19775-5

8. *The Drake Equation: Estimating the Prevalence of Extraterrestrial Life through the Ages*
   Edited by Douglas A. Vakoch and Matthew F. Dowd
   ISBN 978-1-107-07365-4

9. *Astrobiology, Discovery, and Societal Impact*
   Steven J. Dick ISBN 978-1-108-42676-3

10. *Solving Fermi's Paradox*
    Duncan H. Forgan ISBN 978-1-107-16365-2

11. *The Quest for a Universal Theory of Life: Searching for Life As We Don't Know It*
    Carol E. Cleland ISBN 978-0-521-87324-6

ASTROBIOLOGY, DISCOVERY, AND SOCIETAL IMPACT

STEVEN J. DICK
Former NASA Chief Historian
2014 Baruch S. Blumberg NASA/
Library of Congress Chair in Astrobiology

CAMBRIDGE
UNIVERSITY PRESS

CAMBRIDGE
UNIVERSITY PRESS

University Printing House, Cambridge CB2 8BS, United Kingdom

One Liberty Plaza, 20th Floor, New York, NY 10006, USA

477 Williamstown Road, Port Melbourne, VIC 3207, Australia

314–321, 3rd Floor, Plot 3, Splendor Forum, Jasola District Centre, New Delhi – 110025, India

79 Anson Road, #06–04/06, Singapore 079906

Cambridge University Press is part of the University of Cambridge.

It furthers the University's mission by disseminating knowledge in the pursuit of education, learning, and research at the highest international levels of excellence.

www.cambridge.org
Information on this title: www.cambridge.org/9781108426763
DOI: 10.1017/9781108556941

First published 2018
First paperback edition 2020

Printed in the United Kingdom by TJ International Ltd. Padstow Cornwall

A catalogue record for this publication is available from the British Library.

Library of Congress Cataloging-in-Publication Data
Names: Dick, Steven J., author.
Title: Astrobiology, discovery, and societal impact / Steven J. Dick, former NASA Chief Historian, 2014 Baruch S. Blumberg NASA/Library of Congress Chair in Astrobiology.
Description: Cambridge : Cambridge University Press, 2018. | Includes bibliographical references and index.
Identifiers: LCCN 2017057920 | ISBN 9781108426763 (alk. paper)
Subjects: LCSH: Life on other planets. | Extraterrestrial beings. | Exobiology.
Classification: LCC QB54 .D4695 2018 | DDC 576.8/39–dc23
LC record available at https://lccn.loc.gov/2017057920

ISBN 978-1-108-42676-3 Hardback
ISBN 978-1-108-44551-1 Paperback

To those who search for life
and its meaning
in a cosmic context

In what distant deeps or skies
Burnt the fire of thine eyes?
William Blake, *The Tyger*

It is now very close to inconceivable that we could be the only life, and only technological intelligence, in the universe.
David Grinspoon, *Earth in Human Hands, 2016*

Contents

Introduction

When Biospheres Collide

> Over the long term, the psychological and philosophical implications of the discovery could be profound … The discovery of even simple life would fuel speculation about the existence of other intelligent beings and challenge many assumptions that underpin human philosophy and religion.
>
> *World Economic Forum, 2013*

> What has been will be again, what has been done will be done again; there is nothing new under the sun.
>
> *Ecclesiastes 1:9*

In 2013 the Risk Response Network of the World Economic Forum declared the discovery of life beyond Earth one of five X factors – emerging concerns for planet Earth of possible future importance but with unknown consequences. Along with runaway climate change, significant human cognitive enhancement, rogue deployment of geoengineering in the Earth's atmosphere, and the costs of living longer, the authors of the report suggested these were serious issues grounded in scientific findings, but given less attention because they were overshadowed by more immediate concerns like armed conflict, failed states, and economic stability. Giving attention to X factors, they suggested, would lead to a more proactive approach if and when these events actually occurred, resulting in more "cognitive resilience" and perhaps preventing at least some undesirable social consequences. As indicated in the quotation that opens this introduction, such consequences could occur even if simple alien life were discovered.

Speaking on the same subject the following year in testimony on astrobiology before the US Congress, I ventured a bold statement that the authors of Ecclesiastes may have gotten it wrong, that in fact "perhaps there *is* something new under the Sun and the suns of other worlds." Sitting before the full Committee on Science, Space and Technology of the House of Representatives, with NASA's senior scientist for astrobiology and a prominent astronomer from MIT also at the dais, I suggested we need to pay more attention to the societal aspects of astrobiology, in particular what happens when the age-old search for life beyond Earth is successful.

Although there have been false claims of such a discovery in the past, the reaction to an *actual* discovery – now *that* would be something new under the Sun.

Both the concerns of the World Economic Forum and the very fact of the congressional hearings demonstrate that life beyond Earth has become not just a persistent theme but also an active policy issue, not only for funding reasons but also because of the dramatic impact such a discovery might have on society. Astrobiology today is a global and robust discipline that attracts researchers young and old across many fields of endeavor. The discovery of exoplanets by the thousands, of complex organic molecules in giant molecular clouds, of subsurface oceans on multiple Jovian and Saturnian moons, of plentiful water on ancient Mars, of extreme forms of life on Earth, and a variety of other observations, constitute only circumstantial evidence that life may exist. But these developments provide a new scientific basis and enthusiasm for a concept that has captured human imagination over the millennia. The focus of this book is neither the science of alien life nor its history, but the impact should such a discovery be made – cosmic encounters that will surely change our world and our worldviews. In short, this book begins where most other books about life on other worlds end.

In a significant way the idea of extraterrestrial life has already had its impact, for our culture and others have been "captured by aliens" in the felicitous phrase of *Washington Post* reporter Joel Achenbach. In his scintillating book by that title Achenbach examined how aliens have pervaded modern popular culture, from UFO enthusiasts and science fiction to the 39 members of the Heaven's Gate religious cult who in 1997 merrily committed suicide hoping they could then board an alien spaceship following in the wake of comet Hale–Bopp. Interest in aliens dates back much farther than that; as Michael Crowe, a premier historian of the extraterrestrial life debate, has put it, we have for centuries been invaded by aliens – in our imaginations. And as I have shown elsewhere, the scientific roots of the idea are found in the cosmologies of the ancient Greeks, the Copernican and Newtonian transformations of our worldview, the scientific advances of the past century, and a better understanding of our place in 13.8 billion years of cosmic evolution.

Here, however, I focus not on the mere possibility, but the actual discovery of extraterrestrial life. And if anything is clear, it is that the exact nature of the discovery scenario will significantly affect the impact – one of many reasons the word "discovery" appears in the title of this volume. The most startling scenario – celebrated in science fiction film and literature and the subject of sporadic effort in science itself – would be the discovery of extraterrestrial intelligence. The movie *E.T. the Extra-Terrestrial* has become a cultural icon immediately recognizable around the world as symbolizing contact with extraterrestrial intelligence, though not in a very scientific way. Mention E.T. and millions immediately envision a friendly, cuddly creature who enjoys "his" accidental visit to Earth but in the end just wants to go home. UFO reports notwithstanding, such direct contact on Earth is unlikely. But it is not logically impossible, nor is there any guarantee of friendliness should it occur. Indeed, science fiction readers and moviegoers are also familiar with the *Alien* series, which depicts the other extreme in an entire spectrum of possibilities. Even these two extremes, and everything in between, do not cover the possibilities in a universe we are only beginning to understand. In my view direct encounters, either on Earth or in space, are unlikely to be our first experience with E.T.s, but they cannot be ruled out over the long or short term. They are represented in science fiction by movies ranging from *War of the Worlds* to *Close Encounters of the Third Kind*, and are immortalized in the five-year mission of the starship *Enterprise*: "to explore strange new worlds, to seek out new life and new civilizations, to boldly go where no man has gone before." Much more likely in the view of most scientists is Carl Sagan's *Contact* scenario, where initial contact is made via radio telescopes or detectors looking at some other part of the spectrum. But this is possibly a failure of imagination: we do not know how, when, or where it will happen.

Even more likely than the detection of extraterrestrial intelligence will be the indirect detection of more primitive life – the focus of most research in astrobiology today. This could come in the form of microbial life, discovered on Mars or in the oceans of Europa, Enceladus, Titan, or the several other ocean worlds now known to exist around Jupiter or Saturn. The term "habitable zone" has had to be radically redefined in the past few years. A subset of this scenario would be the discovery of fossilized life, a claim already controversially made in 1996 for the infamous Mars rock ALH84001. Yet another subset would be biosignatures in the atmospheres of exoplanets, a field that becomes ever more robust as spectroscopic techniques improve, more planets are discovered, and instruments such as the James Webb Space Telescope and the Transiting Exoplanet Survey Satellite (TESS) come online. The impact of such discoveries is likely to be quite different from the discovery of intelligence. But the reaction could be no less profound, as our worldviews change with the certain knowledge of alien life, and as the search for intelligent life is made more plausible and reenergized. The Europan scenario of life short of sentience is also celebrated in science fiction, ranging from Arthur C. Clarke's novels *2010: Odyssey Two* and *2061: Odyssey Three* to the intriguing film *Europa Report* about a manned mission to search for life. Hints of Europan sentience in *Europa Report*, and its explicit depiction in Kim Stanley Robinson's richly imagined *Galileo's Dream*, are evocative, if unlikely. But intelligence is not at all unlikely on some of the multitudes of worlds beyond our solar system, exoplanets now known to exist around virtually every star. And the impact of that discovery might be considerably greater than microbial life, even given the vast distances.

Our challenge in this book is how to gauge the potential impact of the discovery of extraterrestrial life beyond mere guesswork and facile generalizations. The first three chapters in this volume discuss three possible approaches to the problem, beginning with the reaction to historical episodes where we *thought* such life had been discovered. Such episodes are more common than one might think, and offer a variety of lessons based on real historical data. A second approach is to analyze the nature of scientific discovery itself, now known to be a much more complex process than commonly assumed, and thus an extended affair, a characteristic sure to affect societal impacts in all possible discovery scenarios. A third approach is analogy, frequently used and abused in discussions of alien life encounters without much sophistication. But analogy is a hot topic in cognitive science and philosophy of science, and after examining its validity as a mode of argument, I discuss what analogies might best be applied to our problem, and which to avoid. The promise and problems of analogy are a constant theme throughout this book.

Contemplating the impact of discovering life beyond Earth raises many critical issues. First, it forces us to think about some of our most basic human concepts in more general terms: life and intelligence, culture and civilization, technology and communication on Earth become only a subset of the possibilities in the astrobiological landscape. Chapter 4 analyzes these categories with an eye toward anthropocentric biases, since surely our assumptions about each of them will affect societal impact. The problem here is, how do we get out of our heads to think about the otherworldly? It is not easy, but it is also not entirely impossible. Second, our subject raises the question of whether human knowledge is universal, especially in the event that we discover intelligence. This is the subject of Chapter 5, where I examine how the natural and social sciences might help us determine impact, but also how an extraterrestrial life discovery might affect our knowledge in those disciplines. The questions here are foundational: Could there be a difference between human and nonhuman understanding? Are terrestrial science and mathematics universal? Are our best attempts at "human sciences" universal? The latest research in cognitive science and the philosophy of mind points to some surprising answers. Chapter 6 raises the central dilemma of this book: even applying our knowledge of history, discovery, and analogy, how can we possibly envision impact before it occurs? In other words, isn't this just an exercise in futility? Far from it, I argue. The World Economic Forum is only one among many institutions that constantly try to assess risks, even remote ones, that might affect the human future. And even if life is never found beyond Earth, raising questions we usually take for granted in their terrestrial dimensions makes the discussion worthwhile.

Chapters 7 through 9 of this volume provide the payoff: my best estimate of what the actual impacts will be, for different scenarios over both the short and long term. Any discovery of life beyond Earth, I argue, is sure to affect worldviews ranging from the cosmological to the theological and cultural. In some cases even the *possibility* of life beyond Earth is already giving rise to new worldviews, including astrotheology, bearing on religious worldviews that affect a large percentage of the population. Furthermore, in dealing even with microbial life, and certainly with intelligent life, we will have to come to grips with issues of astroethics. What are our responsibilities to alien life forms, whether microbial or intelligent? What is it like to be an extraterrestrial, and how might extraterrestrials conceive of their responsibilities to us? Given answers to those questions, what does it mean to be human, and what does it mean to be alien? Scholars in the natural and social sciences, as well as the humanities, have begun to discuss these issues, most recently in an important volume of essays, *Encountering Life in the Universe: Ethical and Social Implications for Astrobiology*. How to prepare now, and what actions we might take when the discovery occurs, involves the formulation of what I call astropolicy, part of the embryonic field of astropolitics. This is the subject of Chapter 9, which ranges from protocols and policies for preparing for and managing the discovery of life beyond Earth, to "metalaw," formulating possible universal rules for interacting with intelligence in the universe.

In short, this book is one long argument, to use a Darwinian phrase: that given recent scientific findings, the discovery of life in some form beyond Earth is likely; that because it is a wild card X factor of global concern we need to study the possible impact of such a discovery; that the subject is amenable to study through history, discovery, analogy, and impact models; that we need to think out of the anthropocentric box in our ideas about what extraterrestrial life and its impact might be like; that our usual ideas of life and intelligence, culture and civilization might not be universal in a cosmic context; that our thinking will be transformed, quickly or rapidly depending on the discovery scenario, giving rise to issues in astroculture, astrotheology, astrophilosophy, and astroethics; and that we need to formulate policies to deal with such a discovery. Through all the arguments the reader will see two central concepts arise again and again. The first is the idea of transformation through changing worldviews, which frames my treatment of impact, especially in Chapters 6 and 7. The second is the idea of evolution, which I argue is likely to be universal in both its natural and social science aspects. "Taking Darwin seriously," as philosopher of science Michael Ruse puts it, provides a firm basis for discussing both the biological and cultural nature of life beyond Earth. Evolution's subsidiary ideas of chance and necessity, convergence and progress (or lack of it) frame the possibilities of what we may encounter, which are directly related to impact. Through it all I have also employed science fiction, which at its best represents an important body of thought related to impact. My readers should therefore not be surprised to hear about everything from psychohistory and postbiologicals to Overlords and other exotic aliens, not to mention cosmotheology.

Given how little we know about life beyond Earth, we should not shy away from the potentially negative side of these problems, what we might call "The Andromeda Strain Effect," after the Michael Crichton novel and movie where a microbial extraterrestrial organism wreaks havoc with humans on Earth as a result of back contamination. The *Andromeda Strain* scenario graphically evokes the negative possibilities when biospheres collide, and represents what could happen on other planets if the search for life contaminates the very object of our search. Ramped up to more complex life and intelligence, the Andromeda Strain Effect represents a suite of ethical and philosophical problems: If we decipher a signal from extraterrestrial intelligence, should we reply? Should we initiate messages to extraterrestrial intelligence? If so, who speaks for Earth? What are our moral responsibilities to other beings in the universe? These questions are not hypotheticals; they are real policy questions being asked in the wake of ongoing and proposed programs in the Search for Extraterrestrial Intelligence (SETI) and Messaging Extraterrestrial Intelligence (METI).

Our goals and methods may be seen as part of the broader problem of the impact of science on society. Scientists, historians, and policymakers, not to mention the US Congress, have come to realize the study of impacts is an important part of the scientific endeavor. Already in the 1990s the Human Genome Project set aside 3 percent of its $3 billion funding to study the Ethical, Legal and Social Implications of its work, a serious research program considered an integral part of the scientific project. In 2006 the National Science Foundation began funding the Center for Nanotechnology in Society, including environmental and health risks, as well as economic, technological, and policy impacts. During my days as NASA chief historian I initiated a series of studies on the societal impact of spaceflight, in accordance with the charter of the National Aeronautics and Space Act of 1958 to provide for long-range studies of the implications of NASA's work. And most recently programs are being initiated both in the United States and in Europe to study the problem of runaway artificial intelligence. Such programs and studies are key to the public understanding of science and its integration with the modern world, sorely needed these days more than ever. Surely the impact of the discovery of extraterrestrial life deserves no less attention. Both the World Economic Forum and congressional hearings on astrobiology have highlighted the question "What do we do?" in the event of discovery of life beyond Earth. This book begins to answer that question.

I have been privileged through much of my career to be embedded in the astrobiology community in a variety of forms, participating in astrobiology meetings around the world, attending NASA's small but pioneering "Cultural Aspects of SETI" workshops in the early 1990s, serving as historian of the NASA SETI and astrobiology programs, and undertaking numerous interviews with participants in the field. In the wake of the announcement of possible nanofossils in the Mars rock in 1996, I participated in Vice President Gore's meeting to discuss the implications if it were true – and if it were not. Shortly after Carl Sagan's death I was deposed in the little-known case of *Francis Ford Coppola* v. *Carl Sagan*, having to do with property rights in his now classic book and movie *Contact*. Through all of this and more, the question of what would happen if we actually discovered life has been in the front of my mind. This volume, sprinkled with my personal recollection of these events, is the result of my deliberations on the subject over several decades.

I am acutely aware of the large number of scholars on whom this book depends, not only in the field of astrobiology but also in the many other fields I make use of. In order to think out of the box, one needs to know what is in the box, so in almost every chapter I have often had to describe at least briefly the major themes in subjects as diverse as philosophy of mind, philosophy of the social sciences, cognitive science, theology, and ethics in order to see how they could be applied in an astrobiological context. In the field of astrobiology I wish to thank NASA SETI pioneer John Billingham, who more than two decades ago drew me into the "Cultural Aspects of SETI" workshops, the results of which were later published as *Social Implications of the Detection of an Extraterrestrial Civilization*. A small but growing number of scholars have written on our subject, mostly in the form of articles cited in this volume. Many more have written on subjects indirectly related to the problem, and would no doubt be surprised to see their work used in this context.

I wish to acknowledge several authors who have published book-length works on this subject that have been essential to my work: Albert Harrison's *After Contact: The Human Response to Extraterrestrial Life*, Michael Michaud's *Contact with Alien Civilizations: Our Hopes and Fears about Encountering Extraterrestrials*, Milan Ćirković's *The Astrobiological Landscape: Philosophical Foundations of the Study of Cosmic Life*, and John Traphagan's *Extraterrestrial Intelligence and Human Imagination: SETI at the Intersection of Science, Religion, and Culture*. Similarly, a growing number of edited volumes now constitute a substantial literature on the subject addressed by dozens of authors: Doug Vakoch's four pioneering volumes (all listed in the Bibliography), Connie Bertka's *Exploring the Origin, Extent, and Future of Life: Philosophical, Ethical and Theological Perspectives*, Chris Impey's *Encountering Life in the Universe: Ethical Foundations and Social Implications of Astrobiology*, and David Dunér's *The History and Philosophy of Astrobiology: Perspectives on Extraterrestrial Life and the Human Mind*. Also relevant is Michael Ashkenazi's *What We Know about Extraterrestrial Intelligence: Foundations of Xenology*, published as this volume went to press. Given these volumes and more to come, it is not too much to suggest that a new field of astrobiology and society is developing, part of the history, philosophy, and sociology of astrobiology captured in Dunér's title and systematically outlined in Dick (2012).

This volume was written during my time as the Baruch S. Blumberg NASA/ Library of Congress Chair in Astrobiology, centered at the John W. Kluge Center of the Library of Congress in Washington, DC. The beautiful surroundings of the Thomas Jefferson Building have inspired many writers, myself included. The explicit purpose of the Chair is to look at the humanistic aspects of astrobiology. Barry Blumberg himself deserves my thanks for his support and memorable conversations in my office at NASA Headquarters before his untimely death. I would like to thank Jim Green, director of planetary science at NASA; Mary Voytek, senior scientist for astrobiology at NASA Headquarters; and Carl Pilcher and Edward Goolish, respectively director and acting director of the NASA Astrobiology Institute at the time of this research, for supporting such a forward-looking subject and for their encouragement during my tenure. At the Library of Congress I would also like to thank Kluge Center Director Carolyn T. Brown, as well as Jane McAuliffe, Dan Turello, Jason Steinhauer, Travis Hensley, JoAnne Kitching, Mary Lou Reker, Matthew Hinson, Austin Woodruff, and Nancy Lovas, for their support while I was in residence at the Kluge Center. In the Science Reference Section my thanks go to Margaret Clifton for unearthing many an obscure reference. During my time at the Library I had access not only to the tremendous resources of one of the best libraries in the world, but also to scholars in a variety of disciplines, including David Grinspoon, the inaugural Blumberg Chair who sat in the office next door and produced *Earth in Human Hands: Shaping Our Planet's Future*, a model of good writing on a subject very relevant to the present volume.

For reading chapters or sections, or for discussions relevant to their expertise I would like to thank Gregory Dick (geomicrobiology); Derek Malone-France (philosophy of knowledge); Kelly Smith (philosophy of knowledge and astroethics); Susan Schneider (philosophy of mind); Mark Lupisella (astroethics); Linda Billings (astroculture); Michael A. G. Michaud (SETI protocols and astropolicy); Margaret Race (planetary protection and astropolicy); Lori Marino (intelligence); Michael Meyer (Mars rock); John Rummel (planetary protection); John Traphagan (culture and civilization); Ted Peters (astrotheology); Michael Chorost (communications); and Anthony Dick, Adam Korbitz, Steve Doyle, and George Robinson (all on space law). For Figures 1.5 and 1.6 my thanks to Connie Moore at NASA Headquarters; for procuring Figure 7.5 my thanks to Alexander Geppert in Berlin (as well as for his concept of astroculture discussed in Chapter 7); and for Figure 1.1 thanks to Nasser Zakariya, whose book *A Final Story* has also influenced my thinking on the development of our current cosmic worldview. Finally, thanks go also to my editor at Cambridge University Press, Vince Higgs, for his encouragement and for accepting this, my seventh book for Cambridge University Press, and to Esther Miguéliz Obanos, Samantha Town, and Ami Naramor for seeing it through production.

In addition to human and material resources in the Library itself, in September 2014 I convened scholars from many disciplines around the world to discuss the subject. Their discussion, published as *The Impact of Discovering Life beyond Earth,* has immeasurably enriched my view of the problem. They are acknowledged by the references to their work in each chapter. Readers of that volume will find some similarity in structure with this book, since both were being planned at the same time. This volume, however, is much more of a personal synthesis, and is only a beginning of the work that remains both before and after the discovery of life beyond Earth.

Part I

Approaches

1

History

> Obviously, we have a problem in predicting the nature and behavior of intelligent extraterrestrials, as we have no confirmed information about them. Until we do, we have only two methods of analyzing the possible consequences of contact: analogy with ourselves, and probability based on what we know about human history and behavior.
>
> *Michael A. G. Michaud*[1]

> That men do not learn very much from the lessons of history is the most important of all the lessons of history.
>
> *Aldous Huxley*[2]

It was December 11, 1996, and the vice president of the United States sat at the head of the table to my left, flanked by NASA administrator Dan Goldin and White House science advisor John Gibbons (Figure 1.1). Around the table in the Indian Treaty Room of the Old Executive Office Building across from the White House, where the vice president maintains his ceremonial office and staff, sat luminaries representing a variety of fields: scientists such as David McKay, Stephen Jay Gould, and Lynn Margulis; theologians including John Minogue of DePaul University and Joan Brown Campbell of the National Religious Partnership for the Environment; journalist Bill Moyers; and me, an astronomer and historian of science who had just published a book titled *The Biological Universe*.[3] Astronomer Carl Sagan, one of the pioneers in the search for life in the universe, had been invited but lay in a cancer center in Seattle, nine days from death.

The occasion of this unusual gathering was to discuss the meaning of the announcement four months earlier that possible fossilized life had been discovered in a Martian meteorite that had landed on Earth. Vice President Gore first turned to Gould and asked about the consequences if the discovery turned out to be true. Gould replied that it depended on whether past or present life on Mars represented an "independent genesis." In other words, if life had arisen independently in two places so near each other in the solar system, as opposed to having been seeded from Earth, this would indicate that the universe was filled with life. Gore then

Figure 1.1 In the wake of the claim that the Mars rock ALH84001 contained nanofossils, a high-level discussion of the implications of finding life beyond Earth took place on December 11, 1996, in the Indian Treaty Room of the Old Executive Office Building adjacent to the White House. From right to left: Vice President Al Gore, NASA administrator Dan Goldin, astronomers Anneila Sargent and John Bahcall, historian of science Steven Dick, theoretical biologist Stuart Kauffman, biologist Lynn Margulis, astrobiologist David McKay, theologian Joan Brown Campbell, and NASA associate administrator Wes Huntress. Among those not visible are journalist Bill Moyers, Harvard biologist Stephen Jay Gould, and presidential science advisor Jack Gibbons, who was seated to Gore's left. US government photo.

turned to me and asked what would be the maximum consequences of life on Mars. I replied that humanity's worldview was at stake, as well as the possibility of a universal biology, just as Newton had formulated a universal physics. Around the table we went, with the vice president issuing rapid-fire questions ranging from science to theology and media. Three hours later the meeting was still going on, having overrun its allotted time.

No definitive conclusions were reached that memorable day, and the current consensus is that the purported Martian fossils are not real. But the episode painted a vivid picture of what would likely happen when life was actually discovered beyond Earth. I left the meeting exuberant, but also worried. The discussions revealed we were not prepared for the discovery of life beyond Earth in even the most basic way, and even for the most primitive forms of life, or fossil life. How might we prepare for the implications? For starters, could the history of reactions to purported discoveries of life be of any help? I vowed to find out.

Numerous times in the past four centuries of telescopic astronomy Earthlings believed extraterrestrial life had been detected. And numerous times they were disappointed. Galileo's 1610 landmark telescopic lunar observations had barely been published when Kepler enthusiastically conjectured that one particularly circular crater must be an artificial construction of lunar inhabitants. Two centuries later the famous "Moon Hoax" of 1835 placed lunarians on the Moon, supposedly based on the latest telescopic observations of John Herschel. Sixty years further on, even as H. G. Wells was penning his *War of the Worlds*, astronomer Percival Lowell argued there were canals on Mars, built by a dying civilization managing its water resources. Though largely discredited by the time of Lowell's death in 1916, the idea had not dimmed so much that when Orson Welles broadcast his radio version of *War of the Worlds* on Halloween Eve, 1938, a considerable reaction ensued as some Americans believed a real Martian invasion was under way. Thirty years later, when strange pulses were detected from the heavens (soon dubbed "pulsars"), astronomers briefly but seriously considered the "Little Green Men" hypothesis. Finally, in 1996, NASA announced it had evidence of fossil life from Mars in the form of the now famous and infamous Mars rock dubbed ALH84001, resulting in the scenario described earlier in this chapter.

These episodes not only demonstrate a deeply ingrained human "will to believe" in extraterrestrial life but also provide an opportunity to examine, with some historical rigor, the human reaction to such perceived "discoveries." They represent the first approach, the relevance of history to the problem of impact. Analogy and the nature of discovery are two more approaches that we elaborate in the following chapters of Part I.[4] Keeping in mind that human reactions are always tied to cultural contexts, and mindful also of Huxley's pessimistic judgment that lessons of history are seldom learned, we nevertheless want to ask what lessons we might learn from these episodes about human reactions that might be useful when a real discovery of extraterrestrial life is made. Here history and analogy converge, since the future reaction must remain an exercise in analogy. And while analogy is far from predictive, this form of argument can provide useful guidelines, perhaps even more so when the analogy is another putative discovery of life beyond Earth.

Lunarians: The Great Moon Hoax/Satire (1835)

Passing over Kepler's seventeenth-century claims, which caused little stir because they were not widely known, we begin with the so-called Great Moon Hoax of 1835. Readers of the *New York Sun* for August 25 of that year could hardly miss the headline prominently placed at the top of the front page: "GREAT ASTRONOMICAL

DISCOVERIES LATELY MADE BY SIR JOHN HERSCHEL, L.L.D, F.R.S., etc. At the Cape of Good Hope." Though the article, the first of a series of six that concluded on August 31, purportedly came from the "Supplement to the Edinburgh Journal of Science," readers could not be expected to have known that the journal had ceased to exist three years earlier. A few may have heard of John Herschel, who indeed was the son of the famous William Herschel, had published his *Treatise on Astronomy* in the United States to great acclaim the previous year, and was in fact at the Cape of Good Hope making astronomical observations.[5]

The first three days of the series reported on Herschel's telescope, his geological and botanical observations of the lunar surface, and the discovery of more complex, but not intelligent, life. By the time of the installment for Friday, August 28, readers were primed for the great revelation (Figure 1.2). According to the article, Herschel himself had observed large winged creatures, "wholly unlike any kind of birds, descend with a slow even motion from the cliffs on the western side, and alight upon the plain." The creatures:

> averaged four feet in height, were covered, except on the face, with short and glossy copper-colored hair, and had wings composed of a thin membrane, without hair, lying snugly upon their backs, from the top of the shoulders to the calves of the legs. The face, which was of a yellowish flesh-color, was a slight improvement upon that of the large orang-outang, being more open and intelligent in its expression, and having a much greater expansion of forehead.[6]

And then came an even more important revelation, for:

> these creatures were evidently engaged in conversation: their gesticulation, more particularly the varied action of their hands and arms, appeared impassioned and emphatic. We hence inferred that they were rational beings, and, although not perhaps of so high an order as others which we discovered the next month on the shores of the Bay of Rainbows, that they were capable of producing works of art and contrivance.

The Saturday article, largely devoted to lunar features, ended with the report that Herschel and his team had discovered an immense stone structure, which they pronounced to be a "temple." The series concluded on Monday, August 31, with a description of this find.[7] Oblivious to the fact that no Earth-bound telescope could then – or even now – possess the power to distinguish buildings, much less gesticulating aliens, on the surface of the Moon, the public took the well-written series of articles at face value, eagerly awaiting each day's new revelations.

The articles appeared in the *Sun* anonymously, but within days the author was revealed as Richard Adam Locke (1800–1871), a 34-year-old reporter who had just joined the *Sun* that summer. Within weeks the article was exposed as a hoax, and so it was widely believed to be until 145 years later, when historian Michael Crowe convincingly argued that Locke was actually writing satire.[8] What was he

Figure 1.2 Illustration of winged lunarians in the *New York Sun* for August 28, 1835, part of a series of articles now known as the Great Moon Hoax, though in reality it was probably the Great Moon Satire. Library of Congress Prints and Photographs Division, Washington, DC.

satirizing? According to Crowe, his target was no less than the advocates of inhabited worlds, of which there were many, especially among German astronomers, but none more enthusiastic than Scottish astronomer Thomas Dick. Dick once calculated the number of inhabitants on each of the planets in our solar system, arriving at a figure of more than 21 trillion, not counting those on the Sun! Sincere as he was, such conclusions were ripe for satire, and Locke took the opportunity with aplomb.[9]

Our interest here is not so much in the motivations of the cub reporter Locke, but in the public and scientific reaction to his stories. The *Sun* was a fledgling tabloid newspaper with a circulation of about 8,000. During the Moon episode its circulation reached 19,000, and remained high thereafter. Moreover, the *Sun* sold 60,000 copies of the story in pamphlet form, as well as lithographs of the lunarians. Nor was interest confined to the United States; French, Italian, Spanish, and German editions of the brochure appeared, and numerous other newspapers reported the story both inside and outside the United States.[10] A detailed study of the series of articles concluded, "no other newspaper story of the age was as broadly circulated as Locke's moon series … By the time the series had run its course, the *Sun* had become the most widely read newspaper in the world."[11] Obviously the story was profitable for the *Sun* because it had great popular appeal even if not true, a situation that resonates with tabloid (and some might say mainstream) journalism today.

But it was not only the general public that fell for the story. Crowe again describes the effect: the *New York Times* pronounced the discoveries "probable and possible," while the *New Yorker* credited them as creating "a new era in astronomy and science generally." Religious journals debated the consequences, Yale was "alive with staunch supporters" including students and professors, and astronomy professors Elias Loomis and Denison Olmsted traveled to New York to unearth more details, though they later denied believing the story. Edgar Allan Poe, who had himself just written a fictional lunar account in *Hans Phaall – a Tale*, reported:

> Not one person in ten discredited it, and (strangest of all!) the doubters were chiefly those who doubted without being able to say why – the ignorant, those uninformed in astronomy, people who would not believe because the thing was so novel, so entirely "out of the usual way." A grave professor of mathematics in a Virginia college told me seriously that he had no doubt of the truth of the whole affair!

Poe also called the Moon Hoax series "decidedly the greatest hit in the way of sensation – of merely popular sensation – ever made by any similar fiction either in America or in Europe."[12]

An 1882 *History of New York City* remarked on the graphic details of the series, the ingenious descriptions, and the plausibility of the arguments that even fooled some men of science. Horace Greeley, editor of the *New Yorker*, recalled the "unquestionable plausibility and verisimilitude" of the series, and said that "nine-tenths of us, at the least," were fooled by it. P. T. Barnum declared that "the majestic, yet subdued dignity" of Locke's work "at once claimed respectful attention; whilst its perfect candor, and its wealth of accurate scientific detail, exacted the homage of belief from all but cross-grained and inexorable skeptics."[13]

Not everyone was taken in. John Herschel himself was reported to be "in general amused," though by early 1837 he complained to his sister Caroline (herself a famous astronomer), "I have been pestered from all quarters with that ridiculous hoax about the Moon – in English, French, Italian and German!!" His wife wrote Caroline:

> Have you seen a very clever piece of imagination in an American newspaper, giving an account of Herschel's Voyage to the Cape … & of his wonderful lunar discoveries [of] Birds, beasts & fishes of strange shape, landscapes of every colouring, extraordinary scenes of lunar vegetation, & groups of the reasonable inhabitants of the Moon … it is only a great pity that it is not true but if grandsons stride on as grandfathers have done, as wonderful things may yet be accomplished.[14]

The first thing to note in this fascinating episode in public reaction is that the media played an important role, and indeed was the very vehicle through which the idea originated and was propagated. And, as contemporaries commented, the series was extremely well written, guaranteed to capture the imagination. Second, it mattered not that scientists knew conditions on the Moon did not support such observations; as Crowe puts it, "It was not that Locke lacked the skills of a satirist; it was rather that pluralist preachings and pronouncements had so permeated the thought of his contemporaries that they first failed to see the articles as satire, and failed again as they branded them a 'hoax.'"[15] Third, the episode may have damaged the reputation of a fledgling science in America, causing astronomical news from the United States to be received with great caution abroad.[16] Fourth, it is perhaps relevant that after 76 years Halley's comet was due to arrive in the fall of 1835, and was eagerly anticipated as the first apparition over the United States; by August enterprising businessmen had already set up a telescope near City Hall in New York City. In fact Yale astronomers Loomis and Olmsted claimed the honor of first sighting Halley's comet on August 31, the last day of the Moon series.[17] The public was perhaps primed for celestial news. We hold these lessons in mind as we turn to a quite different episode and a quite different culture, 100 years on.

The Martians Are Coming! The Great Martian "Panic" (1938)

On Halloween Eve, October 30, 1938, an event occurred so famous in radio history that we are still talking about it. Broadcasting coast to coast from Madison Avenue in New York City, 23-year-old American actor, writer, and producer Orson Welles – only three years away from producing and starring in his classic film *Citizen Kane* – directed and narrated *War of the Worlds* as part of the CBS series *The Mercury Theatre on the Air*. Based on H. G. Wells's 1898 novel by the same name in which Martians invade London, Welles's radio version transferred the action from London to Grover's Mill, New Jersey, about an hour's drive south of New York City. The 62-minute broadcast was presented as a series of news bulletins, which (despite warnings at the beginning of the show) many took to be real. It is the reaction to this event that makes it an important part of radio history – and of special interest for studying the reaction to possible extraterrestrial contact in its most extreme and direct form.[18]

There is no doubt that the reaction was considerable. The *New York Times* front-page headline the next day was "Radio Listeners in Panic, Taking War Drama as Fact," adding in the subtitle "Many Flee Homes to Escape 'Gas Raid from Mars' – Phone Calls Swamp Police at Broadcast of Wells Fantasy." It went on to say that

"A wave of mass hysteria seized thousands of radio listeners between 8:15 and 9:30 o'clock last night when a broadcast of a dramatization of H. G. Wells's fantasy 'The War of the Worlds' led thousands to believe that an interplanetary conflict had started with invading Martians spreading wide death and destruction in New Jersey and New York."[19] Many other newspapers carried similar accounts. A study published in 1940 by respected Princeton University professor and public opinion researcher Hadley Cantril, entitled *The Invasion from Mars: A Study in the Psychology of Panic*, confirmed the idea of a widespread "panic" reaction, and the concept was propagated in both scholarly and popular culture over the next 70 years.[20]

It is important to understand that by this time the idea of intelligent Martians had pervaded popular culture for four decades, thanks to the claims of American astronomer Percival Lowell that he had observed canals of artificial construction on the Martian surface. Indeed, according to Lowell, his 18-inch telescope in Flagstaff, Arizona, revealed a network of canals crisscrossing the entire planet. Lowell published his ideas in numerous reports, articles, and three major books – *Mars* (1895), *Mars and Its Canals* (1906), and *Mars as the Abode of Life* (1908) – and spread his ideas at numerous lectures, where his personal charm and enthusiasm usually won the day, at least among the public.[21]

Notably, although Lowell's ideas fired the popular imagination, there was no hint of panic caused by Martians reputed to be active on their home planet tens of millions of miles away. Historian William G. Hoyt, analyzing the press clippings in the Lowell Observatory archives, concluded that the initial public reaction "ranged from simple uncritical wonder through more or less credulous curiosity to skeptical but tolerant amusement." Another historian found that Lowell's theory of Martian canals unleashed a firestorm of controversy. "The public was fascinated, while professional astronomers generally viewed him with suspicion; some were openly hostile." Historians such as Michael Crowe have examined the scientific controversy in detail, but the important point is that a claimed remote detection of intelligent life on Mars resulted in much discussion and interest from both the public and astronomers, but no panic.[22]

This would change with H. G. Wells, or more specifically when Welles took up Wells. Wells first set pen to paper for his *War of the Worlds* in 1897, and it is no accident that Wells's invaders came from the Mars of Percival Lowell, whose Martian canal controversy had by 1897 reached England. Moreover, Wells believed Lowell's theory; as late as 1908, he cited "the work of my friend, Mr. Percival Lowell," as testimony that *War of the Worlds* was not too farfetched. Pronouncing Lowell's case for canals created by intelligent Martians "very convincing," Wells even discussed the forms that such Martians might take.[23] Although Lowell's claims were discredited by observations made during the close approach of the planet in

1909, Mars remained a favorite setting for the alien through many decades. And the idea of a Martian invasion of Earth lived on in popular culture also (Figure 1.3).

Which brings us back to Halloween Eve, 1938. For several decades sociologists, psychologists, and media specialists have questioned the extent of the "panic"

Figure 1.3 Cover from *Amazing Stories* of August 1927, illustrating the Martian invasion in H. G. Wells's *War of the Worlds*, the basis for Orson Welles's infamous Halloween Eve 1938 radio broadcast. Copyright 1927 by Experimenter Publishing Company.

reported in newspapers and supported by Hadley Cantril's study. Sociologist William Sims Bainbridge was among the earliest to criticize Cantril's conclusions, albeit almost a half century later. Writing in 1987 in the context of the sociological concept of "collective behavior," he clearly stated, "the most striking impression conveyed by Cantril's book was a false one – that real mass panic followed the broadcast. By quoting the stories of a few people who claimed to have been very frightened, Cantril implies that there was widespread panic. There wasn't … the whole affair was more a news media craze than a mass panic."[24] Bainbridge (as well as *Time* and *Newsweek* at the time) suspected those who *were* panicked were likely affected by war anxiety in Europe, fanned by the American media. Bainbridge also makes the point that panic is most likely when an ambiguous threat is seen as immediate; in other words, had Welles kept the original location of the invasion in England rather than the environs of New York City, there would undoubtedly have been little effect in the eastern United States. Reaction to Lowell's Martian canals confirms this idea; no one panicked when he suggested Martians on Mars.

The idea of mass panic from the Welles broadcast has been gradually thoroughly debunked. On the sixtieth anniversary of the broadcast in 1998, sociologist Robert Bartholomew reviewed the criticisms of the panic scenario, which included problems with Cantril's estimates of the actual number of people affected, as well as erroneous reports in the media that nevertheless went on to become part of popular culture. He concluded that perhaps tens of thousands rather than millions of people were panicked; in any event his lesson learned is that the mass media significantly influenced public perception of the event.[25]

But such a vivid part of popular culture is hard to excise. On the seventy-fifth anniversary of the broadcast in 2013, two professors of communication criticized PBS for perpetuating the "myth" of the "War of the Worlds" panic in a broadcast in its American Experience series. Among other things they also took Cantril to task for methodology and sloppy terminology, arguing he did not distinguish a state of excitement from a state of panic where people rioted in the streets. They conclude that the newspaper coverage was "dramatic and sensational – but ephemeral." They chalk up the sensational newspaper coverage to competition between newspapers and fledgling radio, which was beginning to draw advertising away from print media. In his book *Getting it Wrong: Ten of the Greatest Misreported Stories in American Journalism*, W. Joseph Campbell comes to the same conclusion, and states that

> the panic and mass hysteria so readily associated with the *War of the Worlds* program did not occur on anything approaching a nationwide dimension. The program did frighten some Americans, and some others reacted in less than rational ways. But most listeners, overwhelmingly, were neither frightened nor unnerved. They recognized the program for what it was – an imaginative and entertaining show on the night before Halloween.

He attributes the myth of the panic to anecdotal news reports that were never confirmed. Why were they not retracted? In Campbell's estimation, they represented an irresistible opportunity to rebuke radio, a rival source of news and advertising, as unreliable and untrustworthy.[26] This view has come to be the consensus among scholars, including Bartholomew, who also places the reporting of the "War of the Worlds" broadcast in the context of other media-driven panics and hoaxes such as the Moon Hoax, the Halley's comet scare of 1910 (due to reports of poisonous gases in its tail), and mythmaking during Hurricane Katrina.[27]

And yet, all this notwithstanding, whether excited or panicked, in small numbers or large, there is no doubt that the "War of the Worlds" broadcast had its effect, both on the night of the broadcast and in contemporary popular culture. And there is no doubt it could happen again on scales large and small. In fact, it did happen again, several times in a rebroadcast in Chile in 1944, one in Quito, Ecuador, in 1949, and on several occasions since. The latter resulted in another front-page *New York Times* headline for February 14, 1949: "Mars Raiders Caused Quito Panic; Mob Burns Radio Plant, Kills 15."[28] Panics related to anniversary rebroadcasts of "War of the Worlds" still occasionally occur, as do other mass panics involving erroneous reports of environmental contaminants, nuclear accidents, and other cultural concerns. Bartholomew's lesson is that only a small portion of the population needs to act on erroneous information over a short period to create large-scale disruptions to society. *Sociological study of the causes and effects of collective behavior are thus critically important to the study of the potential societal impact of discovering extraterrestrial life.*

"A bit of 'scruff'": Pulsars and Little Green Men (1967)

Both the 1835 Great Moon Hoax/Satire and the 1938 "War of the Worlds" broadcast were creatures of the media from beginning to end. By contrast, we now turn to an episode of extraterrestrial discovery that originated with serious scientific research. In the fall of 1967 astronomers at Cambridge University's Mullard Radio Astronomy Observatory made a puzzling discovery. Such discoveries are not unusual in the annals of astronomical history, but this one stands out because it involved a mysterious new class of object and was for a period of weeks seriously considered as a possible signal from extraterrestrial intelligence. It involved new technology, a low-frequency radio telescope that had begun operation only months before. The unusual telescope, built of wooden poles and wires and covering an area near Cambridge the size of 57 tennis courts, was built to observe exotic objects called quasars, located far outside our Milky Way Galaxy in the far reaches of the universe. The telescope was the brainstorm of Cambridge astronomer Anthony Hewish, while graduate student Jocelyn Bell had sole responsibility for operating the telescope and analyzing the data, under his supervision. The data were recorded on long strips of paper that were visually inspected. "Six or eight weeks after starting the survey I became aware that on occasions there was a bit of 'scruff' on the records, which did not look exactly like a scintillating source, and yet did not exactly look like man-made interference either," Bell wrote. "Furthermore I realized that this scruff had been seen before on the same part of the records – from the same patch of sky (right ascension 1919)." In late August, Bell showed the charts to Hewish, and by the end of September, Hewish suspected they had located a flare star similar to the certain dwarf stars under investigation by Bernard Lovell at Jodrell Bank. But by November 28, more observations had been made, and Bell recalled: "As the chart flowed under the pen I could see that the signal was a series of pulses, and my suspicion that they were equally spaced was confirmed as soon as I got the chart off the recorder."[29] The pulses were only 0.3 seconds, separated by about 1.3 seconds, phenomena not only unknown in the astronomical world but also difficult to explain by any natural physical process.

What to make of this observation? Hewish at first thought it must be man-made. Radar reflected from the Moon, satellites in peculiar orbits, and local effects were eliminated when another telescope confirmed the results, and it was established that the source was outside the solar system, but inside the Galaxy. As early as September a "Little Green Men (LGM) hypothesis had been jokingly raised when some of the group dubbed it the 'LGM star.'"[30] As astronomer Alan Penny has documented in his history of the event, Bell wrote in her thesis, "The possibility that the signals were from some intelligent civilization in the universe was not ruled out: hence the unfortunate nickname 'little green men.'" After the November 28 observations, when the regularity and unexplainable brevity of the signal was realized, the LGM explanation became more real. Cambridge astronomer Martin Ryle, the head of the radio observatory, later wrote, "our first idea was that other intelligent beings were trying to establish contact with us." By mid-December, by Penny's account, the LGM explanation was moving up in the list of possibilities. Hewish wrote, "As the days went by excitement rose when we found that the pulses were coming from a body no larger than a planet situated relatively close to us among the nearest stars of the galaxy. Were the pulses some kind of message from an alien civilization? This possibility was also entertained for lack of an obvious explanation for signals that seemed so artificial." He continued, "all kinds of thoughts went through our minds: it was such an artificial signal that I had to seriously consider that the signal was being sent to us … I had a test for the little green men idea though I had to pinch myself to take it seriously."[31] Hewish's test (which he began December 11) was to make timing (Doppler) measures to determine if the signals were on a planet orbiting a star, in which case the signal would shift back and forth in frequency. "Without doubt," he recalled, "those weeks in December 1967 were the most exciting in my life."[32]

By early January 1968. it was clear that the Doppler shifts in the signal showed only the orbital motion of the Earth, not of a planet with extraterrestrials. The discovery on December 21 of similar signals coming from the radio source Cassiopeia A also mitigated the possibility of extraterrestrial life as an explanation, since two civilizations would likely not be signaling at the same frequency. In February 1968 the data were reported in *Nature*, where Hewish, Bell, and their coauthors speculated that the signals could be caused by radial pulsations of exotic stellar evolutionary endpoints known as white dwarfs or neutron stars. Theorist Thomas Gold at Cornell quickly published a model that explained pulsars as rotating neutron stars; though Walter Baade and Fritz Zwicky had predicted neutron stars in the 1930s, their prediction had been forgotten. The rotation of these exotic objects, the aftermath of the collapse of a massive star (though not so massive as to produce a black hole), is the explanation accepted today.

From the first observations of unusual "scruff" and its discussion in August to the realization in late November of the extreme brevity and regularity of the signals, the LGM hypothesis seems to have been taken lightly. But in the wake of the November observations, for three weeks in December 1967, and until another similar source was found at the same frequency, the LGM hypothesis seems to have been considered quite seriously. The hypothesis was taken seriously enough that Martin Ryle expressed concern that the news should not leak out, a concern he amplified a few years later when Frank Drake sent a message to M 13, the Great Cluster of Hercules. On December 21 Hewish and Ryle had discussed between themselves what to do if it was an ETI signal: "you can't just publish it or release it like a news flash; we thought we would inform the Royal Society and get it handled nationally as it was too big a thing to deal with ourselves."[33]

Pulsars were not the first or the last case of unexplained radio signals. Frank Drake's Project Ozma (1960), the first radio telescope search for extraterrestrial intelligence, saw unusual signals, quickly explained as local interference. In the midst of the discovery of the new class of exotic objects known as quasars, in 1963–1964 Russian astronomers Nikolai Kardashev and Evgeny Sholomitskii argued that the unusual spectrum of a regularly varying radio object known as CTA-102 might be generated by artificial signals. Soviet radio astronomer Joseph Shklovskii vividly recalled the announcement: "There was a great uproar over it," he wrote. "I'll never forget the press conference that the Shternberg Institute gave to announce the great discovery. The entire courtyard of the institute was crammed with luxurious foreign cars belonging to some 150 of the leading accredited correspondents in Moscow … The director of the institute, Dmitry Martynov, basked in the limelight." This "CTA-102 affair" remained unexplained for some years, until it was determined to be a type of quasar with an unusually large redshift. But anomalous observations continued: in 1977 astronomers at Ohio State University detected a still unexplained "Wow!" signal; in 1997 astronomers at the National Radio Astronomy Observatory briefly had an anomalous signal; and in 2015 radio astronomers searched for signals after the *Kepler* spacecraft detected unexplained light variations in the star known as KIC 8462852.[34]

These episodes, taken together with pulsars, illustrate several characteristics likely to be present in a real discovery of extraterrestrial intelligence by astronomical observations: confusion, doubt, and uncertainty about what to do. As we see in Chapter 9, these incidents played a role in urging scientists and others to think about consequences if a detection were made. Beginning in the mid-1980s, as NASA was ramping up its own SETI program, the International Academy of Astronautics, in conjunction with the International Institute of Space Law, developed and approved the "SETI Post Detection Protocols," the thrust of which was to confirm the observations, then tell everyone.[35] How such principles would play out in real life is anyone's guess.

Extraordinary Claims: Fossils from Mars? (1996)

Deep in the summer of 1996 a startling announcement came from NASA, the American space agency. Life had been found on Mars! Or at least evidence of past life. On August 7, with little more than a day's notice, reporters descended on a hastily called press conference at NASA Headquarters in Washington, DC, to which the participants themselves had been hurriedly summoned. A carefully planned announcement for the following week had been upstaged by a three-paragraph leak in the industry newspaper, *Space News*, and the exhausted scientists had flown in from around the country. Among the many officials in the audience were the heads of the National Science Foundation (NSF), the National Academy of Sciences, and Gerald Soffen, the project scientist for the *Viking* spacecraft, which had landed on Mars 20 years earlier. First to the podium was NASA administrator Dan Goldin, who had already briefed President Bill Clinton and other top political officials. He waxed eloquent about NASA, American science, and the breathtaking conclusions about to be announced, and reported that the president had asked that the discovery be given top priority.[36]

NASA Associate Administrator for Space Science Wes Huntress then turned the podium over to the scientists, a team of nine led by geochemist David McKay of NASA's Johnson Space Center. Now they presented their evidence to a hushed audience. Organic molecules had been found in a meteorite that was blown off Mars 16 million years ago, had landed in the Antarctic 13,000 years ago, was found there by a meteorite-collecting team funded by the NSF and the Smithsonian

Institution 12 years ago, and had been recognized as Martian only two years ago. Almost three years of exhaustive study had led the researchers to their momentous conclusions. The claim of organic molecules on Mars was already a step beyond the *Viking* results. But there was much more: mineral "carbonate globules" of possible biological origin; evidence of tiny magnetic minerals that on Earth are secreted by certain bacteria; and, finally, pictures of strange, hauntingly wormlike structures that they argued might be microfossils (Figure 1.4). In short, the assembled scientists suggested, life had existed on Mars sometime in the planet's distant past, when Mars was warmer and wetter.

This was not exactly the Martian civilization of Lowell, Wells, and Welles, but compared to the ambiguous results of *Viking* 20 years before, the claim that a Martian meteorite had landed on Earth bearing evidence of past life on that fabled planet was little short of miraculous. Nor was it to be a mere three-week hypothesis as in the case of Little Green Men and pulsars. The very possibility of life on Mars, even past life, set the world afire, igniting media hype, public imagination, scholarly discussion, and scientific curiosity alike. The reaction played out over timescales short and long, and arguably is still strongly felt today, long after scientists reached consensus that the rock most likely does not harbor fossils.

The flurry of events the claimed discovery set in motion, both before and after the announcement, are best analyzed as they unfolded. Before anyone dared make

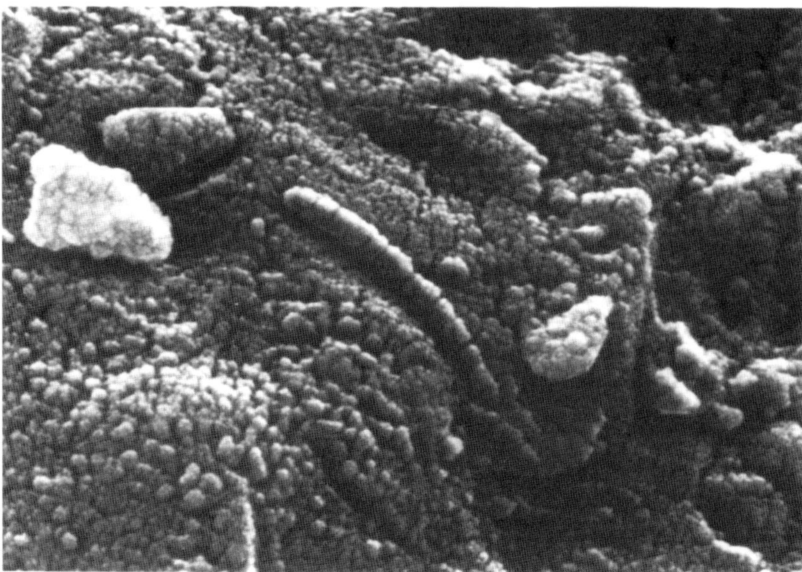

Figure 1.4 Structures interpreted as nanofossils in the Mars rock ALH84001. The consensus today is that they are not biogenic. Credit: D. McKay (NASA /JSC), K. Thomas-Keprta (Lockheed-Martin), R. Zare (Stanford), NASA.

an announcement, almost three years of intense study of the Mars rock had been undertaken, in which the possibility of biogenic origin gradually dawned and was then closely held within a circle of scientists until the evidence was considered compelling enough for publication. That evidence consisted of four parts. None of the parts, the participants pointed out, was conclusive in itself, but taken together they could be interpreted as biogenic. First, the multidisciplinary science team reported, the fractured surfaces of the rock contained large, complex organic compounds in the form of what are known as polycyclic aromatic hydrocarbons, PAHs for short. Even though the NASA team undertook analysis that showed to its satisfaction that the PAHs were not contamination from Earth, this was not proof of life since organic molecules could have originated in nonbiogenic processes on Mars.

But then the plot thickened when inside the fractures the team also discovered carbonates, mineral deposits that may be produced by living creatures on Earth, as in the case of limestone. Within the carbonates they also found magnetite, pyrrhotite, and greigite, minerals that are produced (among other ways) by certain "magnetotactic" bacteria on Earth. Finally, using a high-resolution scanning electron microscope, the team suggested the existence of microfossils in the carbonates and other mineral grains; at only 20 to 100 nanometers they were 100 times smaller than the smallest known bacteria on Earth. All of this work had been done over three years beginning in 1993 when the rock was first recognized as Martian in origin and began to be studied in detail. An article was submitted to the prestigious journal *Science* in April 1996, was accepted in July, and was to be published in mid-August, at which time NASA would make a public announcement. The long gestation time for the claim is notable; no reputable scientist wanted to make a claim that had to be retracted, and this was the mother of all claims.[37]

Another notable point, as historian of science James Strick has observed, is "the JSC [Johnson Space Center] team's unusually secretive behavior during the time the work was being done and even after the paper had been submitted and was under review for publication in *Science*." Everett Gibson, one of the team leaders, stated that the team considered the Clinton administration and NASA Headquarters in Washington a "sieve" that would leak any important story to the press. McKay put it more circumspectly when he said the team members wanted their case to be as strong as possible before they went to their superiors, much less to the public. And they were worried about priorities in discovery – many other groups had samples from the same meteorite and the NASA team did not want to be scooped. A slightly different story comes from Michael Meyer, then the exobiology discipline scientist at NASA Headquarters. He recalls that during the Lunar and Planetary Science Conference in Houston in March 1995, Gibson and McKay "excitedly insisted I come to their lab. Basically they presented the data they had to date and were coming to the life conclusion in ALH84001." After the fall of 1995, when the evidence was piling up, however, the science team likely did not keep NASA managers informed of the details outside of the Johnson Space Center, for the reasons given earlier. These interactions show the complexities that will undoubtedly arise at the participant and institutional level in any announcement of the discovery of alien life.[38]

This semi-secretive situation could not last. Already in April 1996, word had been received at NASA Headquarters, where Wes Huntress decided not to act further until the paper had been accepted by *Science*.[39] When it was accepted in mid-July for publication in mid-August, Huntress immediately went to NASA administrator Dan Goldin. Goldin assembled a team from NASA public affairs, as well as key policy people and scientists to decide how to handle the information. They were concerned with containing the information before it was published in *Science*. But they also had political concerns: the Republican National Convention would take place at about the same time as the *Science* publication, and Goldin considered the timing of the announcement a political decision above his pay grade. On July 30 he and Huntress went to the White House to discuss the matter with Chief of Staff Leon Panetta, who immediately set up a meeting with President Clinton, followed immediately by a meeting with the vice president. In what seems like reverse order to what should have taken place, Goldin then ordered Gibson and McKay to come to Washington. There, on the last day of July, they underwent several hours of Goldin's skeptical scrutiny; Goldin did not wish to have egg on his face (or NASA's collective face) any more than the scientists. But his fears of a premature leak were well founded. The details of the leak are well known, an initialized copy of the galley proofs of the article having gone (according to Gibson) from him to Goldin, Gore, Clinton, White House advisor Richard Morris, and a hooker who tried to sell it to the media. NASA Headquarters began receiving calls from the news media around August 1, inquiring if there was any substance to the story.[40] Goldin attempted damage control by pushing up the announcement by eight days, to August 7.[41]

Late in the afternoon of August 6, the die was cast. NASA sent out a "note for editors," which began: "A team of NASA and Stanford scientists will discuss its findings showing strong circumstantial evidence of possible early Martian life, including microfossil remains found in a Martian meteorite, at a news conference scheduled for 1:00 pm EDT, August 7, at NASA Headquarters, 300 E. St. SW, Washington, DC. The team's findings will be published in the August 16 issue of *Science*." A short time later Administrator Goldin issued an unusual four-paragraph statement: "NASA has made a startling discovery that points to the possibility that a primitive form of microscopic life may have existed on Mars more than three billion years ago. The research is based on a sophisticated examination of an ancient Martian meteorite that landed on Earth some 13,000 years ago." The statement went on to characterize the evidence as "exciting, even compelling, but not conclusive," and to emphasize that this was all about bacteria-like structures, not "Little Green Men."[42]

The rushed press conference on August 7 was a spectacle to behold, with a 1.3 ounce piece of the Mars rock displayed in a place of honor at the front of the room along with the participating scientists, and the rest of the room crammed with reporters, scientists, and onlookers hoping to see history in the making (Figure 1.5). All the major TV stations were broadcasting live. Speakers included not only the scientists making the claim (Figure 1.6) but also J. William Schopf, a UCLA professor and specialist in microfossils on Earth, representing the many scientists who were skeptical of the claim. Schopf had been invited to Houston in early 1995 to look at the evidence, and was skeptical. He remained skeptical at the press conference, describing the entire body of evidence as "circumstantial," saying it did not constitute proof consistent with the normal standards of science. In fact, he had tried to avoid participating in the press conference, citing Carl Sagan's dictum that "extraordinary claims require extraordinary evidence"; having been persuaded by Administrator Goldin to participate, he minced no words about his continued

Figure 1.5 Press conference August 7, 1996, at NASA Headquarters announcing the possibility of fossilized life in the Mars rock ALH84001. The photo, taken by NASA photographer Bill Ingalls, stunningly captures the excitement and media frenzy likely to occur with any announcement of life beyond Earth. Credit: NASA/Bill Ingalls.

Figure 1.6 NASA Administrator Dan Goldin and happy scientists making the claim of the discovery of Mars nanofossils at the August 1996 press conference. Seated left to right: Richard Zare, David McKay, Everett Gibson; standing left to right: Hojatollah Vali, Dan Goldin, Kathy Thomas-Keprta, and Chris Romanek. Perhaps revealingly, skeptic Bill Schopf was part of the press conference, but not in this picture. Credit: NASA/Bill Ingalls.

skepticism, saying that the evidence was not even close. But the others at the table begged to disagree. To some extent it was a debate about what constitutes evidence in science, an argument likely to be repeated in any discovery of extraterrestrial life. NASA's official position on the discovery was summarized in a lengthy but carefully worded press release prepared before the press conference.[43]

The furious reaction to the announcement can be analyzed over both the long and short terms. Even as the press conference was taking place, at 1:15 P.M. President Clinton spoke from the South Lawn of the White House, with science advisor Jack Gibbons at his side. "This is the product of years of exploration and months of intensive study by some of the world's most distinguished scientists," he began.

> Like all discoveries, this one will and should continue to be reviewed, examined and scrutinized. It must be confirmed by other scientists. But clearly, the fact that something of this magnitude is being explored is another vindication of America's space program and our continuing support for it, even in these tough financial times. I am determined that the American space program will put its full intellectual power and technological prowess behind the search for further evidence of life on Mars.

He went on to say he had asked the vice president to convene, before the end of the year, a bipartisan space summit on the future of America's space program, in part to pursue answers to the questions raised by the new discovery. He reinforced what he termed the "aggressive plan" for the robotic exploration of Mars, symbolized by the *Pathfinder* spacecraft launching later in the year. And he concluded by saying "if this discovery is confirmed, it will surely be one of the most stunning insights into our universe that science has ever uncovered. Its implications are as far-reaching and awe-inspiring as can be imagined. Even as it promises to answer some of our oldest questions, it poses still others even more fundamental."[44]

Then, on the scale of hours, days, and weeks, came the reaction in the mass media. Before the press conference was finished CNN quoted a source close to NASA saying it was arguably the biggest discovery in the history of science. Radio, television, and nascent internet media all carried the news as their top story. The next day the *New York Times* again had a front page headline related to extraterrestrial life: "Clues in Meteorite Seem to Show Signs of Life on Mars Long Ago: Startling Find of Organic Molecules from Space." The *Los Angeles Times* editorialized "Maybe It's a Lively Universe: Finding Fuels Speculation, along with Mars Fever," waxing eloquent about the human imagination and future space missions to Mars. Inevitably, the claims would also play a key role in reviving the search for life, as reflected in the headline of the "Science" section of the *New York Times* a few days later: "After Mars Rock, A Revived Hunt for Other-Worldly Organisms."[45]

Over the next few weeks, other newspaper articles speculated about the broader impact of life on Mars, many broaching the question of theological implications. The *Washington Post* for August 11 reported that the consensus among theologians queried was that proof of life on other planets, whether microbial or intelligent, would confirm the expansive nature of God – an old argument from the tradition of natural theology. Some did admit, however, that Christianity would have to rethink its central dogmas of redemption and incarnation, which implied that Jesus Christ came only to this world to save humans from sin. James A. Wiseman, chairman of the Department of Theology at Catholic University, admitted that some people's faith might be shaken, and that changing the "geocentric and ethnocentric mind-set" would take immense amounts of discussion and rethinking, tantamount to Copernicus persuading the world that the Earth revolved around the Sun. A professor of theological ethics at Duke University Divinity School turned things around, saying that the discovery did not challenge Christian and Jewish theism with its expansive God, but "the high humanism that has been characteristic of our lives since the 17th century, namely that the human species is what it's all about, that everything exists to serve us." Some of those queried pointed out that Christian fundamentalists, who take the Bible literally, might have a problem with extraterrestrial life. But James Garrett, a professor at Southwestern Baptist Theological Seminary, a bastion of

Southern Baptist conservatism, said that although the Bible does not mention life on other planets, "we need to be cautious saying that life on other planets is precluded because [the Bible] also says that God is the creator of all." A week later, the Los *Angeles Times* ran an article, "Theologians Find Awe in Possibility of Life on Mars." "We believe a God who is capable of creating one world is capable of creating many worlds," said Rabbi Alexander Schindler, former head of the Union of American Hebrew Congregations. "It does not change our fundamental faith. It doesn't touch it in the slightest." Among others, the *Times* also quoted Richard Payne, dean of the Institute of Buddhist Studies in Berkeley, as saying that Buddhist cosmology puts the Earth "as just one small part of a larger inhabited reality" filled with a variety of sentient beings. The core Buddhist truths, he emphasized, would hold true anywhere in the universe.[46] Further examples of theological reaction could be multiplied, but suffice it to say that around the world, in small-town and city newspapers, on summer beaches and among university scholars, the implications of the Mars rock were debated in a way likely to be echoed by any discovery of extraterrestrial life.

Within five weeks, congressional hearings were held on life on Mars, hearings that mentioned in passing H. G. Wells, Orson Welles, and the blockbuster movie of the summer, *Independence Day*, explicitly acknowledging "the cultural fascination with life in outer space."[47] And on the scale of months and years, the more technical and scholarly examination of the evidence and its implications was carried out. Even as the scientific debate was ongoing, the scholarly discussion of the implications proceeded apace. Such discussions were carried out around the world, in settings both formal and informal. One example is the symposium convened in November 1996 at George Washington University in Washington, DC. Titled "Life in the Universe: What Can the Martian Fossils Tell Us?" the symposium focused not on the science, but on the cultural, intellectual, and societal implications of the discovery. To that end, the speakers included not only scientists, but also a philosopher, two theologians, a historian of science, a policy analyst, a science fiction writer, and the NASA administrator. Again, the questions they addressed are likely to be those addressed in any such discovery: about theory and observation in relation to verified knowledge, religious meaning, the role of the media, the historical context of the discovery, and the place of extraterrestrial life in the imagination.[48]

Among the most consequential events, the one I described at the opening of this chapter, was the vice president's Space Science Symposium held in December 1996, four months after the original announcement and spurred by President Clinton's interest. In preparation for this high-level event, the Space Studies Board of the National Academy of Sciences, at the request of the White House and NASA, first held a workshop to discuss the implications of the Mars rock. Convened in late October within three months of the original announcement, it conveyed the message that the Mars rock findings should be part of NASA's much larger "Origins" program to determine our place in the universe. The more intimate December meeting with the vice president focused on implications of the discovery, with only minimal attention to budget implications.

In keeping with the focus on implications of the discovery, President Clinton had encouraged an exploration of the religious dimensions of the problem because of his interest in religion, and Administrator Goldin later stated it was "crucial that we … have broad consultation with the American people. When you have science – free flying science – funded by tax dollars, you want to avoid crossing ethical boundaries." It was not immediately obvious what ethical boundaries would be crossed by fossils from Mars, but there was reason to be concerned about religious reactions. Some scientists, such as Richard Zare, whose Stanford team had found the organic molecules in the Mars rock, were unsettled by fundamentalists angry about the religious implications during his discussions on ABC's *Nightline* and other television shows. The theologians at the meeting were generally reassuring, arguing that religions would find a way to adapt. The discussions in the Indian Treaty Room lasted three hours, as the participants debated the origins of life (represented by Stuart Kauffman), the astronomical aspects (John Bahcall), and the media reaction (Bill Moyers). The meeting took on the air of an intellectual salon, but with the possibility of real implications for NASA programs.[49]

In fact, on the scale of years and decades, the fallout to the research community in terms of new budgets and new directions of research was considerable. According to David H. Smith, a senior program officer at the Space Studies Board of the National Academy of Sciences deeply involved in astrobiology from a policy perspective, the events of 1996 "had a profound long-term impact on both scientists and policy makers." In particular, he emphasized that when the Clinton administration on February 6, 1997, announced its proposed budget for FY 1998, it contained new funding for the initiative known as "Origins," including initiation of a new program in astrobiology.[50] This was also the beginning of the NASA Astrobiology Institute, which in May 1998 awarded large grants for astrobiological work to 11 institutions, and has been thriving ever since in the sense of funding cutting-edge research. Moreover, the NSF, the principal sponsor of the Antarctic Meteorite Collection Program under which the Mars meteorite was found, initiated new astrobiological activities in early 1997. *The lesson is that any discovery as exciting as finding life beyond Earth will result in increased funding to elaborate that discovery.*

It is notable that these events had a lasting impact even though there was much skepticism about the Mars fossils; indeed, the consensus after a few years was that they were not fossils at all. At the 10-year anniversary of the announcement, aerospace analyst and planetary scientist Jeff Foust characterized the evidence as "inconclusive at best, and outright discredited at worst."[51] Some of the scientists involved with the discovery, including David McKay, at that point still believed their evidence, but admitted they had not convinced the scientific community. With few exceptions, on the tenth anniversary the media had lost interest. Nevertheless, the impact on astrobiology remained palpable because the questions remained compelling. It was once again possible to talk freely about life on Mars; indeed, the robotic exploration of Mars was arguably redirected toward the problem of life directly as a result of the attention given to the Mars rock. Astrobiology became infused with life, helped along by the first discoveries of exoplanets around Sun-like stars. Smith attributes the lasting impact in the field to massive interest among scientists, the public, and the press, combined with the rapid response of the scientific community providing context within the broader exobiology program, and fortuitous timing relative to the budget-planning cycle. Furthermore, he finds that prior plans existed that could quickly be implemented once funding became available. These are perhaps all lessons for a future discovery.[52]

At another level is the question of how the Mars rock discovery affects individuals. When *USA Today* asked readers to share their views five days after the announcement, the responses ranged from "no impact at all" because they already believed there was life in the universe, to "exciting me greatly" because "we start to wonder where our futures are."[53] *This range of responses is likely to represent the reaction to other such discoveries in the same way individuals react to different events in different ways on any subject.* Although it might seem impossible to gauge the "overall impact" on any society, much less the world with its myriads of societies, it is historically apparent that aggregate opinions do affect culture.

The impact of the discovery of microbial life, whether alive or fossilized, is sometimes denigrated as uninteresting compared to the discovery of more complex life. The case of the Mars rock indicates this is far from true, both inside and outside science. Internal to science, such a discovery could be no less transformational than for other societal domains. In the aftermath of the Mars rock announcement two scientists from the Department of Microbiology and Immunology at the Stanford University School of Medicine argued for some "unicellular prejudice," writing that "the detailed description of even one-cell Martian life forms would profoundly alter our understanding of life itself."[54]

The Lessons of History

Unfortunately, one of the great lessons of history is that we do not learn the lessons of history. As a recent author put it while contemplating historical patterns in a much broader context, history "goes unheeded, as it always has and as it always will, because history teaches us that we do not learn from history, that we fight the same wars against the same enemies for the same reasons in different eras, as though time really stood still and history itself as moving narrative was nothing but artful illusion."[55] Even in an optimistic frame of mind, in a world in which we might apply lessons learned if only we paid attention, the problem is determining exactly what those lessons are.

At base, the reactions we have described in this chapter are a question of human behavior, whether as individuals or groups. Studies of the psychology and sociology of discovery, and human reactions to them, are therefore in order to illuminate future possibilities. Meanwhile, we have seen that in our first two cases of claimed extraterrestrial discovery (the Great Moon Hoax/Satire and the "War of the Worlds" broadcast), the media played a precipitating role, the first associated with the rise of tabloid journalism, the second with the concerns of the traditional print media with the rise of the new radio media. The Great Moon Hoax/Satire illustrates a phenomenon all too common even today in tabloid journalism – exploitation in the service of moneymaking, with little regard to the truth. The Orson Welles broadcast of *War of the Worlds* illustrates that the print media can be biased due to internal concerns perceived or real. Although the "panic" was much less widespread than thought, "the famous broadcast suggests that when circumstances are right, the media can create panic and other effects that are unpredictable, disruptive, and wide-ranging."[56]

The media also played an important role in both reporting and sensationalizing the Mars rock claims of nanofossils. The Little Green Men hypothesis of pulsars, on the other hand, generated no media reaction at the time because the anomalous astronomical observations were closely held among the astronomers involved, and in the aftermath the LGM hypothesis was little more than a curiosity. The lesson here is not that such a discovery should be kept secret; with email, the internet, and other social media, news of the discovery of life beyond Earth would spread like wildfire in the twenty-first century.

The case of Mars rock ALH84001 offers the most robust example in the modern era of what might happen following the claim of life beyond Earth. While it is true the Mars rock announcement and subsequent events were relevant to a specific time, culture, and set of circumstances, in broad outline, these events are likely to be mimicked by any discovery of extraterrestrial life. The reaction of government institutions, the media, and the public will occur side by side with the reaction among scientists, who will subject the discovery to their exacting standards. Science is like that; scientific articles, after all, are published only after thorough peer review, then open to withering criticism, which the evidence will either survive or not.

It is our contention, then, that history provides one realistic guide to what will happen when the discovery of extraterrestrial life is announced in the coming years, decades, or centuries. But that discovery will be far from immediate; rather it will consist of an extended process characteristic of all discoveries. In the next chapter we turn to the nature of discovery, and the many possible scenarios of discovery, as another means of illuminating the reaction to the discovery of extraterrestrial life.

2

Discovery

> The real art of discovery consists not in finding new lands, but in seeing with new eyes.
>
> *Marcel Proust*[1]

> [The signal] is the object of intense socially required study for a long period of time. I regard it as much more like the enterprise of history of science than like the enterprise of reading an ordinary message … The data rate will for a long time exceed our ability to interpret it.
>
> *Philip Morrison*[2]

I have a special place in my heart for Pluto. When I arrived in 1978 as a young astronomer at the US Naval Observatory in Washington, DC (better known to many as the home of the vice president of the United States), less than a year had elapsed since astronomers there had discovered Charon, the first and largest of the five moons now known to circle Pluto. Some astronomers denied the discovery, since the evidence came in the form of what appeared to be a small, fuzzy bump on the planet, visible in a series of photographic plates. Only years later did a transit of Charon across Pluto convince most astronomers of its reality, and it was only in 1990 that the Hubble Space Telescope captured an image where the two bodies were clearly separate.[3]

One of the lessons I learned from that experience was that discovery is an extended affair, not only for Charon but also in the case of Pluto, which had been discovered in 1930 at the Lowell Observatory in Flagstaff, Arizona (the very observatory that pushed the idea of Martian canals). Very little was known about Pluto in 1978, but the discovery of a body orbiting around it allowed its mass to be accurately calculated for the first time by Kepler's laws. The determination of its small size was the beginning of Pluto's downfall as a planet. I was among those at the meeting of the International Astronomical Union in Prague in 2006 who voted on Pluto's status. I voted that Pluto should be classed as a dwarf planet, and that a dwarf planet was a subcategory of planet. Alas, others disagreed, and went for the linguistic barbarity we now have that a dwarf planet is not a planet.[4] By international agreement, Pluto is considered one of numerous objects in what is known as the Kuiper Belt. Whatever you call it, thanks to Alan Stern and his New Horizons crew, Pluto is now another world whose features, including possibly an ocean, are well known. It is unlikely that an ocean at the outer reaches of the solar system would harbor life, but stranger things have happened. In any case, my point is that Pluto is continually being rediscovered.

Few would argue that even the discovery of fossil or microbial life, let alone the discovery of extraterrestrial intelligence, would constitute one of the greatest achievements in the history of science. But contrary to common expectations, history shows that scientific discovery is far from the eureka moment it is often thought to be. Archimedes may or may not have run naked through the streets of Syracuse shouting "Eureka!" ("I have found it!") after the discovery of the buoyancy principle that bears his name. But if so, his supposed moment of instant discovery was either legendary or exceedingly rare. Rather, discovery is an extended process involving the efforts of many individuals across multiple stages of detection and interpretation, before true understanding finally dawns, perhaps years or decades later.

We have already seen this in the case of the famous Mars rock, and it is true also of the *Viking* biology experiments, which some believe to this day exhibited tantalizing indications of life on Mars. But it is also true in a much broader sense throughout science; indeed, examining any particular case in detail, we would be hard put to document a pure eureka discovery moment in science. The fine detail always shows structure, and that structure is important to any discussion of impact. In the quotation cited at the opening of this chapter, physicist and SETI pioneer Philip Morrison captures the likely nature not only of an intelligent signal but also of the discovery of microbial life, and everything in between. In a broad sense, the "signal" is the detection of life or past life in any form. We will not "read" the message of alien life; we will have to decipher and interpret it.

Moreover, an analysis of discovery is important because to a large extent the exact nature of any discovery determines its impact. Because the impact of finding life beyond Earth will surely depend on how the discovery is made, this chapter also analyzes in some detail the scenarios under which that may occur. This chapter is informed not only by history and science but also by science fiction, which at its best has laid out thoughtful scenarios of discovery in considerable detail. Science fiction is much more than story; in my experience working for NASA, a large percentage of its employees have been influenced by the science fiction of their youth. Imagination has a way of foreshadowing reality, and the case of alien life is no different. Having examined the anatomy of discovery and the scenarios for discovery of alien life, we are then in a position to look at the structure of an alien life discovery, which in turn will feed into the problem of impact in the post-discovery phase discussed in the remainder of this volume.

The Anatomy of Discovery

If discovery is not a eureka moment, it follows that it must be extended and have some structure. In fact, any scientist can tell you it is indeed a long and complex series of events. "Discovery is a process and must take time," scientist-philosopher Thomas Kuhn wrote in 1962 in his famous *Structure of Scientific Revolutions*. In the half-century since Kuhn's pronouncement, scholars have repeatedly documented the truth of discovery's extended and complex nature. It has been subjected to historical, psychological, and sociological analysis, and philosophers have parsed its conceptual meanings as well. They have also insisted that discovery is not only a technical but also a social process, involving real people with real emotions that have real effects on the final results – and their impacts on society. As a way of making the extended nature of discovery more concrete, let's take a brief look at how it applies in the three specific areas of physics, chemistry, and astronomy. This will help us to recognize the need to take this into account when it comes to the discovery of life beyond Earth.[5]

Eureka Denied: The Extended Nature of Discovery

One of the most famous events in recent elementary particle physics is the discovery of the Higgs boson, the long-sought particle thought to be the source of all mass. In 2012 the journal *Science* wrote that scientists at Fermilab reported seeing hints of the Higgs particle. "The signs are not strong enough to clinch a discovery," *Science* noted, "but they were consistent with earlier reports at Europe's Large Hadron Collider." Such "hints" become real discoveries only after sometimes lengthy interpretation and an even lengthier road to understanding. And even then, consensus is not immediate. Only a few weeks later physicists at the Large Hadron Collider made a more definitive announcement of the Higgs boson, still laced with caution: "To the layman I now say, I think we have it," said Rolf-Dieter Heuer, director general of CERN. "We have a discovery. We have discovered a new particle consistent with the Higgs boson. It's a historic milestone today." While the announcement was met with universal applause, the result was at the lower end of the accepted statistical standard. Others emphasized it was still unknown whether the new particle was the theorized Higgs of the Standard Model, or merely "Higgs-like." Months later *Physics Today* headlined that "The Higgs particle, or something much like it, has been spotted." Properties, such as spin and its interaction with other particles in the manner expected of the theorized Higgs, remained to be determined. Clearly, the process of detection itself may be extended, not to mention its interpretation and understanding.[6]

What is true of elementary particle physics is also true of chemistry, which demonstrates that discovery may extend over years and decades rather than weeks or months, and that a putative discovery may not pan out at all. The history of the chemical elements, for example, is full of cases of mistaken identity or claims of discovery for a given element. What are now considered the first three elements of the Periodic Table illustrate the pattern. The lightest element, hydrogen, was only recognized over a period of 15 years after eccentric British chemist Henry Cavendish first suspected it to be a discrete substance in 1766, and in 1781 found that water was produced when hydrogen was burned. Because of the latter property, two years later Antoine Lavoisier gave the substance the name "hydrogen." Astronomers Jules Janssen and Norman Lockyer independently claimed the discovery of what is now known as chemical element number 2, helium, by observing the Sun in 1868. But they were mocked for years by other scientists who believed no such thing existed. It was 1895 before Sir William Ramsay isolated the element on Earth. An ore containing what turned out to be element number 3, lithium, was discovered in 1800; its identity as a new element was not suspected until 1817, and it was not until 1821 that it was isolated. Every element has its own extended discovery story. One author wryly noted it would be possible to compile a periodic table of 100 elements named in hope but never found. Nebulium, jargonium, virginium, and russium are but a few of these, and they demonstrate the important point that many false alien life discoveries are likely to be announced, just as they have been in the past. The discovery that the real chemical elements could be ordered into a periodic table is another story of extended discovery. Of the four elements predicted in Dmitri Mendeleev's periodic table of 1871, three were found within 15 years, while the fourth (technetium) was only synthesized in 1937 (by Emilio Segre) and was not found to occur naturally until 1952, and even then only in a red giant star. Its existence there was the first proof of nucleosynthesis in stars, the idea that more complex elements were being produced from simpler elements through nuclear reactions. More broadly this discovery showed that cosmic evolution was taking place.[7]

Cases could be multiplied in other areas of science, but let's turn to astronomy, a field closer to our subject that we can parse in more detail for the finer structure of discovery. We might think that with the invention of the telescope, the discovery process would be relatively compressed, as intrepid astronomers turned their increasingly powerful instruments to the heavens and instantly spied new objects just waiting to be revealed. Nothing could be further from the truth. The extended nature of discovery was well illustrated when Galileo turned his telescope toward Saturn on July 30, 1610. It is often said that Galileo discovered the rings of Saturn at this point. But in fact Galileo was totally perplexed: "[T]he star of Saturn is not a single star, but a composite of three, which almost touch each other, never change or move relative to each other; and are arranged in a row along the zodiac, the middle one being three times larger than the two lateral ones." There is no doubt that Galileo detected something new on that date; in fact, he detected what we now know to be the rings of Saturn. But it is equally evident that Galileo had no idea what he had detected. As historians have shown in detail, during his lifetime Galileo treated these bodies as "collateral globes," as satellites such as those he had discovered around Jupiter six months earlier, though strange satellites to be sure. This remained the dominant concept for more than 30 years. Galileo had gone through the phases of detection and interpretation, but as yet had no correct understanding of the most basic nature of the object. Although one commonly sees in both textbooks and historical treatments that Galileo "discovered" the rings of Saturn, it is historically more accurate to say that Galileo contributed significantly to the *process* of the discovery of Saturn's rings. It was only in March 1655 that Christiaan Huygens turned his large telescope toward Saturn and interpreted what Galileo had seen as handles or "anses" – we would say "rings" – that changed appearance depending on the angle of observation from Earth. Lacking knowledge of dynamics, Huygens in turn had no idea how such rings could physically exist and persist around Saturn. That understanding came only in the mid-nineteenth century, when James Clerk Maxwell provided a basic dynamical explanation. We could say the full discovery of Saturn's rings took centuries to make.[8]

Nor is this phenomenon of extended discovery due to a nascent and naïve telescopic astronomy. The search for planets outside our solar system in the last half of the twentieth century, which bears directly on the search for life, again brings the lesson home. The object of that search, now known as "exoplanets," had been a Holy Grail of astronomy for a century or more. Theory had oscillated back and forth between the "nebular hypothesis" of the formation of solar systems and the "close encounter" theory, whereby matter for planet formation was pulled out of stars when two stars had close encounters. The former method was a natural by-product of stellar evolution, and thus predicted abundant planetary systems, while the latter was a rare occurrence. In the 1940s the nebular hypothesis was again gaining the upper hand, and as early as 1943 independent observational claims were made for the existence of two planetary systems around nearby stars, indicating planets might indeed be abundant. By 1963 astronomer Peter van de Kamp made a widely publicized claim for planets around Barnard's star, based on decades of photographic observations of the perturbed motion of the star. All of these observations turned out to be spurious, and it would be decades more before the real discoveries of exoplanets were made. In other words, almost the entire twentieth century may be viewed as a prediscovery phase in the search for exoplanets.[9]

The eventual definitive discovery of exoplanets was closely bound up with the discovery of "brown dwarfs," failed stars whose existence had been predicted as early as 1963. Such stars had a mass of only about 0.08 of the Sun, too small for nuclear fusion. Twenty years earlier, Henry Norris Russell had proposed a mass of 0.05 solar masses as the dividing line between a star and a planet. It is therefore not surprising that the search for brown dwarfs and planets should be closely related, and the historical record demonstrates this was indeed the case. In 1988 two astronomers reported a low-mass star between 0.06 and 0.08 solar masses, which they suggested could be a brown dwarf. It turned out to be what is now known as an "L dwarf," cooler than the well-known M dwarfs but still able to undergo hydrogen fusion. Only in 1995 did astronomers report the first unambiguous discovery of a brown dwarf, photographically observed with the 60-inch reflector at the Palomar Observatory, and confirmed by the Hubble Space Telescope (Figure 2.1).

This discovery of brown dwarfs meant that astronomers were hot on the heels of planetary systems. Indeed, because of the low masses involved, the two searches sometimes became intertwined. In 1989 Harvard astronomer David Latham and his colleagues reported a possible brown dwarf using a very different technique – the change in the line-of-sight "radial velocity" of a star with an unseen companion. The size of the tug indicated a companion to the star that could have a mass as small as 0.001 of the Sun, some 11 Jupiter masses. "Thus the unseen companion of HD 114762 is a good candidate to be a brown dwarf or even a giant planet," they

Brown Dwarf Gliese 229B
Palomar Observatory
Discovery Image
October 27, 1994
Hubble Space Telescope
Wide Field Planetary Camera 2
November 17, 1995
PRC95-48 · ST ScI OPO · November 29, 1995
T. Nakajima and S. Kulkarni (CalTech), S. Durrance and D. Golimowski (JHU), NASA

Figure 2.1 First unambiguous detection of a brown dwarf star, observed (left) in 1994 with adaptive optics using the 60-inch Palomar reflector and confirmed (right) with the Hubble Space Telescope. Credit: left, T. Nakajima (Caltech), S. Durrance (Johns Hopkins); right, S. Kulharni (Caltech), D. Golimowski (Johns Hopkins), and NASA.

concluded, allowing that there was less than a 1 percent chance that this companion could be massive enough to burn hydrogen stably. The uncertainty was due to the unknown orbital inclination of the object with respect to its star as viewed from Earth. Because of this factor Latham and his colleagues cautioned that the object was most likely not an extrasolar planet, but a brown dwarf.[10] Even now, great uncertainty surrounds the nature of this object. It may be a brown dwarf or an L dwarf star, but some astronomers still believe it may be the first extrasolar giant planet discovered. The episode indicates how both theory and observation can cause uncertainties in discovery.

The honor of the first unambiguous detection of extrasolar planets around a Sun-like star is normally credited to a European team. Also applying the radial velocity technique, in 1995 the Swiss team of Michel Mayor and Didier Queloz unambiguously detected the first extrasolar planet around a Sun-like star, 51 Pegasi. In this case the orbital inclination factor was only half a Jupiter mass, 20 times less than Latham's object, and so could not substantially affect the planetary status claim. Their observation was confirmed, and soon supplemented, by the American team consisting of Geoff Marcy and Paul Butler.[11] They became the two primary planet-hunting teams over the following two decades prior to the *Kepler* spacecraft, though many others joined them as more discoveries were made. Over the next 20 years thousands of planets were found circling other stars, many of them in systems with more than one planet.[12] The radial velocity technique was supplemented by the transit technique, whereby starlight is observed to dim as it passes across the face of its parent star as viewed from Earth. This was the technique used by the *Kepler* spacecraft, which found most of those thousands of planets and indicated by extrapolation that planets surround most stars in the universe.

Again, exemplars could be multiplied in these cases, all of which fall in the category of the discovery of new classes of astronomical objects. You might well argue that the discovery of life beyond Earth is not a new class of object, but more akin to the discovery of new astronomical phenomena. But such discoveries show the same extended pattern, whether the expanding universe in the 1920s, the three-degree cosmic microwave background in the 1960s, or the accelerating universe in the 1990s.[13]

The General Structure of Discovery

The descriptions just given – and many more that could be elaborated – demonstrate not only the extended nature of discovery but also its general structure. Tantalizingly, the history of discovery indicates that there is very often a prediscovery phase, during which the true nature of an object, a signal, or a phenomenon goes unrecognized or unreported, or during which theory only indicates the phenomenon *should* exist. This is true not only for pre-telescopic astronomy and pre-microscopic biology but also after these technologies have reached advanced stages. We are perhaps not surprised that Galileo did not recognize his observations of the apparent protuberances of Saturn as rings, or that the intriguing nebulous objects in the sky were not known to be external galaxies until the twentieth century. But the same is often true today, as objects are not immediately recognized for what they truly are, or are theorized long before they are found (think black holes). Needless to say, the phenomenon of life beyond Earth is still in a prediscovery phase. Whether anything has been observed that turns out to be an indication of life, or life itself, remains to be seen. In my view, and that of history, it is entirely possible.

The discovery process itself consists of detection, interpretation, and understanding, as vividly seen in the examples discussed earlier. At each stage there are technical, conceptual, and social elements, which play out in different ways depending on the community and the environment in which the discoveries are made. The details of these discoveries are the bread and butter of historians of science, who have illuminated just how this plays out. Their work should have a large role in analyzing just how a discovery of life beyond Earth might occur.

Finally, if there is a prediscovery phase, it makes sense that there is also a postdiscovery phase, the period of most interest in this volume because it is the one that deals with reaction and impact. In this final phase, as shown in Chapter 1, the media will play a crucial role. Numerous disciplines will be involved. In our case a basic but essential insight about the postdiscovery phase is that the reaction to the discovery of alien life will be very much scenario-dependent. That is why I spend so much time on it in the next section.

Enough has been said to indicate that the discovery of extraterrestrial life, whether microbial in our solar system or intelligent in the far reaches of our Galaxy, would be an extended process with the general structure displayed in Figure 2.2. As in other astronomical discoveries, it will likely be preceded by a prediscovery phase in which the microbe, biosignature, or electromagnetic signal is not recognized as extraterrestrial life. The discovery will then go through phases of detection and interpretation before understanding is achieved. These phases could be long or short, depending on the evidence, and might result in success or failure. But in all cases the discovery will be an extended process, especially in the case of extraterrestrial life, where the postdiscovery societal implications are so great. And how society reacts to the discovery will be affected by its extended nature, as evidence is presented, interpreted, and understood. In short, the basic structure of discovery is essential to any understanding of its impact.[14]

The Anatomy of Discovery:
An Extended Process

Discovery

Detection **Interpretation** **Understanding**

Technological, Conceptual and Social Roles at Each Stage

Pre-Discovery

- Theory
- Casual or Accidental observations
- Classification of Phenomena (Harvard spectral types)

Post-Discovery

- Issues of credit & reward
- How do discoveries end?
- Classification of "The Thing Itself" (MK spectral types)

Figure 2.2 The extended structure of discovery, showing its three stages of detection, interpretation, and basic understanding, as well as the prediscovery and post-discovery phases. Discovery also has a microstructure consisting of conceptual, technical, and social roles. From Dick (2013c), where this image was used to illustrate the structure of discovery of new classes of astronomical objects, including their classification. In the long run, the discovery of life beyond Earth may also require classification.

Scenarios for Alien Life Discovery

In addition to understanding the structure of discovery, the most fundamental insight into the impact of discovering life beyond Earth is that it will most certainly depend on exactly how the discovery is made. Luckily, human imagination has been working overtime on this problem. Numerous scenarios for the discovery of life beyond Earth have been imagined in science fiction literature and film; six of the American Film Institute's top 10 science fiction films feature aliens.[15] The top three on the list are *2001: A Space Odyssey* (1968), *Star Wars* (1977), and *E.T. the Extra-Terrestrial* (1982), with the remainder including *The Day the Earth Stood Still* (1951), *Alien* (1979), and *Invasion of the Body Snatchers* (1956). All deal with encounters with intelligence rather than microbes, undoubtedly for dramatic effect, although Michael Crichton's *Andromeda Strain* microbial invasion (novel 1969, film 1971) of Earth was pretty dramatic also. All involve close encounters, either on Earth or in space, rather than indirect contact by radio transmission. Again, this was undoubtedly for dramatic effect, though Carl Sagan's *Contact* (novel 1985, film 1997) was no sleeper. And beyond the list of the American Film Institute there are plenty of aliens in science fiction literature and film illustrating all possible modes of contact.

How to make sense of this bewildering array of alien life encounter scenarios? Perhaps the simplest way of parsing possible scenarios, at least in terms of impact, is according to the degree of complexity of the life encountered, ranging from microbial to intelligent. The human mind seems naturally, perhaps irrationally, to think in terms of complexity. I say irrationally because microbes arguably have a huge effect on humans and the environment, despite their small size. This goes without saying in an era when both the beneficial and pathological effects of microbes, including the Ebola and Zika viruses, are well known. Without slighting microbes, which microbiologists naturally love, we can agree that they are simpler than humans, even while recognizing that microbes are a major part of the human ecosystem, not to mention the so-called microbiome of the human body.[16] Such microbes, which include bacteria, viruses, and fungi, are essential to human existence, but we are all aware that some microbes can also wreak havoc. We need hardly be reminded that the most brutal effects of the European culture contacts with the Americas were wrought not by warfare, but by the Columbian exchange, including the diseases brought on by "alien" microbes fatal to the natives.[17] And H. G. Wells's *War of the Worlds* ended not by Martians conquering the Earth, but by simple microbes destroying the invaders. All this notwithstanding, microbes are inarguably very different from intelligence, and despite any symbiotic relationships between the two, surely constitute a distinct difference in complexity.

A very close second to the complexity of life discovered is its proximity – how near or far the discovery is from Earth. Andromeda strain microbes delivered to Earth, or microfossils similar to those claimed in the famous Mars rock, will have a very different impact from microbes discovered on a planetary surface like Mars, in the ocean of a planetary satellite like Europa, or in water spouts emanating from such an ocean. The discovery of biosignatures in a planetary atmosphere beyond our solar system is very different again. The discovery of intelligence offers even more hair-raising possibilities, again depending on the mode of contact, whether direct or indirect. If the mode of contact is by signal, the nature of that signal is surely paramount when contemplating impact: a "dial tone" of unambiguous intelligent origin raises one set of questions, an undeciphered signal bearing information raises another, and a signal with a deciphered message is on another scale entirely. Each of these scenarios will certainly affect the nature of our encounter with alien life and our reaction to it.

Other means of parsing alien life discoveries are surely possible: the physical nature of the life, whether biological or postbiological; the moral nature of the life, ranging from inferior to superior (loaded terms to be sure); and an altruistic spectrum, ranging from competitive to collaborative. One might further characterize intelligent encounters according to energies harnessed, technology utilized, or communicative ability. But arguably in the most basic sense, the two variables of complexity and proximity provide a first-order approximation to discovery scenarios, which may then be parsed more finely as the situation demands.[18] Keeping in mind these variables of complexity and proximity, discovery scenarios can be characterized to a first approximation in Tables 2.1 and 2.2.

Table 2.1 *Microbial Life Discovery Scenarios*

	TERRESTRIAL	**EXTRATERRESTRIAL**
	(Earth or Vicinity of Earth-Moon System)	(Beyond the Earth-Moon System)
DIRECT (In the Form of Extant Life or Fossil Life)	**Encounter Type 1** Accidental Contamination by sample or astronaut return Panspermia or interplanetary matter transfer (ALH 84001)	**Encounter Type 2** Human Exploration of planets, comets, satellites
INDIRECT (In the Form of Extant Life or Fossil Life)	**Encounter Type 4** Shadow Alien Biosphere on Earth, Unknown Alien Microbes on or near Earth	**Encounter Type 3** Robotic Exploration of planets, comets, satellites of outer planets Biosignatures

Table 2.1 depicts direct and indirect encounters with alien microbes in terrestrial and extraterrestrial environments, where terrestrial is broadly defined as the Earth or the vicinity of the Earth-Moon system. Direct encounters of Types 1 (terrestrial) and 2 (extraterrestrial) are the most unlikely to occur in the near future, at least to the extent they are under our control. A Type 3 indirect encounter is the most likely scenario, or at least the most anticipated, since remote detection of microbes beyond Earth by robotic space exploration is one of the main goals of NASA's astrobiology program. A Type 4 encounter raises interesting possibilities not often discussed – that unknown alien microbes actually exist on Earth or in its vicinity, but are having an indirect or passive effect on the biosphere. In its static form Table 2.1 lays out scenarios, but in its dynamic form, indicated by the arrow proceeding clockwise from Encounter Types 1 to 4, it arguably depicts impacts from their strongest to their weakest, a dynamic to which I return in Chapter 6. Direct contact with microbes on Earth is likely to have a greater effect than direct contact beyond Earth or indirect contact by robotic exploration or biosignatures.

Direct Contact with Microbial Life: Contamination to Human Exploration

Each of these types, beginning with a direct terrestrial encounter, has its variations and raises its own problems and promise. Direct contact on Earth could occur, for example, if soil or rock samples are returned to Earth from another planet or satellite. The avoidance of direct contact between the terrestrial biosphere and alien microbes, the "back contamination" problem, was the main concern of the quarantine protocols of project *Apollo*, during which more than 800 pounds of rock samples were returned from the Moon. Because the potential also existed for contamination by returning astronauts, the first three *Apollo* lunar landing flights (11, 12, and 14) were subject to strict quarantine protocols in their Mobile Quarantine Facilities, much to the dismay and discomfort of the astronauts. The same precautions apply to any sample return missions. The Stardust mission, which collected samples from comet Wild 2 and returned to Earth in 2006, was subject to the highest "level 5" planetary protection protocols, because of possible contamination by alien life. Prevention of back contamination of the Earth by extraterrestrial microbes in general remains one of the primary goals of NASA's robust planetary protection program. But it is a striking fact that the planetary protection programs of other nations such as Russia and China are not so robust, not to mention the uncertain policies of private companies with the explicit goal of landing humans on Mars and presumably returning them to Earth.[19]

Encounter Type 1 also encompasses the "Andromeda Strain" scenario vividly displayed in Michael Crichton's novel and film by that name, exploring how an extraterrestrial microorganism could have devastating effects no less than the Columbian exchange of microbes did five centuries ago. In the novel the organism, code-named "Andromeda," arrives on Earth on a military satellite that had been collecting atmospheric samples of microorganisms for bioweapons research. The result is lethal blood clotting in humans, resulting in the death of almost everyone in the town near its landing place. The "discovery" of an alien microorganism is thus evident from its effect on humans, a scenario that planetary protection programs hope to prevent.

Finally, Encounter Type 1 encompasses the interplanetary transfer of material from space to Earth, whether in the form of extant life or fossil life. The unexpected claim in 1996 for nanofossils in the Mars rock ALH84001 is an example. As seen in Chapter 1, that rock landed on Earth some 13,000 years ago before being discovered in the Antarctic ice in 1984, only one of many that have been discovered. We cannot dismiss delivery of microbes by such rocks, a process known as *panspermia* and what scientists have called "exogenous delivery." The term is usually used in connection with the delivery of organic molecules to the early Earth, but might also apply to the delivery of life to Earth. The development of primitive life so soon after the late heavy bombardment of Earth 3.8 billion years ago gives rise to suspicion that life on Earth, or at least the organics leading to life on Earth, might have originally been delivered from space in this way. And while some current theories of the delivery of actual microbes from space are not widely accepted, the logical possibility remains.[20]

A Type 2 Encounter, direct exposure to microbes in an extraterrestrial environment, could occur in the event space exploration brings humans into contact with microbes in a planetary or satellite setting. This was a fear of the *Apollo* program, but since no humans have returned to or gone beyond the Moon since 1972 (much to the dismay of some of us), so far this remains the stuff of science fiction. Over the next century, however, this scenario may not remain fiction, as NASA and other space agencies plan missions to Mars, and eventually to comets and the satellites of the outer gas giants, some of which are believed to harbor vast oceans beneath their surfaces. David Brin and Gregory Benford dramatized this scenario in their 1986 novel *In the Heart of the Comet*, and it was featured as well in the 2013 film *Europa Report* and astronomer Dirk Shulze-Makuch's *Alien Encounter*.

Indirect Contact with Microbial Life: Robotic Exploration to Biosignatures

This brings us to the indirect categories of Encounters Type 3 and Type 4. The discovery of microbial life in a Type 3 Encounter has been the main goal of NASA's exobiology and astrobiology programs almost since NASA's founding in 1958.[21] Given the interest in life beyond Earth, it is not surprising that as soon as spaceflight was possible, plans were laid to search for life on Mars, the planet most similar to Earth. The *Viking* biology experiments, though now four decades in the past, still represent the most robust attempt at this mode of microbial discovery. And, though little known to the general public, some still claim that the *Viking* biology experiments discovered life. Notably, the principal investigator of the so-called labeled release experiment, Gilbert Levin, remains adamant that his experiment demonstrated life on Mars. For a long time this was Levin's private crusade, but after the *Phoenix* lander discovered perchlorates 30 years later, reputable astrobiologists have pointed out the ambiguities and uncertainties of the experiments. That controversy has been very drawn out and continues today, a vivid example of extended discovery, or extended failure to discover.[22]

The possibilities for indirect discovery of microbes is not limited to Mars. One of the most surprising results of the robotic exploration of the outer solar system has been that some of the satellites of Jupiter and Saturn are prime suspects for life, including Titan with its organics and Enceladus, Europa, and others with their oceans. Europa is a particular target for the search for life, being the first satellite discovered to harbor an ocean, and NASA is actively planning missions to that enigmatic target. The discovery of its ocean was itself extended, beginning in 1979 with *Voyager* observations, confirmed in the year 2000 with *Galileo* orbiter observations, and still the object of intense research. The existence of microbes or more complex life is purely conjectural, but has driven studies of possible science probes that could land on Europa and drill down below the ice looking for life. The recent discovery of water spouts emanating from the Jovian moon could make the process much easier and cheaper, since a spacecraft in orbit around Europa could run through a plume of water and analyze it for life.

Europa has been the subject of science fiction ever since the first hints of an ocean. Arthur C. Clarke's *2010: Odyssey Two* (1982) features aliens transforming Jupiter into a star to kick-start the evolution of microbial life believed to be in the Europan ocean. By the time of *2061: Odyssey Three* (1987), humans are banned from Europa, being warned by the aliens to "attempt no landing there," presumably so as not to interfere with the ongoing evolution of lifeforms. Much of the fiction related to Europa is fanciful, including Kim Stanley Robinson's *Galileo's Dream* (2009), in which Galileo is transported to twenty-ninth-century Europa where sentient organisms live in the subsurface ocean. For dramatic effect, the literature tends to prefer sentience to microbes, or at least to place microbes in a lesser role, as in the movie *The Europa Report*, which ends with the discovery of a more complex bioluminescent creature.

The discovery of microbial life in a planetary atmosphere in the form of a biosignature also falls in the category of a Type 3 Encounter. The atmosphere of Mars is less than 1/100th that on Earth, and in any case one does not expect microbes in such a thin atmosphere. But some of the numerous exoplanets being discovered at an increasing pace do have atmospheres, and can be probed by various biosignature techniques. Oxygen, methane, and ozone are the most obvious gases to search for from our perspective, but a recent study emphasized that thousands of molecules could be biosignatures given the likely diversity of planetary atmosphere masses and compositions. The problem is that they could have originated in ways other than biology, giving rise to false positives. As a result, any claim of life based on biosignatures is also likely to be the subject of extended controversy.[23]

This brings us to Encounter Type 4, indirect terrestrial contact with microbes. What could this mean? It could be a null set, an empty box. But it could also imply the existence of alien microbes on Earth or in its vicinity that we are unaware of, thus "indirect" but possible in the sense that some have raised the idea of a "shadow biosphere."[24] That term is usually used in connection with terrestrial microbes that have not been detected because they have a different and entirely unexpected biology compared to the rest of the biosphere. This seems far out, but remember that extremophile life was under our very noses until discovered a few decades ago. Could alien microbes already exist on Earth, either in a shadow biosphere or having been delivered long ago and incorporated into our "normal" biosphere? Like empty boxes on the Periodic Table in the nineteenth century (which admittedly had a stronger physical basis!), a microbial Type 4 Encounter raises tantalizing possibilities.

Direct Contact with Intelligent Life: UFOs to Human Exploration

Not surprisingly, the scenarios multiply for more complex life, but we can envision the same kinds of cases for direct and indirect discovery in terrestrial and extraterrestrial environments (Table 2.2), now also taking into account some twists peculiar to intelligence, such as the possibility of artifacts. Also not surprisingly, in contrast to microbes, each of these scenarios has triggered a large number of science fiction novels (Table 2.3), beginning substantially with H. G. Wells's *War of the Worlds* more than a century ago, accelerating in the second half of the twentieth century with the rise of the Space Age, and continuing unabated today. The literature in this "alien encounter" or "first contact" genre is very large, but I need examine only a few classic representatives to highlight my points about the four encounter types.

Encounter Type 1 with intelligence represents direct contact on Earth or in its vicinity, defined here as the Earth-Moon system. This encounter scenario, of course, has been the subject of much speculation, science fiction, and controversy. It encompasses the entire realm of astronomer J. Allen Hynek's 1972 classification

Table 2.2 *Intelligent Life Discovery Scenarios*

	TERRESTRIAL (Earth or Vicinity of Earth-Moon System)	**EXTRATERRESTRIAL** (Beyond the Earth-Moon System)
DIRECT	**Encounter Type 1**	**Encounter Type 2**
(In the Form of Extant Life, Fossil Life, or Active Artifact)	Alien Space Exploration	Human Space Exploration
	Resulting in: UFO Phenomenon Alien Landing Active Alien Artifacts	Resulting in Discovery of: Extant Aliens Active Alien Artifacts
INDIRECT (In the Form of Extant Life, Fossil Life, or Passive Artifact)	**Encounter Type 4** Extant Life	**Encounter Type 3** Radio/other Remote Contact Via SETI or METI
	Fossil Life Passive Artifacts	Passive Artifacts Dyson Spheres and other Macro-engineering Artifacts

Table 2.3 *Alien Life Discovery Scenarios in Science Fiction*

	TERRESTRIAL (Earth or Vicinity of Earth-Moon System)	**EXTRATERRESTRIAL** (Beyond the Earth-Moon System)
DIRECT (In the Form of Extant Life, Fossil Life, or Artifact)	**Encounter Type 1** Wells, *War of the Worlds* *The Day the Earth Stood Still* Clarke, *Childhood's End* Crichton, *The Andromeda Strain* Spielberg, *E.T. the Extra-Terrestrial* Spielberg, *Close Encounters of the Third Kind* *Clarke, *2001: A Space Odyssey* Brin, *Existence* *Liu, *Three Body Problem* trilogy Chiang, "Story of Your Life" (Filmed as *Arrival*)	**Encounter Type 2** *Clarke, *Rendezvous with Rama* Bradbury, *Martian Chronicles* Lem, *Solaris; His Master's Voice* Scott, *Alien* (and sequels) McDevitt, *Engines of God*
INDIRECT (In the Form of Extant Life, Fossil Life, or Artifact)	**Encounter Type 4** *Clarke, *2001: A Space Odyssey* McCollum, *Lifeprobe* Hoyle, *The Black Cloud*	**Encounter Type 3** Hoyle, *A for Andromeda* Clarke, *2010: Odyssey Two* Clarke, *2061 Odyssey Three* Gunn, *The Listeners* Brown, *The Cassiopeia Affair* *Sagan, *Contact* *The Europa Report* *Liu, *Three Body Problem* trilogy

*More than one mode of contact takes place

of close encounters with aliens, made famous by Steven Spielberg's film treatment, *Close Encounters of the Third Kind.* It includes as well the entire UFO debate and an addition to Hynek's classification, alien abductions, which are often referred to as "Close Encounters of the Fourth Kind." While speculative, direct terrestrial contact is based on the analogy that advanced aliens would undertake not only local space exploration as do humans, but also have the ability to undertake interstellar travel. Interstellar travel is often seen as unlikely from the point of view of the SETI community, in terms of both technology and economics. But it remains a logical possibility. The problem of all past claims, whether involving UFOs or alien abductions, is evidence. In my view UFO claims have been greatly undermined in an era when a large percentage of the population has cell phones with cameras, yet all we have for putative UFOs are blurry photos. The future is another matter, but solid evidence must remain the bedrock principle. Encounter Type 1 also includes an actual alien spacecraft landing, as opposed to UFO phenomena in the skies. The fictional prototype for this kind of encounter is H. G. Wells's *War of the Worlds* (1898), and it has been immortalized in such classics as the 1951 movie *The Day the Earth Stood Still*, in which the alien Klaatu comes to planet Earth to warn humanity about the dangers of nuclear war, and Arthur C. Clarke's *Childhood's End*. Most recently Ted Chiang's short story "Story of Your Life" was transformed into the cerebral and critically acclaimed film *Arrival* (2016).

With intelligence we also have the possibility of a new kind of Type 1 contact not present with microbe scenarios: the discovery of what I call active alien artifacts, rather than extant life, in the Earth-Moon system. By "active" I mean those that are transmitting information or have a direct and active effect on Earth, potentially making their impact almost equivalent to that of extant life. Examples in science fiction are the iconic black monolith in Arthur C. Clarke's *2001: A Space Odyssey*, or the crystal alien artifacts in David Brin's *Existence*. Clarke's monolith is implied to have affected the course of human history, while Brin's artifact is filled with the uploaded personalities of nearly 100 aliens and carries an enigmatic message for Earthlings: "Join Us." Clarke's artifact is planted millions of years ago even before the arrival of modern *homo sapiens*, while Brin's crystal artifacts are discovered serendipitously in the course of searching for space garbage in Earth orbit – just the first of millions eventually discovered. The possibility of active artifacts in science fiction reflects a rather new trend in the scientific literature – a new field of "interstellar archaeology," the Search for Extraterrestrial Artifacts (SETA), implying that we should search for them no less than radio or optical signals. The existence of alien artifacts is certainly a logical possibility no more and no less reputable than SETI itself. Reputable scientists such as Paul Davies have suggested searches for them.[25]

Encounter Type 2 encompasses contact with aliens or alien artifacts in the course of human space exploration beyond the Earth-Moon system. Again, this is the stuff of science fiction, although the recent discovery of a mysterious elongated interstellar object entering the solar system, dubbed Oumuamua, has greatly raised the profile of this scenario. Arthur C. Clarke's *Rendezvous with Rama* depicts humans trying to unravel the mysteries of an alien spaceship that has entered the solar system. Astronomers discover the 30-mile-long cylindrical spaceship serendipitously while searching for near-Earth asteroids. At the time of discovery it is beyond the orbit of Jupiter, but its speed and trajectory betray its interstellar origins. The Ramans prove harmless. By contrast, Ridley Scott's treatment in his *Alien* movie series (1979–1997) depicts the other end of the spectrum for a Type 2 Encounter, with evil aliens perfectly designed for action movies. The commercial spacecraft *Nostromo* is returning to Earth from an unspecified location in the solar systems when it detects a signal from a nearby asteroid. Upon investigation the crew finds an abandoned alien spacecraft, a dead alien, and a room full of eggs, one of which hatches – and the chase is on. Stanislaw Lem's novels *Solaris, His Master's Voice*, and *Fiasco* all present Type 2 discoveries of intelligence beyond Earth's vicinity, and are well known for conveying the theme of unknowability of the alien.

Given the long timescales over which life could have evolved in the universe – billions of years by some accounts – direct extraterrestrial Encounters Type 2 also need not involve extant aliens. Arguably, the more likely scenario in such an old universe where civilizations have limited lifetimes is the discovery of alien artifacts. The abandoned alien spaceship as in *Rendezvous with Rama* would be a very large artifact indeed. Although this might seem a kind of indirect-direct encounter since the aliens might be long gone, the possibility of an active artifact – say in the form of a beacon transmitting information – would place this scenario more firmly in the direct category. More passive artifacts would have indirect impacts, and so fall in our next category of encounters.

Indirect Contact with Intelligent Life: SETI, METI, and Artifacts

Encounters of Type 3 encompass indirect contact beyond the Earth-Moon system due to alien signaling, the discovery of macro-engineering alien artifacts, or the discovery of active or passive artifacts. Signaling is the tradition begun in the modern era by the advent of radio telescopes, beginning with the first Search for Extraterrestrial Intelligence (SETI) search in 1961. Undertaken by young astronomer Frank Drake at the nascent National Radio Astronomy Observatory in Green Bank, West Virginia, it consisted of a 200-hour search targeting only two nearby Sun-like stars, Tau Ceti and Epsilon Eridani, around the 21 cm line of neutral hydrogen. Although it ended in failure, it captured the imaginations of scientists and public alike. The Ozma search (though independently conceived by Drake) followed the landmark publication by Giuseppe Cocconi and Philip Morrison arguing on theoretical grounds that such a search should be undertaken. Numerous but sporadic radio searches have been undertaken over the half century since Drake's pioneering observations, still covering only a small portion of the possible search space. The search has since been expanded to other regions of the spectrum, particularly optical SETI using lasers, and efforts are now under way to revitalize SETI by thinking out of the box and formulating a new Roadmap similar to the Astrobiology Roadmap, which focuses on microbial life. The $100 million infusion of funds from Russian entrepreneur Yuri Milner is also revitalizing SETI observational programs – all the more reason to begin considering societal impacts in the event of success.[26]

Encounters of Type 3 would also encompass Messaging Extraterrestrial Intelligence (METI), a relatively new endeavor that might have a quite different impact from a SETI discovery. In this effort Earthlings send deliberate messages to specific targets, as opposed to the less detectable leakage radiation already emanating from Earth over the past century of our Radio Age. This activity is extremely controversial in some quarters, since it raises all kinds of questions about who speaks for Earth, and whether revealing our position in space might trigger harmful effects when the aliens reply – or arrive. The latter scenario is played out in Chinese science fiction writer Cixin Liu's *The Three Body Problem* trilogy, in which a military project sends signals to the stars, resulting in alien plans to invade Earth. As a Type 3 Encounter gradually transforms into a Type 1 Encounter, a "Trisolarian fleet" with 1,000 ships approaches Earth from the triple-star Alpha Centauri system four light-years away. They are motivated by the unstable situation of their planet with a chaotic orbit around three stars. Liu's universe, similar to that of David Brin in *Existence*, is premised on the pessimistic survival of the fittest theory that "the universe is a dark forest" where every civilization will eliminate any others it comes into contact with. The difference between METI and SETI is that a message has already been sent, so the aliens know something about us depending on the message, or at a minimum know our place in space. Such messaging has a short history, beginning with Frank Drake's missive, sent from the Arecibo Observatory in 1974 to M13, the Hercules globular cluster of stars some 25,000 light-years distant. It was a relatively strong signal and caused little controversy at the time; 25 years later, Cornell University even put out a press release praising the effort. Other attempts, notably by the Russians, have been sporadic, but there is now an institution known as METI International, with the express purpose of sending messages to the stars. As a member of its board of trustees, I am fully aware of the controversy this has caused. Suffice it to say there is no evidence whether aliens are good or evil, a subject I take up in Chapter 8 in the context of astroethics and extraterrestrial altruism. Policy issues for SETI and METI are discussed in Chapter 9.[27]

As in direct encounter modes, we need to take into account the possibility of an alien artifact discovered remotely. For Encounter Type 3, this would be the discovery of an alien artifact detectable from Earth in the form of macro-engineering projects. Dyson spheres – large structures hypothesized to have been built around a star by very advanced aliens in order to capture the star's total energy output – are an example of such a macro-engineering indirect encounter. In 1960, physicist Freeman Dyson proposed they might be detected through their infrared radiation. Like more standard SETI projects, observations of infrared excess have been attempted, but none has been unambiguously successful, despite tantalizing observations such as the strange dips in brightness of "Tabby's star" that some have interpreted as caused by an artificial structure.[28] Discovery of smaller passive artifacts in the course of human space exploration might also be considered "indirect," since we do not come face to face with extant alien life or its active artifacts. This scenario has also been represented in science fiction, notably in Jack McDevitt's *Engines of God* (1994), which takes place in a universe where most life is dead, leaving behind only artifacts.

Encounter Type 3 remains the most robust area of research in terms of the search for intelligence, largely because the technology exists for the search, both in the form of the detection of electromagnetic signals and in the form of infrared excess radiation that would be produced by structures such as Dyson spheres. It is thus not surprising that encounters with intelligence of the Third Type have generated the most science fiction because they were stimulated by real life events such as Ozma. In 1962 British cosmologist Fred Hoyle and John Elliott wrote *A for Andromeda*, where alien life is discovered serendipitously while sweeping through the Andromeda Galaxy with a radio telescope looking for other astronomical phenomena. Geochemist Harrison Brown penned another variation in *The Cassiopeia Affair* (1968), and the radio contact genre reached a classic form in 1972 with James Gunn's *The Listeners*, although with a twist. In Gunn's *Listeners* the SETI project utilizing the largest steerable telescope on Earth – the "Little Ear" – has not been successful for 50 years when the true discovery is revealed in data from the "Big Ear," the five-mile-diameter network orbiting 1,000 miles above the Earth's surface. But the Big Ear was not in itself a SETI project in the sense that rather than observing, it only received data tapes related to other astronomical observations so scientists could search for any signals of intelligent origin, a kind of piggyback operation. And in doing so, the first signals detected were Earth radio transmissions being beamed back from the direction of the star Capella. So the discovery was again in a sense serendipitous. In both cases, part of the drama is the shock when the signal comes in, followed by doubts, elation, confirmation, hard decisions about who to tell, and the political and social consequences.

The indirect extraterrestrial encounter form reached its greatest popularity with Carl Sagan's novel *Contact* (1985) and its film version in 1997. I became all too familiar with its plot when I served as an expert witness in the intellectual property rights battle of Francis Ford Coppola versus the Sagan estate and Warner

Brothers, filed six days after Sagan's death in December 1996.[29] Again, the discovery is made as part of a deliberate search, although a lonely one constantly struggling for adequate funding. Sagan's plot has some similarities to Gunn's treatment, but features multiple modes of contact. It too depicts astronomers discovering a signal, this time sent from the Vega star system. On the surface it consists of prime numbers, but this is only the beginning. Deeper in the signal scientists find a retransmission of Adolf Hitler's speech at the 1936 Summer Olympics in Berlin, indicating the extraterrestrials want the Earthlings to know they have received this undirected transmission from Earth. Finally, another message is decoded with instructions for building a machine. The machine turns out to be a spacecraft that transports five humans through a series of wormholes to the center of the Milky Way Galaxy. There (in a Type 2 Encounter) the astronauts meet extraterrestrials who assume the shapes of their loved ones. The genre continues to be developed in creative ways, and I say more about it in discussing impact scenarios in Chapter 6.

Encounter Type 4 involves indirect contact in the vicinity of Earth, in the form of extant aliens, fossilized aliens, or passive artifacts. Again, science fiction writers have given us several scenarios. An early treatment was astronomer Fred Hoyle's *Black Cloud* (1957), in which intelligence takes the form of interstellar matter, some half billion years old. The discovery is first made during a sky survey search for exploding stars known as supernovae. Appropriately for that era, the search is undertaken with photographic plates taken with the 48-inch Schmidt of Mt. Palomar Observatory in California. Using a "blink microscope" to compare plates taken on different nights – the same method Clyde Tombaugh used to discover Pluto in 1930 – the astronomer notices a circular dark patch with some of the surrounding stellar images blinking, indicating a change has taken place during the period between the plates. After consulting colleagues and holding a dramatic but purposely private meeting, the most likely explanation offered is that some kind of dark cloud is approaching Earth, blotting out more stars as it approaches. Based on the size changes of the cloud over a period of a month, astronomers calculate it will arrive in the vicinity of Earth in less than 20 months. Meanwhile, astronomers in England have noticed discrepancies between observations and the calculated positions of Jupiter and Saturn, and conclude that a cloud, with a mass two-thirds that of Jupiter and presently at a distance of 21.3 astronomical units, is approaching Earth at a speed of 20 kilometers per second. The cloud approaches Earth and threatens chaos and destruction, but is convinced to retreat. This, then, is discovery as serendipitous anomaly rather than as deliberate search.

The idea of an indirect Earthly encounter would seem to have been brought to classic form with Arthur C. Clarke's *2001: A Space Odyssey* series, which begins with the discovery of an alien artifact in the form of a black monolith, first on the Earth and then on the Moon. But as I argued earlier, this encounter might more appropriately fall in the Type 1 direct category, since the monolith was an active artifact that had a direct effect on human history and set off a whole chain of events including the famous voyage to Jupiter and adventures with HAL the computer. Similarly Michael McCollum's *Lifeprobe* features a space probe sent by the "Maker" civilization that arrives in the vicinity of Earth after a 100,000-year journey. If it were just a passive probe, even one reporting back to its Makers, it would fall squarely in the Type 4 indirect terrestrial contact mode. But the Lifeprobe is actually very active, asking humans to repair the probe in return for all of the Makers' advanced science. The impact is almost as if the Makers themselves had arrived.

Again, considering Tables 2.1 and 2.2 in a dynamic way, the potential impact decreases in a clockwise direction from Type 1 direct contacts on Earth to Type 4 indirect contacts. Obviously, a terrestrial encounter of Type 1, whether in the form envisioned by H. G. Wells's *War of the Worlds* or by Arthur C. Clarke's *Childhood's End*, would have a strong impact affecting the entire population of the Earth. Type 2 would affect far fewer people, but the astronauts it did affect would feel a strong impact, and when relayed back to Earth would morph into an encounter of Type 3 – indirect for Earth's population. The impact of Type 3 Encounters by radio or other remote electromagnetic means would depend on various sub-scenarios. A so-called dial tone that appears to be intelligent in origin based on defined characteristics of the signal, would be very different from the impact if a message were obviously present. If the message were deciphered, the impact could be different still, depending on the content of the message. The discovery of a Dyson sphere would be more like the discovery of a dial tone in its impact, since no communication is possible. Finally, the impact of a passive artifact found on Earth, or a spacecraft in the vicinity of Earth, might have the least impact. An artifact with a dead message and no communication possible would have some impact, but would not require immediate attention. An alien spaceship or creature in the vicinity of Earth would also have impact, depending again on the scenario. Hoyle's *Black Cloud* had a devastating impact before the creature passed on.

As philosopher Clément Vidal has pointed out, there are many ways to parse the discovery of life beyond Earth other than the direct/indirect and terrestrial/extraterrestrial matrix presented here. Vidal proposes a multidimensional impact model with 26 parameters, of which 10 relate to the discovery process, with our essential criteria of proximity and complexity (microbial or intelligent) comprising only two. For example, the *intent* of the extraterrestrials could be neutral, benevolent, malevolent, or mixed. Their *size* could range from microscopic to so-called

Kardashev Type III civilizations that utilize energies on a galactic scale. They could be *dead* or *alive*, or revealed in the form of the artifacts I have discussed. All of these factors and more would surely affect the impact of discovering extraterrestrial life at some level. But as I have argued, distance and complexity as parsed earlier are two of the most important, the top of a hierarchy of traits that might make us despair in laying out scenarios of contact. But people ranging from hard-nosed economists to leaders in the business world do this all the time, separating the most important factors of their particular endeavors from secondary effects. Vidal himself does not despair; the remaining 16 of his 26 dimensions relate to impact rather than discovery, and when it comes to impact, Vidal believes we can intelligently discuss reactions ranging from universalizing knowledge to the impact on religion and world philosophy.[30]

We have our work cut out for us, even if we limit ourselves to the ranges of complexity and proximity. For example, one possibility not often considered among the intelligent life scenarios is whether it will be biological – as in our most frequent and most likely erroneous projections of ourselves onto the cosmos – or postbiological. By postbiological, I mean intelligence that has evolved beyond flesh and blood to artificial intelligence. Far from science fiction, this possibility is in my view actually more likely than not. It is a product of considering cultural evolution as seriously as we do astronomical and biological evolution. I discuss this possibility more in Chapter 4.

The Structure of Alien Life Discovery

Our discussion of the anatomy of scientific discovery, the variety of discovery scenarios, and possible impacts viewed through the imaginations of science fiction writers puts us in a much better position to examine the general structure of alien life discovery under a variety of conditions. We can look at this from at least two levels, what I call the "level 1" general structure described in the first section, which I consider a pretty good bet to occur based on past discoveries in astronomy, and the more specific level 2 structure, for which we can provide guidelines, but not predictions.

Level 1 Structure: Prediscovery, Discovery, Postdiscovery

First, let's look at the most general "level 1" structure (Table 2.4). For any scenario, there will likely be a prediscovery phase, during which theory or evidence leads scientists to believe there could be life beyond Earth, or during which the true nature of a phenomenon already found is unrecognized. My detailed study of 82 classes of astronomical objects discovered over the past 400 years

Table 2.4 *The Anatomy of Alien Life Discovery*

Discovery Phase	Activities	Comments
Prediscovery	Ideas based on assumptions and theory (Copernican, Darwinian, etc.) Casual or accidental observations not yet properly interpreted Analysis of historical arguments	Almost always occurs in scientific discovery
Discovery	Detection of microbe, signal, biosignature, or artifact Interpretation over a long period Understanding over an even longer period	Technological, conceptual, and social roles at each stage
Postdiscovery	Issues of credit and reward Issues of short-term societal impact Issues of long-term societal impact Classification	Role of media crucial Numerous disciplines involved Assimilation of implications and new worldview

of telescopic astronomy illustrates that prediscovery phases occurred in almost every case.[31] In the same way that the first quasar declared in 1963 had already been detected optically as a 13th magnitude star in 1887, it is entirely possible that a signal or other phenomenon has already been detected that has not yet been interpreted in the proper way as alien in origin. It may have been misinterpreted, or may still lay unsuspected in some database awaiting discovery of its true nature.

Also in the prediscovery category is the web of assumptions and circumstantial evidence indicating there *should* be alien life. As philosopher Iris Fry has argued, the possibilities of life beyond Earth are underpinned by the Copernican presupposition that the Earth is not special, and by the Darwinian presupposition that life on Earth evolved by natural processes and that it might do so wherever biogenic conditions prevail throughout the universe.[32] While these presuppositions are not proven, astrobiologists continually test them, and, she argues, there are grounds for optimism. Others do not share these assumptions, and indeed have their own, as in the Intelligent Design movement, which sometimes drives opposition to the search for life. If a person believes a deity uniquely created life on Earth, the assumption of many terrestrial religions, that person might not be optimistic about life beyond Earth because it would undermine the uniqueness of the relationship between God and humanity. This is indeed the argument seen in books such as *The Privileged Planet*, but we see it is not necessarily true in the context of astrotheology discussed in Chapter 7.[33] It remains a religious view underpinned by a different set of presuppositions that are difficult to test.

Broadly speaking, the entire history of the extraterrestrial life debate is part of the prediscovery stage, including the historiographic analysis of its arguments and their success or failure. We can assess, for example, how successful arguments from analogy have been, since they have played an important part in the history of the debate extending even to today. We can analyze the general arguments that have been made in the debate, including the uniformity of nature, the Principle of Plenitude, the Principle of Mediocrity, the Goldilocks argument for rare Earth, large number arguments, and the Fermi paradox. We can see how popular culture has reacted to the possibility of alien life in the form of the UFO debate and science fiction. From this point of view, history becomes relevant to societal impact in a much broader way than the specific cases of putative life discoveries described in Chapter 1. Of course, the strengths and weaknesses of some of these arguments, whether in science or popular culture, will only become fully apparent once a discovery is made – or not.[34]

After prediscovery comes the discovery phase, which in astronomy and probably all of science most often consists of detection, interpretation, and understanding – as I have demonstrated. An anomalous phenomenon – whether in the form of a microbe, electromagnetic signal, biosignature, spacecraft, or artifact – will be detected. Interpretation of the anomalous phenomenon will likely take years, during which ideas are discussed in the scientific community, reported and hyped in the mainstream and social media, and in the end accepted or rejected. If accepted, understanding the true nature of the discovery may take even longer. Comets, for example, were known in antiquity, but only in the sixteenth century were discovered to be astronomical in nature (prior to that they were believed to be atmospheric). And only in the 1950s was the "dirty snowball" model put forth, subsequently proven by ground observations and up-close spacecraft, the beginning of true understanding of comets. What is true of astronomical objects and phenomena will also be true of the discovery of life beyond Earth. The historical cases of putative extraterrestrial life discovery discussed in the previous chapter bear out this extended nature of discovery, ranging from a period of weeks for pulsars – from the detection of the pulsing signal to its interpretation as a natural rather than an artificial phenomenon, to an understanding in terms of the theory of neutron stars – to years for the supposed fossils from Mars, and even decades if one takes seriously the reinterpretation of the *Viking* biology results.

Finally, while short-term impact will have been occurring during the detection and interpretation stages of discovery, an extended postdiscovery phase will occur. It is here where human reaction and longer-term impact become dominant. Chapters 6 through 9 detail this long-term impact in the postdiscovery phase. This phase might be considered to terminate when the implications – whether in the form of theology, worldview, or more practical effects – are fully assimilated into culture and the new worldview replaces the old. While it might seem this would be an open-ended process, there is precedent for this structure as expressed, for example, in Thomas Kuhn's *Structure of Scientific Revolutions* and its aftermath. Here the old worldview – whether in science, politics, or some other area of human endeavor – ends over generations as a new one takes its place. I discuss this in more detail in Chapter 6.

Level 2 Structure: History, Imagination, Analogy

Beyond this general level 1 structure of alien life discovery, at least three sources exist for fleshing in what we may call the "level 2" details of societal impact in ways that are suggestive, but not predictive. The first is the history of past reactions to putative alien life discovery, which suggests three dominant influences: media, government, and popular culture. As argued in Chapter 1, the media will play an important and confusing role. It played a precipitating role in both the 1835 Great Moon Hoax/Satire and in the 1938 "War of the Worlds" broadcast. It played only a minor role in the pulsar "Little Green Men" hypothesis, but only because scientists held events closely within their circle over the short period of time it took to resolve the issue – an unlikely scenario in our current age of information and social media. In the case of the real possibility of even fossil life in the Mars rock, the media played a decisive role in the way the public received and reacted to the news. Studies suggest that in the current age of mainstream and social media, the media will play a large role in determining the short-term impact, and that this dynamic might be illuminated not only by history but also by theories of mass communication.[35]

The Mars rock episode indicates that governments will also play a decisive role. It is true that in this case the discovery happened to be centered around a government institution in the form of NASA, but it is likely that governments would get involved under any scenario. The desire to have a strong case before going public, or even before informing institutional superiors, was combined with fears inside the NASA team of a scoop by other groups working on the same samples. The pressure to make an announcement was strong, but external political considerations also had to be taken into account. This involved discussions with the White House, and in the end the announcement was made at a rushed press conference at NASA Headquarters. Even as the press conference was under way the president spoke from the White House lawn with his science advisor by his side. Within a few weeks, congressional hearings were held. Within a few months, the White House had arranged the Space Science Symposium on the societal impact of the finding, with funding implications for NASA. And within a few years, the NASA Astrobiology Institute was founded, making permanent the impact of the finding with robust new research directions.

Finally, the Mars rock episode also clearly demonstrates the public firestorm that will be generated as scientists attempt to clarify the evidence. The NASA scientists were subject to withering attack. They presented four independent lines of evidence to indicate possible biogenic origin for the purported nanofossils found in the rock. Taken singly, each one did not clinch the case. But taken together, the scientists argued, it was a strong argument for past life on Mars. Others pointed out that four weak arguments do not add up to one strong one. It all added up to an interesting argument about evidence and inference, a case study in the philosophy of science. "Extraordinary claims require extraordinary evidence," the critics intoned. Even as this internal fight was going on within the scientific community, scientists were under extreme public scrutiny. Scientists argued about the science, while scholars, the media, and the public argued about the implications both on an individual level and on broader cultural levels such as theology. The resulting range of responses is likely to be repeated in any discovery of alien life.

The second source of level 2 insight on societal impact is human imagination, especially when still tethered to history and science. As suggested in this chapter, the best science fiction explores in vivid terms the process of discovery and possible impacts for each discovery scenario over the short and long terms. We cannot say whether any particular scenario will play out in the way it is imagined in fictional accounts, but most likely these accounts contain many of the elements that will play out in one combination or another. Again, I have more to say about this in Chapter 6.

A third source of level 2 insight on societal impact is related to the historical episodes of Chapter 1, which turn out to be just a subset of the broader argument known as analogy. In comparing past episodes of putative alien life discovery with what might happen in the future, we are engaging in a form of analogical argument, in which what happened in the past may illuminate what may happen in the future. Some may well question the validity of this argument. But it is a type of argument so important that it deserves a treatment of its own in the next chapter.

Whether our first encounters with alien life are direct or indirect, in the vicinity of Earth or in distant space, the general structure of discovery laid out in this chapter is very likely to occur. Although history offers some guidelines, envisioning the second-level details of impact falls more in the category of laying out possible scenarios that may or may not occur when a discovery is made – although some of them such as the media firestorm are a pretty sure bet. Envisioning such possibilities is stimulating, but also dissatisfying in the sense that we cannot at present choose which will occur in reality. But history shows it is better to be prepared to deal with a variety of well-thought-out scenarios than to have to deal – perhaps rather quickly – with events that have not been foreseen. Whether in the context of pandemics such as Ebola, defense strategies prepared well before battle, or near-Earth asteroid detection, early actions are often the most important. If we are not careful, the road taken in haste might turn out to be the road that should not have been taken at all.

3
Analogy

One should not think of analogy-making as a special variety of *reasoning* (as in the dull and uninspiring phrase "analogical reasoning and problem-solving," a long-standing cliché in the cognitive-science world), for that is to do analogy a terrible disservice ... To me analogy is anything but a bitty blip – rather, it's the very blue that fills the whole sky of cognition – analogy is *everything*, or very nearly so, in my view.

Douglas Hofstadter[1]

The problem with analogies is that they are highly persuasive, inherently limited, and easily overextended.

Kathryn Denning[2]

In such a difficult and multifarious study as history, it is all too easy to find evidence to "prove" almost any proposition.

William H. McNeill[3]

The *conquistadores* came, 168 strong, and swept across the Peruvian highlands, dealing death and destruction along their path. They were, one historian wrote, "the sixteenth century equivalent of today's astronauts – emissaries from a distant and alien civilization." In 1532 they entered the Inca empire, stretching some 2,500 miles along the Pacific coast down into Chile, home to 10 million souls controlled by 100,000 ethnic Incans. In a striking stroke of luck, the invaders arrived at a moment when the empire was in crisis, a civil war in which two brothers had fought to the bitter end for power, and one, the Sapa Inca Atahualpa, had just emerged victorious. But the empire was weakened, and through a series of remarkable circumstances, the Spaniards captured and eventually killed Atahualpa. Despite some resistance, his followers largely laid down their arms, and within a few years the entire Inca empire, the largest of the four pre-Columbian civilizations in the Americas and one of the largest in the world, was no more. By 1572, all resistance had ceased when the last Inca emperor died, disgraced in a corner of his own former grand empire.[4]

When I arrived some 450 years later in Cuzco, the former capital of this once glittering empire situated at 11,000 feet above sea level, plenty of traces of the former civilization were still to be seen: the remains of the palace of the Incas, the Temple of the Sun, streets lined with buildings undergirded by exquisitely sculpted and chiseled Inca stone, all now appropriated by the new culture for its own purposes. Descendants of the Incas were still in evidence, most visibly in the form of the colorful textile weavers who demonstrate their skills and sell their wares. But there was no doubt that the culture of the invaders had completely appropriated the old culture, and that the natives had been assimilated into the new. Not far away, past Ollantaytambo and the roiling Urubamba River in the Sacred Valley of the Incas, the mountain citadel of Machu Picchu stands, one of many ruins across the former empire that are mute testimony to the past. A once great civilization had been replaced by the descendants of a distant and alien one.

The analogies of the rapid destruction of the Incas – as well as the Aztecs, and many other cultures throughout history – are ubiquitous when scholars contemplate culture contact with extraterrestrials. Everyone from Stephen Hawking to the proverbial man in the street has used them to urge caution in our search for extraterrestrials, and certainly in any attempt to send messages to our celestial counterparts. Films from the *War of the Worlds* to *Independence Day* reinforce the images of power and destruction that would supposedly overwhelm us as they did the Aztecs and Incas. But just how valid are such analogies? And wouldn't other analogies prove the opposite? Even as cognitive scientist Douglas Hofstadter emphasizes the ubiquity of analogies in human reasoning, and anthropologist Kathryn Denning cautions they are highly persuasive, inherently limited, and easily overextended, historian William McNeill bluntly observes that history can be used to prove almost anything. Since I am arguing that analogy is an important approach in determining the impact of discovering life, it is essential that we probe just when analogy is useful – and when it is not.

Our first two chapters have already illustrated the use of analogy, employing past reactions to claimed discoveries of extraterrestrial life to illuminate possible future reactions, and investigating the nature of discovery and its scenarios. We now take a hard look at analogy itself – its strengths, weaknesses, and applicability to the problem of the discovery of life beyond Earth. Although analogy must be used with caution, cognitive science and philosophy of science demonstrate it is not the weak form of argument some seem to think. Once past that misconception, history offers the opportunity to study analogs in a variety of forms. For present purposes, I concentrate on four analogies, corresponding to different discovery scenarios and timescales: the microbe analogy, the culture contact analogy, the transmission/translation analogy, and the worldview analogy. I find that despite the obvious need for caution and the lack of predictive value, analogies can indeed serve as solid guidelines to cosmic encounters with alien life.

The Promise and Problems of Analogy

For those who consider analogy a fuzzy form of reasoning, it is comforting to know that it has been the subject of considerable scholarly research. This does not mean analogies are always correctly used, but it does mean we understand their use more than one might think. Analogy has been defined as "a structural or functional similarity between two domains of knowledge," more precisely, "a mapping of knowledge from one domain (the base) into another (the target) such that a system of relations that holds among the base objects also holds among the target objects." At its best, analogy is a cognitive mechanism that allows us to learn and solve problems by drawing a map from the known to the unknown.[5] In a broader sense, analogical reasoning is helpful in discovering, developing, explaining, or evaluating scientific theories. As philosopher Clément Vidal summarizes its potential, "such reasoning enables us to propose new hypotheses, and thus discover new phenomena. These new hypotheses trigger us to develop new experiments and theories." And, I hasten to add, new ideas. All researchers on analogy caution that analogy is not a proof, but a heuristic, a method that can serve and has served as an aid to learning and discovery. They also caution that the relation between any two systems needs to be examined in detail before an analogy or dis-analogy can be tested. Scholars have also offered principles of analogical reasoning meant to avoid its many pitfalls; these principles include the structural consistency of what is being compared, the depth of any relation, avoidance of mixed analogies, and the realization that analogy is not causation.[6]

Almost 50 years ago, philosopher of science Mary Hesse penned her classic *Models and Analogies in Science*. Here she argued that models and analogies are integral to scientific practice and advancement, and she distinguished positive analogies, negative analogies, and neutral analogies. She pointed to exemplars of analogy in science ranging from German organic chemist Friedrich Kekulé's dream of a snake with its tail in its mouth that (by Kekulé's own account) helped him arrive at the structure of benzene, to wave models for sound and light, and the use of billiard balls in random motion as a model for the behavior of gases. She emphasized that, while no account of the snake appears in textbooks of organic chemistry, models and analogies have been essential to the logic of scientific theories. Her book is largely a debate about just how essential analogies and models are to science, and she comes down strongly on the side of their overwhelming importance.[7]

Since Hesse's work, scholars have enumerated many more examples, ranging from the physical domain such as Kekulé's dream, to the biological (chromosome/beaded string) and the cognitive (mind/computer) domains. Sometimes the analogies are grand, as Darwin's analogy between artificial and natural selection, which arguably propelled one of biology's most important insights. In other cases, they are much less grandiose. Historians and philosophers have documented many such cases of both.[8] Most recently philosopher Paul Bartha analyzes analogy as a hot topic in artificial intelligence research, psychology, and cognitive science. He lays out a widely accepted process of analogical thinking as follows:[9]

(a) Retrieval or access of a relevant "source" analog;
(b) Mapping that sets up systematic correspondences between the elements of the source and "target" analogs;
(c) Analogical inference or transfer of information from source to target;
(d) Learning of new categories or schemas in the aftermath of analogical reasoning.

In their book *Surfaces and Essences*, Douglas Hofstadter and Emmanuel Sander go even further, describing analogy in their subtitle not just as an important mode of argument but as "The Fuel and Fire of Thinking."[10] Hofstadter, who has a PhD in theoretical physics and is the author of the widely acclaimed *Gödel, Escher, Bach*, was convinced of the importance of analogy long before he wrote *Surfaces and Essences*; in 2001 he wrote that "analogy is … the very blue that fills the whole sky of cognition – analogy is *everything*, or very nearly so, in my view."

Having at least established analogy as a respectable scholarly topic, let me step back and recall how it has been used at many levels both internal and external to science and technology. In the broadest terms, historian John Lewis Gaddis has argued that "science, history, and art have something in common: they all depend on metaphor, on the recognition of patterns, on the realization that something is 'like' something else." The problem is, just how alike are the two things to be compared? Cullen Murphy's colorful volume *Are We Rome?*, comparing the fate of Rome to persistent claims of the decline of the West, appropriately indicates with a question mark the skepticism with which all such global comparisons should be undertaken. One constantly sees political analogies fraught with ideology: is Vladimir Putin like Hitler, is Donald Trump like Putin, is Obama giving in to Chamberlain-like appeasement by not reacting to Russia's grab of Crimea in 2014, and so on.[11] Analogy is part of the discourse of everyday life, but one person's casual analogy is another's casual dis-analogy; the prime consideration (all too often lacking) is informed knowledge of both the source and target analogy. In general, polemic is usually not fertile ground for the enlightened application of analogy.

At a lower level of generality, 50 years ago, historian and MIT professor Bruce Mazlish produced an important study, *The Railroad and the Space Program: An Exploration in Historical Analogy*. This volume, populated with well-known scholars, addressed the problem of analogy in considerable detail. Mazlish himself spoke of "attempting to set up a new branch of comparative history: the study of comparative or analogous social inventions and their impact on society." Although originally suspicious of parallels with the past, present, and future, contributors found analogy a useful tool; historian Thomas P. Hughes saw "the possibility of moving up onto a level of abstraction where the terrain of the past is suggestive of the topography of the present and its future projection." The authors cautioned that as much empirical detail should be used as possible and that analogies drawn from vague generalities should be avoided. Confident in the use of historical analogy as suggestive but not predictive of the future, Mazlish and his coauthors went on to elaborate their analogy with the railroad and the space program with such a degree of success that their work is still used today and has inspired continued use of analogy in the space arena. Fifty years after the railroad and the space program study, former NASA chief historian Roger Launius, fully aware of the use and abuse of analogy, penned the exemplary *Historical Analogs for the Stimulation of Space Commerce*. The bottom line is that, despite the misuses of analogy, the proper analysis of history can and should affect current and future programs. We discount human experience at our own peril.[12]

Analogies, then, are never perfect, but can be useful and illuminating as guides for thought, suggesting the topography of the future. They can also be overstated and misleading, as in the case of the "frontier analogy" so prominent in American space exploration. Many historians do not agree with the classic "frontier thesis" that the western frontier was the primary source of the American characteristics of inventiveness, inquisitiveness, and individualism. And yet space as a new frontier has been, and continues to be, a driver of the US space program. Some have argued (implausibly in my view) that the comparison is actually harmful; in any case, it is an analogy that needs to be used with qualification and caution.[13]

Turning to astrobiology, analogies are immediately obvious, and in more than just "the grand analogy" used in the past and even today arguing that since life exists on Earth, it must exist on other planets. Certainly astrobiologists dealing with the microbiological parts of the subject make heavy use of analogy in the most general sense: since they have no extraterrestrial microbes at hand, they use terrestrial microbes as analogs. Anthropologist Stefan Helmreich spoke of the "novel methodological strategy to those who would scout for extraterrestrial life. Astrobiologists treat unusual environments on Earth, such as methane seeps and hydrothermal vents, as models for extraterrestrial ecologies. Framing these environments as surrogates for alternative worlds has made marine microbes like hyperthermophiles attractive understudies – what scientists call analogs – for aliens." Although astrobiology is far from "a novel methodological strategy," Helmreich is correct in suggesting that it would not exist without this most general use of analogy. Put another way, withdrawal of analogy as a form of argument would be an existential threat to the survival of astrobiology as a discipline.[14]

In addition to microbes themselves, astrobiologists employ geographic conditions as analogs to other planetary conditions: to use the language of Bartha, Lake Vostok in the Antarctic is deployed as the source analog to the target analogs of ocean worlds such Europa and Ganymede, while the Atacama desert is the source analog to the target analog Mars, to name only two specific cases.[15] Analogs may also be two-dimensional, combining both microbes and conditions: biogeochemists have recently used observations of microbial mats in sinkholes in Lake Huron (which thrive and persist today under low oxygen concentrations) as models for the low-oxygen early Earth prior to the Great Oxidation Event 2.4 billion years ago, as well as for the pre-Cambrian, when oxygen concentration was still relatively low.[16] That is an example of analogy working in the backward direction – a current condition (the source analog) is used to illuminate the past (target). Analogs can also work in the forward direction: climate scientists use past climate change records to predict, controversially to be sure, what may happen in the future. In short, without analogical reasoning, not only astrobiology but also much of science would come to a standstill.

I have dwelled at some length on the foundations of analogical argument because these insights indicate that the systematic application of analogy to the problem of the societal impact of extraterrestrial life is a field that holds much promise, if properly used. I return to this again and again in this volume. Serious precautions are indeed in order, since analogical reasoning can be misleading. Examples are attempts to show that religion is analogous to science, or to spaceflight, or to SETI, or to Eastern mysticism and quantum mechanics. The often-heard analogy of the book of Genesis compared to the details of the Big Bang theory has been thoroughly debunked.[17] These are often what I would call "polemical analogies," and the goal of analogical argument should not be to polemicize but to illuminate. In addition, as anthropologist Kathryn Denning has warned, it is easy to get carried away with analogy, descending to a level of detail unlikely to be useful or relevant, if not downright harmful. But others see great promise in analogy cautiously used. In the words of anthropologist John Traphagan, "One of the most potent ways the social sciences in general and anthropology in particular can contribute to SETI is through analogy, using an analysis of anthropology's own history of contact as a framework for thinking about potential contact with an extraterrestrial civilization."[18]

Taking all this into account, I offer what I term the "Goldilocks Principle of Analogy": *Analogy must not be so general as to be meaningless, nor so specific as to be misleading. The middle "Goldilocks" ground is where analogies may serve as useful guideposts.* On the one hand, it does little good to argue that science and religion are both searching for humans' place in the universe, when one addresses the natural world and the other invokes the supernatural – differences so great as to swamp any comparison whatsoever. And it does little good to argue that because there is life on Earth there is life out there – the very thing we seek to demonstrate. On the other hand, it is hopelessly naïve to expect that contact with extraterrestrial intelligence will change our worldviews in ways precisely mirroring past discoveries, culture contacts, decipherment efforts, or revolutions in thought, leading us to reiterate that under no circumstances will analogy predict the future.

The Microbe Analogy: Discovering the Microcosmos

I now turn to the central problem of this chapter: how can analogy illuminate the reaction to the discovery of life beyond Earth? I examine several cases to demonstrate the promise and peril of analogy under some of the different scenarios discussed in the previous chapter: the discovery of microbial life, physical culture contact, remote contact, and changes in worldview. In doing so we invoke the Goldilocks Principle, attempting to avoid wishful thinking, all the while keeping an eye on useful guideposts.

I begin with the discovery of microbial life, and ask what analogies might best represent the reaction to the discovery of extraterrestrial microbes, whether alive, fossilized, or indicated in a more ambiguous biosignature from a planetary surface or atmosphere. Whatever else it may be, the discovery of microbes beyond Earth in any form potentially represents a revolution in biology, depending on the nature of the microbes found. Taking this as a starting point, we can therefore ask what discoveries in the history of biology might approximate such extraterrestrial discoveries. More specifically, assuming that such a discovery would change our view of life, what discoveries in biology have changed biologists' view of life? A number of candidates immediately come to mind, including Darwinian evolution by natural selection, the role of DNA in genetics, and the discovery of extremophile microorganisms. But perhaps no biological discovery closer approximates the potential fallout than the discovery of terrestrial microbes themselves, representing a new world of life on Earth.

The details of this amazing and unexpected discovery are well known, and rather surprising. In a parallel to the development of the telescope, the discovery of microbes was enabled by the invention of the microscope in the early seventeenth century, or even earlier in its embryonic forms. But, unlike the telescope, early instruments were so crude they yielded no immediate results. It was only later in the century that two remarkable geniuses, using simple microscopes of their own construction magnifying 25 to 250 times, first spied what we now call microbes. English polymath Robert Hooke published the first drawing of a normally invisible microorganism (the microfungus *Mucor*, the common bread mold) in his *Micrographia* in 1665. Inspired by Hooke (and impelled by the practical need to grade the quality of warp and weft in fabric), Antonie van Leeuwenhoek, a Dutch draper and city official in Delft, undertook much more extensive observations. His discoveries included what we now call bacteria, protists and yeast, which he described in a series of often lengthy letters to the Royal Society of London beginning in 1674. Today, most historians and biologists consider both Hooke and Leeuwenhoek "the major discoverers of the microbial universe."[19]

Several things stand out about the initial discovery of what some term today, in direct comparison to the astronomical universe, the "microbial universe" or "microcosmos."[20] Leeuwenhoek himself was astonished at the large number of "animalcules," bacteria as we now know them, in a single drop of pepper water that had been stagnating for three weeks (Figure 3.1). "This exceeds belief," he wrote in a letter of October 5, 1677. In his letters to the Royal Society, he was therefore careful to include testimonials from "eight credible persons" who affirmed his observations. Hooke himself was also astonished despite his earlier observations, but confirmed Leeuwenhoek's observations of animalcules, writing in his 1678 *Microscopium* "much to [my] wonder I discovered vast multitudes of these exceeding small creatures which Mr. Leeuwenhoek had described … and some of these so exceeding small that millions of millions might be contained in one drop of water." Hooke was now the secretary of the Royal Society, and the authority of that body (later to be presided over by Sir Isaac Newton) left no doubt about the validity of the observations.[21] It is surprising, therefore, that although these observations excited considerable interest among both scientists and the public, they led nowhere during the lifetime of Hooke or Leeuwenhoek and for several generations thereafter. It is true that Hooke's *Micrographia* was very popular, one of the most influential books in the history of the microscope, and that it "stimulated for the first time a wave of public interest in the microscopic world." The book was the first to use the term "cell," in connection with the appearance of sections of cork. But as one historian of biology wrote, while there were occasional observations, "scant research was done in the study of microbial life during the next century."[22] In another parallel to the telescope (a technology to which Hooke also contributed), progress in microscopy was slow, and in any case no one quite knew what to make of the new discoveries.

Once microscopes had been considerably improved, however, the impact was swift and profound. That was not until the nineteenth century, and not until then did microbiology even begin to develop into a science, with the names of Christian Ehrenberg, Louis Pasteur, Thomas Huxley, Ferdinand Cohn, and Robert Koch particularly notable.[23] If Leeuwenhoek and Hooke were the Columbuses of the new world of microbes, Christian Ehrenberg has been characterized as its Humboldt, exploring its broader landscape. In 1838 he published his famous book in German, translated as "The Infusoria as Whole Organisms," in which he described thousands of new species of animalcules, or "infusoria," as he called them. The

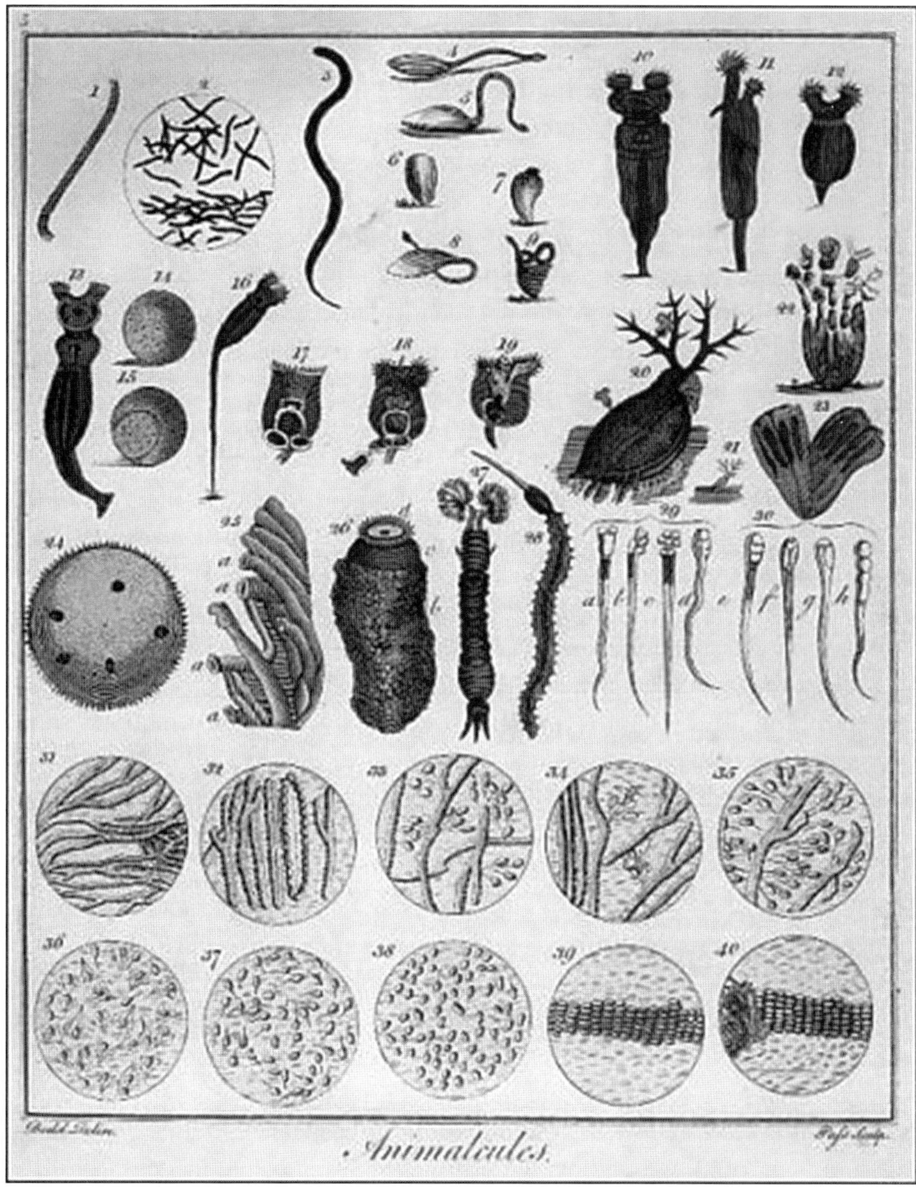

Figure 3.1 Leeuwenhoek's depiction of animalcules, the beginnings of the discovery of the microbial universe. The stages of discovery, including not only detection, but also interpretation and understanding, were very extended, as in Figure 2.2. This image, published c. 1795–1798, via Wikimedia Commons.

infusoria became so numerous they demanded classification, and in this Ehrenberg and Cohn were leaders. With his studies of the role of microbes in fermentation, and his "pasteurization" process demonstrating how knowledge of microbes could be put to practical use by destroying pathogenic microbes in milk, Louis Pasteur showed how microbes could be an eminently practical part of science. With T. H. Huxley's address "Biogenesis and Abiogenesis" (1870), the subject became linked to the problem of the origin of life in Darwinian terms. And in the 1870s Robert Koch showed that anthrax was linked to a bacterium, validating the germ theory of disease. Once it was realized microbes were responsible for many diseases, their study spun off the field of bacteriology, beginning with Cohn's landmark studies in the 1870s. Even then (as microbiologist Thomas Brock has remarked), microbiology did not become a true field of study until the twentieth century, in the sense that nineteenth-century workers did not identify themselves as bacteriologists or microbiologists, but as chemists or physicians. The implications of microbes were profound, and are being felt today at an accelerating rate.[24]

It is notable that both pure and applied effects stemmed from the new realm of life. The discovery of the microbial universe greatly enlarged our view of life, carrying what had been a macro-science into the domain of the very small, the microcosmos. Classification of microbes became a substantial endeavor, an expansion of the species and genera system in effect for zoology, but replete with controversies about what characteristics should be used for classification. Nevertheless, attempts were made to classify objects still at the limits of resolution, often lacking any knowledge of their internal structure. Some argued the bodies observed were the same species in different forms, complicating classification even more. In short, the science of biology was expanded in ways never before seen, but often with one step forward and two steps back. Even today the number of unknown bacterial species is estimated in the millions, and the microbial universe remains in many ways largely unexplored.

Nor does the story of microbes end with their first discovery and their elaboration in pure and practical science. Three centuries after Hooke and Leeuwenhoek, microbes once again radically changed biologists' view of life with the discovery of extremophiles, microorganisms functioning under conditions far beyond what had been considered the limits of life on Earth. Beginning in the 1960s and 1970s such organisms were found thriving near the boiling point of water in hot springs, in conditions of high salinity, acidity, radioactivity, deep underground, and perhaps most surprising of all, around hydrothermal vents deep below the surface of the ocean at extremely high pressures and temperatures, and with no sunlight as an energy source.

These discoveries in turn led to something even more surprising – a complete reclassification of life. In 1977 Carl Woese and his colleagues announced what historians of science generally agree was one of the most remarkable discoveries of twentieth-century biology. Studying the 16s ribosomal RNA of many different microorganisms, they found that methanogens (methane producers), halophiles (salt lovers), thermophiles, and hyperthophiles, all previously classified as bacteria, were as different from them as the bacteria were from eukaryotes (all plants, animals, and fungi). Woese and his colleagues claimed this was a new "Third Kingdom" of life, later called Archaea, one of three discrete divisions of life, along with Eubacteria and Eukarya. Although very controversial for decades, this new classification system for life is now increasingly accepted. Microbes not only enlarged our view of life, they also resulted in an entirely new view of life.[25]

Taking the long view, the first lesson of the microbes must be the importance of new technology in revealing the microbial universe. Hooke and Leeuwenhoek could not have done their pioneering work without their microscopes, crude as they were by modern comparisons. One is tempted to say that the second lesson is that the discovery of extraterrestrial microbes may also see a scenario in which there is great initial interest following the first observations, but no scientific progress for a long time. This is indeed a possible scenario, but unlikely. One needs to take into account that the scientific enterprise has changed a great deal since the seventeenth century. In particular, if technology is perceived as holding back the elaboration of such a fundamental and popular discovery as extraterrestrial microbes, in the modern world government funding and other resources are much more likely to be made available than in centuries past, when science was largely the endeavor of individuals and their private patrons. A third possible lesson comes from the discovery of extremophiles and the resultant reclassification of life: the discovery of extraterrestrial microbes may also lead to a reclassification of life; indeed, it may lead to a universal biology, in which biological principles are elaborated that apply to life with many different origins, genetic bases, and fundamental constructions. A fourth lesson is that the discovery of new microbes has in the past often had practical applications in medicine and other fields, and therefore has had immense economic value. As Aaron Gronstal has pointed out, most recently extremophiles "have provided chemical and molecular products that have completely changed areas of medicine and industry." He suggests the same may be true of the discovery of microbes beyond Earth.[26]

Finally, we should not overlook possible negative effects of the discovery of extraterrestrial microbes. We recall that in H. G. Wells's *War of the Worlds* the Earthlings were saved from the Martians by terrestrial microbes that destroyed the invaders. This was all to the good from the point of view of the Earthlings. But the reverse can also happen, in ways large and small, as demonstrated in the Columbian exchange during the European culture contacts with the Americas. It was not the Spaniards that wiped out the native populations, but their microbes. This *Andromeda Strain* scenario is not mere science fiction, but a real possibility that must be guarded against. Indeed, NASA's planetary protection program exists precisely for the purpose of ensuring Earth is not contaminated by extraterrestrial microbes, with possible catastrophic effects, and vice versa.[27]

While the discovery of microbes on Earth may represent a particularly apt analogy for the reaction to the discovery of extraterrestrial microbes, other revolutions in biology may also prove apt. The Darwinian theory of natural selection and the discovery that DNA carries the genetic code are possibilities, among others. A detailed look at such discoveries may provide a more robust guide to the potential reaction to the discovery of microbes beyond Earth. Similar analogies may apply to the discovery of microbial fossil life and microbial biosignatures, but their implications should be examined with a view toward differences inherent in the peculiar nature of their discovery. In whatever form they are discovered, the importance of microbes should not be underestimated merely because of their small size. Like it or not, we should embrace the possibility that microbes, not intelligent beings, rule the universe, just as in a sense they rule the Earth. It follows that the implications of finding microbes should receive no less attention than the implications of finding intelligence.

The Culture Contact Analogy: Direct Contact with Intelligence

As intimated at the beginning of this chapter in connection with the downfall of the Incas, terrestrial history abounds with contacts between cultures, often with unhappy effects. The idea that such culture contacts might be used as analogies in the context of the Search for Extraterrestrial Intelligence (SETI) dates back at least to the beginning of the Space Age. The new era was only a few years old, and Frank Drake's project Ozma only one year old, when the NASA-sponsored Brookings Institution Study on the Implications of Peaceful Space Activities for Human Affairs warned in connection with the discovery of extraterrestrial life: "Anthropological files contain many examples of societies, sure of their place in the universe, which have disintegrated when they had to associate with previously unfamiliar societies espousing different ideas and different life ways; others that survived such an experience usually did so by paying the price of changes in values and attitudes and behavior." Thirty years later the authors of NASA's Cultural Aspects of SETI (CASETI) report took a more nuanced view, pointing out that there were both pessimistic and optimistic views on culture contact, which they termed "catastrophist and millenarian," with some arguing that an optimistic view of extraterrestrial contact might usher in "Childhood's End" for Earth.[28]

Both of these reports had in mind remote electromagnetic contact, not physical culture contact. The former is indeed the more likely, but here we look at what might be learned from physical culture contacts, not only because it remains a logical possibility, but also because it illuminates the problems of remote culture contact I discuss in the next section. In doing so, we move beyond catastrophist views of cultural destruction brought about by such contacts – historically frequent, valid, and vivid as they are in some cases – to more general effects involving cultural problems of communication, behavior, attitudes, and categories of thinking and understanding. This approach resonates with the suggestion of anthropologist Ian Lowrie that we shift our approach to culture contact in this context from the material aspects to their symbolic and cognitive dimensions.[29] The problem here is not the lack of data, but too much data, which must be analyzed carefully in a balanced and useful way. As another anthropologist has put it, "If terrestrial analogues are to be employed in relation to SETI, then we should explore the wide range of human experience around the globe and not focus solely on familiar cases that appear to reinforce our most earnest hopes."[30] Or, I might add, our greatest fears. Here I can only provide a sampling of issues and provisional results that arise by taking this approach.

One of the most trenchant criticisms of using historical terrestrial culture contacts as analogs to extraterrestrial contacts is that, different as terrestrial cultures may be, they all involve *homo sapiens*, sharing a common ancestry and thus common mental and behavioral patterns in the most general sense. This warning is well taken. One prehistoric terrestrial analog partially transcends this criticism: the overlap of modern *homo sapiens* with other human species. As the Smithsonian's Human Origins program makes clear, during its 200,000-year career *homo sapiens* has shared the planet with at least three other humans of the same genus, but different species: *homo erectus* (extinct about 143,000 years ago), *homo neanderthalensis* (extinct 28,000 years ago), and *homo floresiensis* (extinct 17,000 years ago). The latter, however, are only known to have inhabited Indonesia, and interactions with *homo erectus* are unknown.

That leaves Neanderthals, which overlapped with *homo sapiens* for tens of thousands of years in both space and time. Indeed, though Neanderthals became extinct about 28,000 years ago, modern DNA evidence demonstrates that the two species interbred between 50,000 and 80,000 years ago, to the extent that all of us of European descent have 1 to 3 percent of Neanderthal genes in our genome.[31] The mind reels at what such an encounter might have been like (Figure 3.2). Unlike other animals modern humans may have hunted, domesticated, or otherwise interacted with, evidence indicates Neanderthal cognition was more on a par with humans'. And they were the first "humans" known to have buried their dead, even with ceremony and flower offerings by the evidence of the Shanadar cave in modern Iraq. Yet paleoanthropologist Ian Tattersall has argued that the two species are different enough that we are not justified in using modern humans as ethnographic analogs to make sense of Neanderthals. "When we look at *homo neanderthalensis*," he observed, "we are looking at a creature possessed of another sensibility entirely."[32] While his point is well taken as an ethnographical analog,

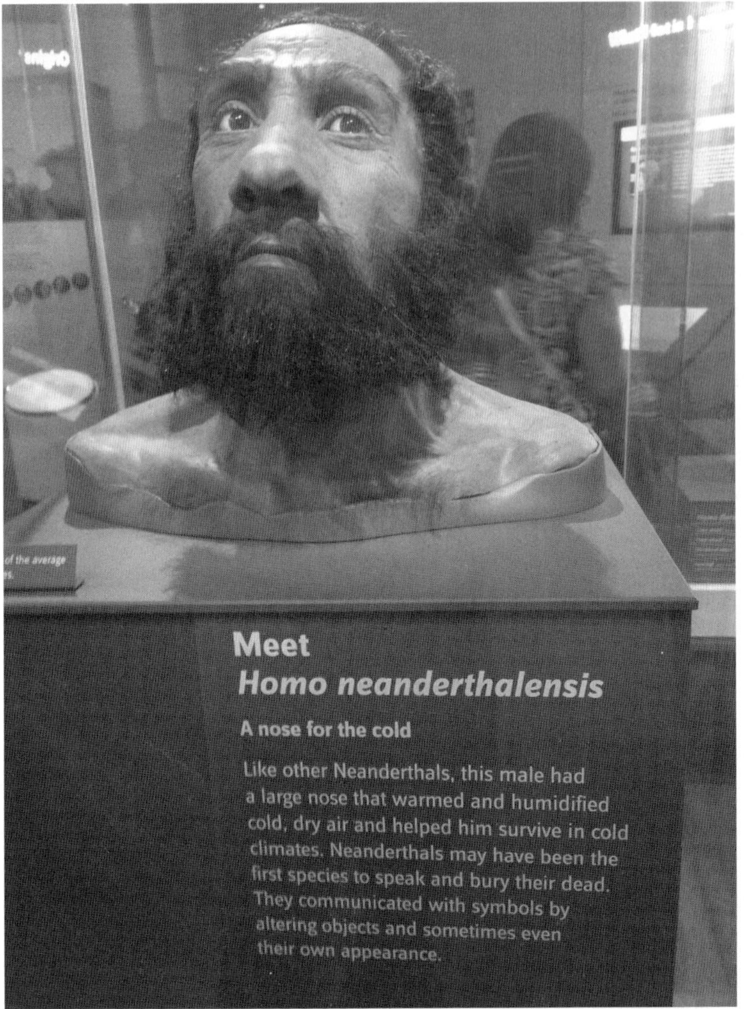

Figure 3.2 The Neanderthal Analogy: Humans encounter Neanderthals at the Smithsonian's National Museum of Natural History, Hall of Human Origins. Understanding of what a real human–Neanderthal interaction would have been like is hampered by lack of evidence, aside from one hard fact: the non-African human genome contains up to 3 percent Neanderthal genes. Credit: Smithsonian Institution, photo by Steven Dick. Art by John Gurche.

this particular "culture contact" among different species provides a rather unique analog for contact with extraterrestrials.

The problem is the lack of data on any such interaction. Nevertheless, one of the first articles of anthropological interest to SETI, published in *Nature* 50 years ago in the wake of the modern beginnings of exobiology, invoked just this comparison. The authors of the article, an anthropologist and a mathematician at Cornell, argued that models founded on our knowledge of human evolution might contribute to our understanding of the impact of success in SETI. In particular, the authors suggested an "analogy between prehistoric contact and exchange, and hypothesized extraterrestrial contact and exchange." In early prehistory, when biologically distinct hominid populations existed, they pointed out, contact "occurred between technologically similar but biologically diverse populations. In later prehistory contact was usually initiated by those populations with advanced techniques and equal exchange was rare."[33] This history, they suggest, might shed light on the nature of contact with extraterrestrial civilizations.

More recently Paul K. Wason has attempted to elaborate the interaction between *homo sapiens* and *homo neanderthalensis*, and in ways that indicate that much more research could be done.[34] Lacking empirical data, he approached the problem from the point of view of models of cognitive evolution. One such model is Merlin Donald's four stages of cognitive evolution: episodic culture, mimetic culture, linguistic or mythic culture, and theoretic culture, all based on Jean Piaget's work with cognitive development in children. These stages in turn represent, according to Wason, primate cognition, nonverbal communication as in Neanderthals, early *homo sapiens*, and modern *homo sapiens sapiens*. Another model that Wason rather prefers over Donald's and Piaget's is the modular view of the human mind developed by Steven Mithen. In Mithen's model, the three major domains of intelligence are social, technical, and natural history intelligence, and all three were well developed among Neanderthals, including the technical skill of tool-making, and extensive knowledge of the environment. Modern humans, however (dating to about 200,000 years ago), also have a "generalized intelligence" that allows a fluidity among the other modules allowing them to think thoughts and solve problems that Neanderthals were incapable of thinking and solving.

Drawing on Mithen's insights of the structure of the prehistoric mind, Wason concludes Neanderthals might have been amazed at modern humans in any encounter, but would not have had the capacity to know exactly what to make of them. Among other things Neanderthals differed from humans in their tool traditions, symbolic behavior, and aesthetic sensibilities such as art and music. "Altogether, we might well expect any encounter between Neanderthals and Cro-Magnons [early modern humans, *homo sapiens sapiens*] to have been a difficult and ineffective affair, fraught with misunderstanding," he says.[35] And if two such closely related terrestrial species have this problem, how much more difficult might an extraterrestrial contact be? As an analog for physical first contact this description seems likely to be quite accurate. "We can expect there to be large areas of incomprehension, relieved by small elements that seem to make sense," but that mask deeper differences. Although we have few data for the meeting of Neanderthal and *homo sapiens*, aside from the DNA evidence in our current genome that they actually did meet, cognitive models hold out hope of making these analogs more useful. To put the situation in Bartha's terms of analogical analysis, even poorly understood source analogs may illuminate target analogs in unexpected ways. And perhaps elements common to any intelligence will enable some type of communication or understanding.[36]

Much more common than data-poor paleoanthropological contacts among species are studies of culture contacts throughout recorded history. Indeed, culture contact is an entire subfield of archaeology and anthropology, whose studies are often stimulated by the desire to understand culture change and refine theoretical and methodological approaches to that problem.[37] The field embraces a variety of cases, ranging from Cortez and the Aztecs, Pizarro and the Incas, and assorted Europeans engaging the American Indians, to European clashes with the Ottoman empire in the sixteenth and seventeenth centuries, the British Raj in India, the attempts of Peter the Great at the Westernization of Russia in the eighteenth century, and Matthew Perry and the opening of Japan in the mid-nineteenth century. And this is to name only a few examples from the Western tradition of expansionism.[38] So diverse are these culture contacts that almost any lesson could be drawn from their history of devastation, cooperation, or assimilation. We must therefore seek the higher level of common elements.

With that precaution in mind, and recognizing that all such comparisons represent contacts among the same species, let us nevertheless examine a few exemplars for any universal elements. Certainly one of the most vivid and well-documented events in world history is the contact between the Old World and the New, that is, between Europe and what is now known as the Americas. Between the early fifteenth and the mid-seventeenth centuries, what is often called the Age of Discovery resulted in dramatic culture contacts as Europeans sailed the Atlantic and Pacific Oceans. The process began in the fifteenth century with Portuguese sailors hugging the west coastline of Africa, then sailing outward to increasingly distant islands. At the end of the century, after the Portuguese denied Columbus the funding he requested, it was the Spanish who financed the first plunge across the Atlantic Ocean, followed by many others, in a remarkable story that is a centerpiece of world history no matter what the cultural perspective. A "Stories of Encounters" kiosk at the National Museum of the American Indian in Washington, DC, depicts several dozen such contacts, beginning in the fifteenth century. By about 1650, the first Age of Discovery was over, as Africa, Asia, and the Americas had become routine destinations.[39]

It is also sobering to note that even before European contact with the New World, Europeans had "imagined the other," just as we are doing now before any contacts have been made with extraterrestrials. It is a matter of historical record that the European imaginings did not accord with reality (Figure 3.3). The dangers of anthropology at a distance were still apparent in Ruth Benedict's

Figure 3.3 Medieval European depictions of "the other" featured monstrosities that might live in unexplored countries, from Gregor Reisch, *Margarita Philosophica* (Basel, 1517). Europeans found "others" different and inferior, while native American Indians tended to view "the other" as similar or superior. Are our imaginations about aliens any better?

mid-twentieth-century study of Japanese culture only through its literature and arts, summarized in her book *The Chrysanthemum and the Sword* (1946). And in recent years, anthropologist John Traphagan, drawing on lessons from Benedict and others, has warned about the implications of anthropology at a distance in the context of SETI. Our "imaginary" of extraterrestrials and their cultures, he argues, may be no better than those of the Europeans or of twentieth-century American anthropologists trying to determine how they would interact with the Japanese following World War II. Even with a SETI signal, the data are likely to be so time-lagged and incomplete as to allow our imaginations to run wild: "even if social scientists, philosophers, and historians are involved, there is a good chance that humans will create an imaginary about ETI that is really a reflection of ourselves and our theories about how we organize ourselves culturally, politically, economically, and socially." Even that insight, however, shows how valuable it will be to have the social sciences and humanities at the table when discovery occurs.[40]

Although Columbus had already wreaked havoc with the Tainos in the Caribbean during his second voyage, arriving in 1493, no culture contact was more dramatic than that between Hernán Cortés and the Aztecs in Mexico a quarter century later (except perhaps that of the Incas already mentioned). I choose it here not only because it is one of the best-documented and most notorious culture contacts from the Age of Discovery but also because it is often used to imply that the impact of extraterrestrial contact will be disastrous, often by people who know nothing about history. The general outline of the events is well known. Cortés left Cuba in February 1519 with three ships and 530 men, supposedly on a voyage of trade and exploration. He reconnoitered the Yucatan and set off in late April for the Mexican mainland, where he found the deeply divided Mexica empire of Montezuma. By August, the Spaniards and thousands of allied natives they had recruited were heading inland toward the city of Tenochtitlan, and on November 8, they made contact with Montezuma and 4,000 "gentlemen of the court." The initial contact was friendly, but Cortés soon took Montezuma into custody, and the latter capitulated, at least in the accounts of the Spanish. Fourteen months of bitter fighting ensued, and in August 1521, Tenochtitlan fell after a long siege, and with it the Mexica empire, soon to be transformed into the viceroyalty of New Spain (Figure 3.4).[41]

The lesson often learned here is that culture contacts can be destructive, especially considering other culture contacts between the Old and New Worlds such as the Inca experience. Indeed that is the indisputable case, but what is important to us

Figure 3.4 A European culture contact analogy, the conquest of the Aztec capital Tenochtitlan (present-day Mexico City) in 1521. From the Conquest of México series, Mexico, second half of seventeenth century, oil on canvas. Jay I. Kislak Collection Rare Book and Special Collections Division, Library of Congress (26.2). Unknown artists.

is not the conquest itself, but the broader outlines of this dramatic culture contact. First, the communications were fraught with difficulties, beginning with the fact that the Spaniards may have thought the Aztecs (also known as the Nahua) were surrendering when in fact they were not. Second, there are always two sides to any story. While the Spanish side is represented most robustly by Bernal Diaz del Castillo's *Historia verdadera de la conquista de la Nueva España* [*True History of the Conquest of New Spain*], published more than a half-century after the fact, there was another side to the story, that of the Nahua who were conquered.[42] Third, culture contacts work both ways. Not only were the Americas affected by Europe, Europe was also affected by the Americas. As historian J. H. Elliott has shown, there were intellectual, social, and economic consequences not only for the Aztecs but also for the Europeans. And as Nobelist Baruch Blumberg has emphasized, aside from their negative impacts, massive amounts of scientific and cultural data were collected from the voyages of exploration.[43]

Significant culture contact in the Americas took place dozens of times over hundreds of years, sometimes with destruction, sometimes with more sensitivity, often a mixture of both. An example of moving beyond physical destruction of cultures to their intellectual and spiritual interaction is the encounter between French Jesuit missionaries and native Americans living in what is now eastern Canada, detailed in 74 volumes over 150 years beginning in the first half of the seventeenth century. As Jason Kuznicki points out, the Huron, Iroquois, and Montagnais peoples of the northern Great Lakes region had developed their technologies and worldviews in almost complete isolation from any outside contact, so in a sense they were truly alien except in a long-ago common ancestry. Despite the advanced Jesuit training in languages, philosophy, the sciences, and theology, Jesuit hopes for establishing communication based on what they considered universal ideas of God and the soul were dashed. The native Americans had no such ideas, at least none that matched those of the Europeans. While expressing caution, Kuznicki suggests any extraterrestrial contact will likewise experience difficulty "grounding any conversation in foundational ideas shared by both communicating groups." More specifically, "these foundational difficulties may well come at precisely the points that we now take to be the most unproblematic: basic concepts of language, ontological categories, supposedly universal knowledge, and the assumed relations among them, may be presented to us in configurations that are presently unknown – and these new configurations may be precisely what give meaning to extraterrestrials' concepts of self, spirituality and ethics."[44]

Outside the tradition of Western expansionism, the less well-known great voyages of the Chinese treasure fleets 80 years before Columbus offer yet another scenario. It is a matter of historical fact that from 1405 to 1433, the Ming Yongle emperor and his grandson sent seven massive expeditions into the Indian Ocean and beyond, all commanded by fleet admiral and Hui court eunuch Zheng He.

The first expedition alone may have included 62 "junks" three or four times larger than Columbus's flagship, 225 support vessels, and 27,000 men. Any single expedition was larger than the Spanish Armada, and with much larger ships. It is also well known that following a maritime tradition stretching back to the eleventh century, these ships had plied the seas of Southeast Asia. But then over the course of the seven voyages over almost 30 years, they sailed to India, the Persian Gulf, the Red Sea, and down the east coast of Africa. And the sudden end of this distant voyaging is indisputable: with changing internal political conditions and the external threat of the Mongols, the fleet was withdrawn in 1433 and its records burned. The subsequent inward turn, it is often argued, set China back centuries. Indeed, in another analogy the fifteenth-century Chinese treasure fleet has often been used as a lesson for those who would withdraw from the Space Age to seek shorter-term goals on Earth.[45]

The contentious questions have always been why the Yongle emperor mounted these expensive expeditions, and why they ended. Hypothesized motives of exploration and naked sea power have given way in modern scholarship to a very different explanation. Zheng He's largest ships, one noted scholar concluded based on Chinese sources, "were not well suited for either exploration or for combat with other ships, but their sheer size was awe-inspiring, and they were intended to ferry Chinese troops around the Indian Ocean in order to impress, or if need be to overpower, the local authorities. The goal of this effort was neither conquest nor the promotion of trade or exploration, but the enforcement of the Chinese tributary system on the countries of the Indian Ocean." In other words, the voyages were a kind of power projection, to use modern terminology.[46]

The detailed nature of the culture contacts made by the Chinese during these amazing voyages is hampered by lack of extensive records. But enough is known, from writings on stelae constructed by Zheng among other sources, to conclude that the contacts were quite different from the case of Cortés. As historian Edward Dreyer puts it, this power projection and tributary system "required a much greater naval presence than any amount of exploration would have needed. Zheng He's armada was frightening enough that it seldom needed to fight, but being *able* to fight was its primary mission."[47]

While it is known that during these voyages the Chinese intervened more than once in the internal affairs of the cultures they visited, the early Ming expansionist style was very different from its Mongol predecessors, and from its later Western counterparts: "It was no longer a question of undertaking mere conquests for the sake of economic exploitation but of securing the recognition of the power and prestige of the Ming empire in South-East Asia and the Indian Ocean," another historian wrote. And in this goal Ming China largely succeeded: "China acquired great prestige in all the seas of East Asia and in the islands and peninsulas of

South-East Asia, and trade in the shape of tribute from all the states of those areas expanded rapidly."[48] Under the fifteenth-century Chinese tribute system, foreign states were left very much intact, their rulers or ambassadors conveying continuous tribute in the form of local products to the Chinese emperor, "thus acknowledging his unique status as the Son of Heaven and ruler of the Middle Kingdom of lands directly under Heaven."[49] This was no small achievement, considering Yongle had assumed power by armed usurpation. In contrast to the case of Cortés and the Aztecs, Chinese culture contacts resulted in new embassies in the Near East, deification (to this day) of Zheng He and his cult in Southeast Asia, and the publication of works enlarging Chinese knowledge of the oceans and landmasses visited.

There is another lesson here about communication and symbolism. It was during the fourth voyage of 1414–1415 that part of Zheng He's fleet visited Bengal, and saw for the first time a wondrous creature with a long neck and horns. It had come from the African city-state of Malindi, and the Bengal king offered it to Zheng He as a gift for the emperor. It arrived in Beijing in 1415 with another giraffe Zheng He acquired directly from Africa (Figure 3.5). As anthropologist Samuel Wilson described it, the animals were a sensation, not only for their odd appearance but also because the Chinese interpreted it to be a unicorn, "whose arrival, according to Confucian tradition, meant that a sage of the utmost wisdom and benevolence was in their presence." The emperor was pleased and the high cost of the expedition vindicated. Trade with Africa was heightened, with fabulous gifts exchanged, including zebras, ostriches, and other exotics. The emperor's giraffe may be seen as symbolic of culture contact quite different from that of Western expansionism in the Age of Discovery. At the same time it is sobering to think what might have occurred if the symbolism had been accidentally bad.[50]

Examples could be endlessly multiplied of culture contacts stemming from the Age of Discovery, much less other traditions such as the Chinese. A final, and peculiar, type that deserves mention because it is occasionally raised in the context of SETI, are the relatively modern "cargo cults." Early in the twentieth century the Australians, British, Dutch, and Germans made contact with the long-isolated inhabitants of the Melanesian island of New Guinea, now part of Indonesia. At first they engaged in trade, and even participated in working foreign-owned plantations. But not for long. Suddenly the New Guineans began to engage in elaborate rituals, building faux radio stations and crude landing strips meant "to attract ships and planes laden with 'cargo,'" items the Westerners had brought and of which the New Guineans wanted more. "Once these cults gripped the people, the colonial officials found it virtually impossible to talk them out of their beliefs," according to anthropologists.[51] Such behavior may seem bizarre to us, but it represents both a massive cultural misunderstanding and the kind of cults that develop even in modern times, and that are difficult to rationally

Figure 3.5 Symbolism in the Chinese culture contact analogy. A giraffe from the African city-state of Malindi, the gift from the Sultan of Bengal to the Chinese Ming Dynasty Yongle Emperor in 1415, resulting in increased relations and more trade, symbolic of a different kind of culture contact. Chen Zhang's copy of "Tribute Giraffe with Attendant"(1415), attributed to Shen Du, a calligrapher in the Ming dynasty.

dissuade by any means. Societies under the stress of novel occurrences are particularly susceptible to such behavior.

What, in the end, is the lesson from such a broad panoply of culture contacts? Is it that a supposedly "more advanced" society will overpower and supplant a supposedly inferior one? Far from it, as indicated by both the case of the Jesuits among the native Americans and the Chinese voyages. Though destruction, power projection, tribute systems, and acculturation are indeed sobering historical facts, we cannot conclude based on terrestrial analogy that destruction is the probable outcome when an advanced extraterrestrial civilization makes contact with a less advanced one, even if it is physical contact. More illuminating are the communication and cultural interactions, which include both positive and negative aspects. Taking any one interaction as an exemplar, almost any lesson could be learned. Far better to learn from the entire set of contact experiences, or better yet, to learn collective lessons common to all contact experiences. This entails knowing something about history, and nuanced history at that.[52]

To bring the point home, anyone who thinks sixteenth-century European exploitation of the West is typical should read Dirk Hoerder's *Cultures in Contact: World Migrations in the Second Millennium*. Among the hundreds of documented culture contacts over the past 1,000 years, the sixteenth-century European exploits cover only a few of the book's 800 pages. Moreover, the contacts were arguably an equal mix of good, bad, and indifferent. Today culture contacts under the name of globalism are also such a mix, as were the culture contacts in the first era of globalization dating from the fifteenth to the twelfth centuries BC in the Aegean, Egypt, and Near Eastern civilizations. While it is understandable for Westerners to focus on our own recent history from the Age of Discovery, to use the Aztecs and Incas as the sole or chief analogies for contact with extraterrestrial intelligence is to ignore the bulk of history. While we cannot draw definitive conclusions even based on the small sample of culture contacts mentioned in this chapter, the determination of characteristics common to all contacts is a respectable research program that could pay significant dividends.[53]

As Samuel Wilson has noted, while some contact was indeed brutal and immediate, "other processes of culture contact took place over many generations, so gradually that individuals might not notice the kinds of cultural change that were slowly taking place." Moreover, although "other aspects of Renaissance thought – the belief in the value and ultimate compatibility of different systems of thought and the innate dignity of humankind – do not at first sound like they had much to do with the period of colonialism and conquest that followed … they did play a key role in this period, which in the end was much more than a story of imperial conquest." As both Wilson and John Elliott have emphasized, contrary to popular belief, the interactions over the past 500 years have changed all the participants, not just those in the New World. And in this change there was always the element of chance. As in biology, so in history: play back the tape of history and things would have gone differently, a tribute to the importance of preparation, diplomacy, and getting contact right the first time.

The Transmission/Translation Analogy: Remote Contact with Intelligence

It is widely assumed that first contact with intelligence beyond Earth will not be physical, but remote, in the form of an electromagnetic signal in the radio, optical, infrared, or an even more exotic region of the spectrum – Encounter Type 3 in Table 2.2. A "dial tone" signal proven to be of artificial extraterrestrial origin might have one set of implications, such as a change in worldview, as I discuss in the next section. But if a SETI signal is deciphered and significant information is transmitted, the flow of information between civilizations across time may be a more appropriate analog. The historians in the CASETI meeting in 1991–1992 saw one such analog in the transmission of Greek and Arabic knowledge by way of the Arabs to the Latin West in the twelfth and thirteenth centuries. This vivid historical analog has since been elaborated by others, and is but one example of what historian Arnold Toynbee, in his massive *Study of History*, called in the terrestrial context "encounters between civilizations in time."[54]

The details of this twelfth- and thirteenth-century transmission activity are well known. After the collapse of the classical Roman Empire, the Greek knowledge that formed much of its basis was lost, but maintained first in the Byzantine and then in the Arab worlds. In fact, the Arab encounter with Greek knowledge provides an embedded example of culture contact across time, but one too complex to unravel here. Suffice it to say that for 500 years prior to the revival of learning in the Latin West, Islamic scholars translated the works of classical Greeks into Arabic, and sought to reconcile Islamic values with secular Greek values. Thus they preserved those Greek works in the same way that the scholars of Byzantium had been doing for centuries. Between the two cultures many works were preserved that would otherwise have been lost.

As the European West revived learning with the founding of its universities, it turned to those works preserved in Greek and Arabic. The problem was one of translation, and Spain provided a center of translation. Especially at the libraries of Toledo, the first great Muslim city conquered by the Christians in 1085, scholars worked tirelessly to uncover the ancient treasure of knowledge. The result was profound. "First a trickle and eventually a flood," one historian of science wrote about the twelfth-century endeavor, "the new material radically altered the intellectual life of the West." Western Europe, which had been struggling to keep the intellectual flame from being extinguished, now had to assimilate a torrent of new ideas (Figure 3.6). It took virtually all of the thirteenth century to absorb the new science,

Figure 3.6 The transmission/translation analogy. Following the transmission of Greek knowledge via the Arabic civilization to the Latin West, new knowledge had to be rapidly assimilated beginning in the twelfth and thirteenth centuries. This analogy of the transmission of knowledge across time is symbolized here by this 1352 fresco by Tommaso da Modena of Albert the Great, thirteenth-century teacher of Thomas Aquinas and leader in the spread of the new knowledge. The painting is located in the Chapter House of San Nicolo, Sala del Capitolo, Seminario di Treviso. Wikipedia Web Gallery of Art.

followed by detailed elaboration and alteration, which reached a peak in the early fifteenth century. The result of the newly recovered knowledge is a matter of record. Thomas Aquinas and other scholars, often with agendas of their own, attempted to reconcile the new Greek and Arabic knowledge with Christianity, and with current knowledge – such as it was. "Without the valiant labors of this small army of translators in the twelfth and thirteenth centuries," another historian of science wrote, "not only would medieval science have failed to materialize, but the scientific revolution of the seventeenth century could hardly have occurred."[55] The result was the European Renaissance, which spread gradually through the continent.

While we do not fancy our civilization analogous to the Middle Ages, the torrent of new ideas might well be analogous to a significant flow of information from an extraterrestrial civilization to one probably less knowledgeable but eager to learn. The army of translators involved in the recovery of lost learning in the Middle Ages may find its analogy in the legions of scientists, cryptographers, linguists, and others sure to participate in any attempt to decipher an extraterrestrial signal. While one cannot guarantee a global terrestrial renaissance based on extraterrestrial knowledge (it might have an opposite and depressing effect), one can project with some certainty that personal and institutional agendas would play a role in deciphering and spreading the information.

The Greek-Arabic-Latin transmission analog is a seductively uplifting scenario. But as anthropologist Ben Finney and historian Jerry Bentley have since emphasized, such an analog fails to take into account the complexity of the task, since Greek, Latin, and Arabic are closely related languages by comparison to any extraterrestrial language. Theirs was a *translation* problem, not a *decipherment* problem, as any extraterrestrial communication is likely to be. Instead, they suggest, a better analog would be "the lengthy and troubled efforts of western scholars to decipher ancient Mayan inscriptions, and just as difficult, to infer from this decipherment the nature of an ancient civilization."[56] The translation of the Mayan hieroglyphs, they point out, had been delayed by the same fallacy that had delayed the decipherment of Egyptian hieroglyphs, namely the assumption that the glyphs represented ideas independent of any language. It was only when scholars "approached the glyphs as symbols for the phonemes and morphemes of speech, studied the modern languages descended from ancient Egyptian and Mayan, and discovered translation keys (such as Egypt's famous Rosetta Stone) that they were able to decipher the hieroglyphic texts." Just how problematic the decoding process could be for alien communication is well illustrated in the recent movie *Arrival*.[57]

As Finney and Bentley emphasize, one often-proposed way out of this dilemma is to assume that extraterrestrials will communicate via some supposedly shared logic, using mathematics or recognizing physical constants.[58] The transmission of prime numbers was one of the earliest proposals made by SETI pioneers. This scenario assumes aliens civilizations will have convergent mental perceptions and conclusions about nature, a very large assumption indeed. Even if other civilizations were able to communicate following decipherment, Finney and Bentley argue, the Maya case does not bode well for full understanding of the civilization. The early decipherment of Mayan numerical notation and calendars did not lead to the translation of the bulk of Mayan texts. In fact, they contend, that understanding may have actually retarded it, leading researchers to mistake dynastic histories for mathematical and numerical matters.

In any case, a study of the decipherment of Egyptian hieroglyphics, Minoan Linear B, and other dead languages likely hold lessons for the remote contact scenario.[59] The transmission/translation analogy would be better termed the transmission/decipherment/translation analogy, which involves multiple analogies representing each of its stages: the transmission of Greek knowledge, the decipherment of ancient dead languages, and the translation of Greek text by way of the Arabs.

Yet another step is involved in this multilayered process. Assuming that a message is received, translated, and understood, vast amounts of information may be released into terrestrial cultures. The transmission of Greek knowledge to the West in the twelfth and thirteenth centuries again applies. But another analogy is also worth considering: Gutenberg and the printing press. The transmission of Greek knowledge was truly novel for the Latin West, while the printing press was primarily a method for spreading knowledge new and old. The effect of the printing press is well known. For example, it is difficult to imagine the Reformation without the printing press to spread the word. And that was just the beginning of a process echoing down through history. And we need not go back that far. The ongoing information explosion with the advent of the internet provides a current example of the effects of an information explosion. With this real-time example in which we are all participating, we can sense firsthand the complexity of the problem of addressing the impact of the discovery of ETI. Nate Silver, famous for his political predictions, remarks in his book *The Signal and the Noise* that "We face danger whenever information growth outpaces our understanding of how to process it."[60] This is one reason to study such effects, past, present, and future, in both terrestrial and extraterrestrial contexts.

In the end the translation of Egyptian hieroglyphics and Mayan glyphs added a great deal to our understanding of the history of those cultures. But the new revelations of these dead civilizations did not greatly impact the common person, rather the intellectual elite was most affected. A message from the stars has the potential to affect a much broader population, particularly with respect to such personal areas as religion and other ideologies. Here analogies fail, with one possible exception. Over the short or long term, individual and collective worldviews would likely be changed by the new knowledge. I end this chapter with a brief look at the promise of worldviews as analogy.

The Worldview Analogy: Copernicus, Darwin, Hubble

Even if an extraterrestrial message is not deciphered, and perhaps even in the case that "only" microbial life is discovered constituting a second Genesis, a change in worldview would likely gradually take place. But what does that mean? In general, individuals have worldviews, and so do cultures and nations. "We see the world differently, in ways that are determined by our culture," anthropologist Sam Wilson has argued.[61] Can one really speak about a change in a cultural or national, indeed even a human, worldview? Surely the national worldviews of the United States, Russia, China, India, and other such nation states are different from each other, and different now than they were a century ago. And surely the integrated human worldview is different now than it was in the seventeenth century. In general, we might postulate that the larger the entity, the longer it takes to change its worldview, individuals doing so on a timescale of years (if their thinking is progressing), cultures and nations on a scale of decades and centuries, humans on an even longer scale. Surely the human worldview has changed since the medieval world centuries ago, perhaps even since pictures of the Earth, the fragile "pale blue dot" and "Blue Marble," have been beamed back from space (Figure 3.7). Although that change would seem to be subtle since it has certainly not brought

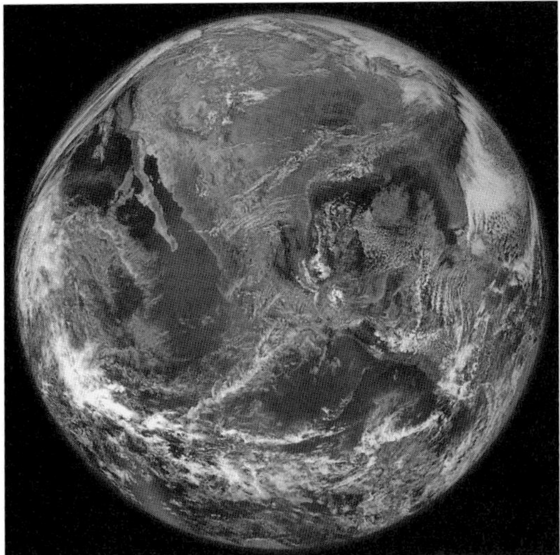

Figure 3.7 Human worldviews generally change over long timescales. Arguably, the "Blue Marble" and other spacecraft images are gradually changing our worldviews in this way. This 2012 Blue Marble image of Earth was taken from NASA's Earth-observing satellite Suomi NPP, the National Polar-Orbiting Partnership, from about 500 miles above the Earth's surface. The original Blue Marble was taken from *Apollo 17* in 1972. Credit: NASA/NOAA/GSFC/Suomi NPP/VIIRS/Norman Kuring.

the world together, such an "overview effect" should not be underestimated over the long term. Indeed (as I show in Chapter 7 in connection with astroculture), as inspiring spacecraft images are beamed back to Earth, the raising of cosmic consciousness may be the most important long-term change in worldview. The gradual construction of such worldviews, and their influence on our thinking, is a deep philosophical problem requiring more research.[62]

Recognizing these difficulties, we argue that the discovery of microbial or intelligent life beyond Earth might be analogous to grand changes in scientific worldviews, exemplified in the Copernican worldview that originated in the sixteenth century, the Darwinian worldview of the nineteenth century, or the Shapley–Hubble worldview of the twentieth century, in which our solar system was demonstrated to be at the periphery of our Milky Way Galaxy, which was itself only one of billions of galaxies. The gradual acceptance of the heliocentric theory of Copernicus (Figure 3.8), followed by its triggering of the Scientific Revolution and indeed its impact in all areas of human thought, has been studied extensively. The Copernican theory eventually gave birth to a new physics, caused wrenching controversy in theology, and made the Earth a planet and the planets potential Earths in ways we are still unraveling today. Gradually, and more broadly, it changed the way humans viewed themselves and their place in the universe.[63]

The Darwinian revolution (Figure 3.9) provides a compelling analog of the trajectory of a biological worldview that bears directly on humanity's place in nature. There are, of course, differences in the epistemological status of a theory of evolution by natural selection and a discovery of life beyond Earth. Yet, like the Darwinian theory and in line with the idea of extended discovery laid out in Chapter 2, the interpretation of any discovery of life is likely to be ambiguous and debatable, and the diverse reaction to such a discovery may therefore be comparable. Like the Copernican reaction, the details of the Darwinian revolution are well studied. From the early general historical treatments of Darwinism to recent historical, philosophical, and scientific analyses, the Darwin industry itself provides a model of scholarship likely to be precipitated by a discovery of extraterrestrial life. These studies show that although there was a long period of prediscovery of the idea of evolution, the reaction to Darwin's *On the Origin of Species* was immediate, widespread, and felt at many levels of society. Debates raged, and historians have demonstrated how Darwin's theory had distinctive impacts over the short term and the long term, and among scientists, theologians, and other segments of the population. Many of these characteristics are likely to be mimicked by the discovery of life beyond Earth: an immediate strong reaction, the short-term heated controversies, the spur to scientific research, the diversity of opinion among and within groups, the widespread effect on areas of society we cannot now predict, and above all, the transformation of the way in which we view our place in nature.[64]

Figure 3.8 The Copernican worldview analogy: The discovery of life beyond Earth may gradually usher in a new worldview, just as the Copernican heliocentric theory did by making the Earth a planet and the planets potential Earths. This image is a modern reworking from Pierre Gassendi's 1654 biography of Copernicus.

Like the Darwinian worldview, but in contrast to the Copernican, the Shapley–Hubble revolution, associated with astronomers Harlow Shapley and Edwin Hubble, is still playing out. Astronomers celebrated the discoveries in the second and third decades of the twentieth century, the press routinely reported them, and the general population went about its business as usual despite humanity's slide from the center to the edge of the Galaxy in the midst of billions of galaxies.[65] That revolution continues today, with the revelations of the Hubble Space Telescope and other spacecraft, which have demonstrated our place not only in space but also in

Figure 3.9 The Darwinian worldview analogy. The work of Charles Darwin on natural selection gave rise to a new biological worldview that is widely accepted by almost all scientists but that has proven very controversial among the general public. The trajectory of this worldview may help us understand how the discovery of a biological universe might play out. This image showing Darwin late in life is attributed to British photographer Julia Margaret Cameron.

time, part of a universe 13.8 billion years old. The long-term implications of the Shapley–Hubble revelations have not been absorbed into culture; many religions, for example, have not come to terms with our seemingly insignificant place in the 13.8 billion years of cosmic evolution. Humanity is clearly not physically central in any sense in the universe. The question remains whether we are biologically central. That is the very question astrobiology is trying to answer – and we can see how high the stakes are for religion, philosophy, and the future of humanity. I have much more to say about worldviews in Chapters 6 and 7.

The Lessons of Analogy

In the end it must be emphasized that these analogies – from microbial discovery, culture contacts, the translation/transmission of knowledge, and changes in worldview – remain just that, analogies, not predictions. At the same time, we have also argued that analogy is a powerful method not to be underestimated. In a little-noticed but strongly argued essay on analogy in science, historian John North hammered home this point: "One often reads the platitude that analogical arguments are inevitably limited in their scope, because what is radically new is precisely that which cannot be accounted for in familiar terms. The whole purpose of analogies, however, is to *explore*, and to explore in the hope that what *seems* to be radically new will have unsuspected elements in common with what is familiar."[66] I could not agree more, and the "Goldilocks Principle of Analogy" enunciated earlier must be our guide: *analogy must not be so general as to be meaningless, nor so specific as to be misleading. The middle "Goldilocks" ground is where analogies may serve as useful guideposts.*

Some may argue that the discovery of life beyond Earth will be an event unlike any other, especially if it is intelligence capable of communication. That may be true to some extent, but taken in its strongest sense such a view amounts to throwing up our collective hands and ignoring the problem – a problem in which the US Congress, astrobiologists, and the general population have expressed considerable interest. In contrast, I argue that just as scientists have profitably employed analogies throughout the history of science, just as historians have investigated analogies to the impact of spaceflight and other human endeavors, and at a time when cognitive scientists have come to see analogy as the "fire and fuel of thinking," so may we cautiously deploy analogy in order to illuminate the impact of discovering extraterrestrial life, even as fundamentally novel aspects may also exist. The human mind, the basis of all behavior, remains the constant factor in our capacity to react to novel events.

Part II

Critical Issues

4

Can We Transcend Anthropocentrism?

That is the exploration that awaits you. Not mapping stars and studying nebulae, but charting the unknown possibilities of existence.

Q to Captain Jean-Luc Picard, "All Good Things..."
Series finale of Star Trek: The Next Generation

Man is not the measure of all things … and his mental models cannot be usefully projected onto the universe.

Darko Suvin[1]

Tralfamadorians, of course, say that every creature and planet in the Universe is a machine. It amuses them that so many Earthlings are offended by the idea of being machines.

Kurt Vonnegut Jr.[2]

If a lion could talk, we could not understand him.

Ludwig Wittgenstein[3]

In 1884, British schoolmaster Edwin Abbott famously wrote the satirical novel *Flatland: A Romance of Many Dimensions*. In this book the author ingeniously explores what the world would be like if it were only two-dimensional (Flatland), one-dimensional (Lineland), or zero-dimensional (Pointland), as seen from the point of view of its inhabitants. The reputed author of *Flatland*, written by "A Square," has a dream about Lineland, in which Lineland inhabitants cannot fathom what an exotic two-dimensional Flatland might be like. In the other direction "A Sphere" then visits Flatland, and its inhabitants have similar difficulty envisioning a world of three dimensions. And when, warming to the idea of many dimensions, A Square tries to convince A Sphere of the existence of even higher dimensions, he is immediately sent back to his native Flatland in disgrace for raising such a ridiculous idea. The story at the time was meant as a commentary on Victorian culture, but for us the point is that Abbott has tried to get out of his head and imagine a "what if" scenario: what if we did not live in three dimensions? Almost all science fiction begins with a "what if," but we are about to embark on a nonfictional journey where we must attempt to strip ourselves of our assumptions, just as Abbott and his characters A Line and A Square tried to do. It is notable that Abbott did not entirely succeed in his goal: after all, he still assumed a world with points, lines, and squares. We cannot entirely succeed either, but in dealing with the possibilities of alien life we must make the attempt.[4]

How do we move beyond our own preconceptions and "get out of our heads" to study concepts in astrobiology that are by definition otherworldly, perhaps entirely divorced from our everyday experience on Earth? The question is far from academic, and indeed becomes very practical when searches for extraterrestrial life and intelligence require a specific idea of the object of the search. The $100 million *Viking* Mars lander biology package could not have been built without some specific strategy and technology to search for a preconceived type of life (in that case life that metabolizes). Search for Extraterrestrial Intelligence (SETI) programs assume a certain mode of intelligence that involves technology, usually technology that will reveal itself in the electromagnetic spectrum. High technology in turn assumes a culture and civilization sophisticated enough to produce it, which assumes in turn that aliens have a social organization in some way similar to ours. And communication with such intelligence makes assumptions about technology, language, metaphysics, universals, and the nature of knowledge.

Fortunately, both scientists and philosophers have given some attention to these astrobiological difficulties, though the difficulties are far from resolved, and seldom discussed in concert. Both the scientific and philosophical approaches are useful: scientists, intimately familiar with the technical aspects of their subject, need to keep philosophers grounded in reality. Philosophers, while all too often living up to their reputation as esoteric and divorced from real-world problems, nevertheless can perform a service essential for astrobiology: they step back and analyze the most basic categories of thought, reminding scientists that their long-held assumptions may not apply universally.

In this chapter I examine the assumptions behind the crucial concepts of life and intelligence, culture and civilization, technology and communication, as exemplars of concepts that need to be critically examined in an extraterrestrial context. The Greek philosopher Protagoras notwithstanding, humanity may be the measure of all things when it comes to its own knowledge, but not when it comes to projecting its own mental models onto life beyond Earth. The process of examining our assumptions, a necessary prelude to any discussion of the impact of discovering life – and perhaps necessary to discovering it at all – is all too often marginalized in astrobiology. Our goal is not to assess the likelihood that life or intelligence will arise in any given circumstance, but to assess how our parochial ideas about these concepts might be expanded in a cosmic context. Although the length of this chapter is an indication of the importance of the problem, we can give only the briefest overview of the subject, some of which has generated a large and often contentious literature.

But the effort will suffice to highlight the questions that natural and social scientists alike must ask if we are to make a fair attempt to transcend anthropocentrism.

Life and Intelligence

Any discussion of the fundamental concepts of life and intelligence must begin with what we know happened on Earth. If we are to think out of the box, first we must know what is in the box. Table 4.1 shows the fundamental types of life on Earth from a top-level (one might say alien) view, arranged in terms of complexity moving clockwise from microbes to intelligence, which also corresponds to the evolution of biological and cultural complexity. We know single-celled microbial life arose between 3.5 and 3.8 billion years ago, or even earlier according to some latest results, strikingly fast after the Earth's formation, cooling, and the "late heavy bombardment" by rogue objects. In one form or another microbial life ruled the Earth for 3 billion years. In some ways, it still does, in terms of both numbers and impact, including our own microbiome. The rapid development of complex multicellular life began between 1 billion and 600 million years ago in the pre-Cambrian period, resulting in life such as the Ediacaran fauna. But the rapid evolution of life with hard parts is associated with the "Cambrian explosion," so called because of the fossil beds dating to the Cambrian period beginning about 540 million years ago, exemplified in the famous Burgess Shale in Canada. This is when nature first experimented with the basic body types we know today.

Intelligence is more difficult to define, and its first appearance is correspondingly more difficult to pinpoint. Depending on whether we are talking about human intelligence (itself difficult to define) or intelligence in animals such as dolphins,

Table 4.1 *Fundamental Types of Life on Earth*

Microbial	**Complex**
Single-celled Originated 3.8–3.5 billion years ago Biochemistry Originates	Multicellular up to proto-intelligence Originated about 600 million years ago Body Types Originate Contingency and Necessity Issues
Postbiological Intelligence	**Biological Intelligence**
(Products of Cultural Evolution) Artificial Intelligence Brain Emulation (ems) Human–Machine Interfaces	A very diffuse category, but about 200,000 years ago in *homo sapiens,* earlier in other species, depending on definition

chimpanzees, and elephants, or at an even earlier period in the evolution of life, the origins of intelligence on Earth might be dated to a few hundred thousand years ago or a few million years ago. Only by the broadest definition would the existence of the phenomenon we normally call "intelligence" extend back further than 10 million years on Earth, at least in the context of the evolution of humans. To round out this typology of life, postbiological intelligence, of which one form is artificial intelligence (AI)), is a long-sought goal not yet achieved on Earth at the human level except in the most rudimentary way. But some specialists in the field believe it may not be long in coming, and it may well already exist beyond Earth. This postbiological category also covers a variety of intelligence thus far only imagined in science fiction, most often in robotic form as in Isaac Asimov's novels, but occasionally in more exotic form such as Fred Hoyle's "Black Cloud" scenario discussed in Chapter 2.

Microbial Life: Scaffolding, Genetics, and Solvents, Oh My!

Microbial life immediately presents us with an opportunity to explore concepts of life because it raises the age-old questions "how did life originate?" and "what is life?" Is it based on a vital principle, an *élan vital* inexplicable in physical terms, or can it be reduced to physics and chemistry? If the latter, is its physical basis protoplasm, protein, the virus, the gene, the cell, or something even more fundamental? All these options have been considered throughout the history of biology over the past century and more amid much controversy. Today the naturalistic approach to life is accepted by almost all scientists, and debates about the origin and evolution of life are at the level of the molecular assembly of life. Experiments have demonstrated Darwinian evolution in the laboratory at the level of the nucleic acids to the extent that a widely accepted definition of life is a "self-sustaining chemical system capable of Darwinian evolution," where Darwinian evolution at the molecular level refers to descent with modification by natural selection in a replicating genetic system such as RNA or DNA on Earth.[5] The "scaffolding" for this replication system on Earth is based on carbon, hydrogen, oxygen, and nitrogen (CHON), with carbon as the anchor and water as the solvent. With the discovery decades ago of the extremophiles, microbes thriving under extreme conditions of temperature, pressure, acidity, radioactivity, and other parameters, our conception of the environment in which life might emerge has significantly broadened. New views of the diversity of metabolism, such as chemosynthetic ecosystems like hydrothermal vents deep in the ocean, have broadened our conceptions of life even more.

Given our knowledge of life on Earth, how might our concepts of life be expanded in the universal context of astrobiology? How might astrobiology change our core conceptions of life? Triggered by the search for life beyond Earth, heroic efforts have been made to take a non-Earth-centric approach. Fifteen years ago astrobiologists at NASA's Jet Propulsion Laboratory defined life as a phenomenon incorporating a structure that converts physical or chemical energy to a biologically useful form; a unique chemistry that carries out this conversion; high-fidelity replication; evolution by natural selection; energy consumption and metabolic production; and (eventually) the ability to move.[6] Given these general properties, scientists hope they can contemplate measuring biosignatures in any alien environment without the preconception that life will be similar to that on Earth other than having these general properties. And identification of such biosignatures must be done over various spatial and temporal scales, not assuming they will be Earth-like. At the same time life detection techniques must be ground-truthed by being capable of detecting any form of life on Earth.

Another attempt at describing the characteristics of life was based on enumerating even more general criteria believed to be necessary under any minimum definition for the origin and persistence of life: a fluid medium, a source of energy, and constituents and conditions compatible with polymeric chemistry. Under this system, inspired by the possible new habitats being discovered both in our solar system and on exoplanets, scientists developed a system with categories ranging from I to V, sensibly giving the Earth a plausibility rating of I (highest), and the Sun, Moon, and gas giant planets V (lowest). Based on evidence for water, energy sources, and possible organic compounds, they rated Mars, Europa, and Ganymede at II (Enceladus would certainly now join this group); Mercury, Venus, and Io at IV based on harsh conditions, while other bodies in the solar system were rated accordingly. The attempt is notable for trying to imagine all possibilities for life. Europa, for example, might use some non-terrestrial kind of chemosynthesis to provide metabolic energy. Instead of water as a solvent, the sulfur-rich volcanic Io might use hydrogen sulfide, and Titan might use hydrocarbons. And who knows what might be possible in the exotic environments of planets beyond the solar system? If locales such as hydrothermal vents on Earth are any guide, the natural environment will experiment with producing life based on whatever materials are available.[7]

Taking a step back, astrobiologists have also asked why terrestrial life should be based on carbon for its scaffolding. The reasons for carbon are well known, and often expressed something like the following:

> Carbon can form stable polar and non-polar single bonds with many other non-metals, and double and triple bonds with itself and a few other, non-metallic elements, forming complex functional groups that can bond to all other elements (with the possible exception of the noble gases). Carbon can also form polymers, bonded to itself or to other elements, that can be decorated with an immense variety of side groups to provide information storage and catalytic effect.[8]

Some astrobiologists, such as University of Colorado biochemist Norman Pace, have gone even further, suggesting that many specifics of terrestrial biochemistry are to an extent "universal," in that peptides, sugars, and nucleic acids are the most likely biological components to be built from CHON. Given this conclusion, the practical lesson is that the search for life should be limited to carbon biochemistries.[9]

Nonetheless, alternate alien biochemistries have long been a staple of science fiction and now real scientific research. In 1894 H. G. Wells conjured up "visions of silicon-aluminum organisms – why not silicon-aluminum men at once? – wandering through an atmosphere of gaseous sulfur, let us say, by the shores of a sea of liquid iron." Forty years later Stanley Weinbaum was elaborating silicon life in his short story "A Martian Odyssey" (1934). Crystalline or silicon life was also the stuff of scientific thought early on. A number of authors speculated on the idea of silicon life in the first third of the twentieth century, and one, historian of science and science popularizer Desiderius Papp, broke new ground on the subject in his 1931 book *Was lebt auf den Sternen?* [*What Lives on the Stars?*].[10]

With the rise of astrobiology the idea of alien biochemistries has come into its own, not just as a thought experiment but as a necessity in the search for life. In 2004 Cambridge University's William Bains argued against Pace and others that many chemistries might be possible for living systems, that the universality of carbon biochemistry is not necessarily the case. He argued that an "ammonochemistry" might arise in the internal oceans of Jupiter's Galilean satellites, or a silicon biochemistry in the presence of liquid nitrogen, and that this should be kept in mind in the search for life on other planets with those liquid environments. The liquid environment, he suggested, was the key to which biochemistry life would employ in any given locale. He concluded that searches for life beyond Earth should "look for a self-sustaining ability to perturb a local environment from thermodynamic chemical equilibrium into complex, patterned, non-equilibrium chemistry, rather than for narrower measures of terrestrial CHON biochemical output."[11] In a 2007 overview, paleontologist Peter Ward and biochemist Steven Benner looked at alternative biochemistries in terms of possible scaffolding elements (carbon and silicon), genetic material (DNA, RNA, proteins), and solvents (water, and various cryosolvents), tying these to particular bodies in the solar system where they might have occurred. Such seemingly far-out thinking is not merely academic. Not only should these possibilities be kept in mind in our search for life in the solar system, even as a purely intellectual exercise they would help put terrestrial life in perspective.[12]

Taking one more step back, it is important to note that aside from the metabolism concept adopted in the *Viking* experiments, definitions of life are possible based on other concepts such as energy and thermodynamics, complexity theory, and cybernetics. And while some concept of life needs to be adopted for operational life detection purposes, some scientists and philosophers have concluded that all attempts to define life are fundamentally misguided. In 1939 British biologist Norman W. Pirie, in a widely cited article, reviewed the definitions of life and argued that the terms "life" and "living" were meaningless and that the transition from nonliving to living was like the transition from green to yellow in the spectrum or from acid to alkaline in chemistry. More recently University of Colorado philosopher Carol Cleland has argued repeatedly that a definition of life is not possible in the absence of a general systems theory of biology, something we cannot have with a sample of one (Earth). Others have argued this "definitional pessimism" has gone too far, and that definitional phobia must be balanced with pragmatism. Still, there is some agreement that definitions of life are inseparable from the theories that give them meaning. There is no doubt that scientists must adopt one or more definitions of life for operational purposes such as the search for life in the solar system. But there should equally be no doubt that they may be choosing the wrong definitions. This would argue for a strategy that incorporates more than one definition, but that must be balanced with the realities of funding and weight limitations all too common in spaceflight.[13]

Ultimately, these considerations give rise to the idea of a universal biology, a program being actively pursued by several research groups, including the Institute for Universal Biology at the Institute for Genomic Biology of the University of Illinois, funded in part by the NASA Astrobiology Institute. The prospects for universal biology – a field that seeks to determine what properties must be present in any form of life, with the idea of producing a living systems theory – are so important that I discuss them as part of the next chapter on universal knowledge.

Complex Life: Chance and Necessity for Body Plans

Following the eons that microbes ruled the Earth, the next major development was the origin of "complex life." Whereas microbes like bacteria and Archaea consist of relatively simple non-nucleated cells, around 2 billion years ago much more complex nucleated "eukaryotic" cells arose, followed by multicellular eukaryotes about 1 billion years ago, in the form of plants. Complex life, including ourselves, has these eukaryotic cells as its basis, and what they accomplish in an extremely small space is nothing short of amazing (Figure 4.1). Whereas a pinhead is about 2 mm in diameter (2,000 microns), bacteria and Archaea cells are generally 1,000 times smaller (1,000th of a millimeter, or one micron), and even the eukaryotic animal cell is only 10 to 100 times larger than that. These are very small volumes considering the functions they encapsulate, including power generation carried out by the cell organelles known as mitochondria.[14]

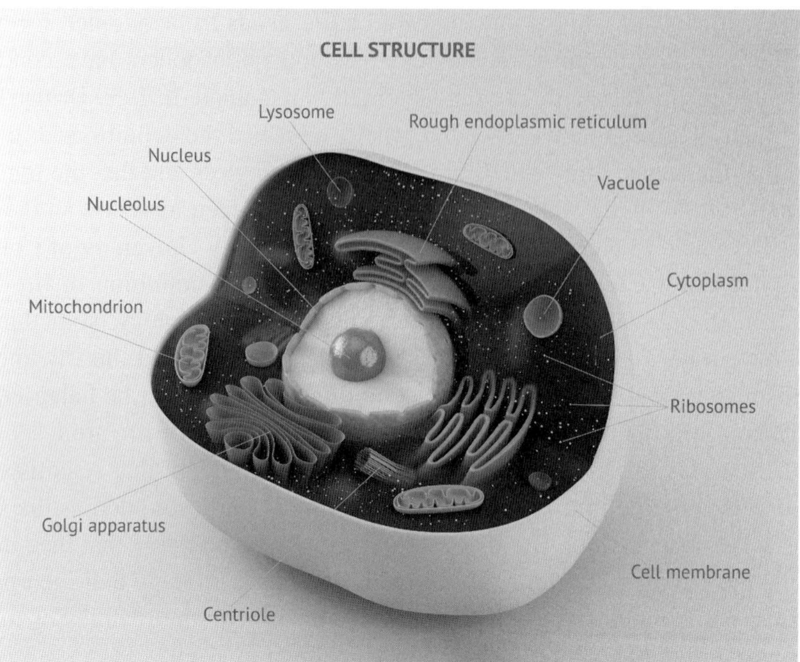

Figure 4.1 The eukaryotic animal cell, the basic building block for all complex life on Earth, is about 100th of a millimeter across. In this small but complex encapsulated space, the functions of life are carried out, including power generation in the mitochondria, the enzymatic breakdown of biomolecules and waste disposal in the lysosomes, and protein synthesis in the ribosomes. Ribosomes, seen here as small dot-like points, are about 0.03 micron (0.00003 millimeters) in diameter. Credit: Eranicle/iStock/Getty Images Plus/Getty Images.

Some moderately complex organisms originated beginning 1 billion years ago, but it was only with the Cambrian explosion some 540 million years ago that the pace of evolution of complex life picked up in a dramatic way. Whether this was due to increased oxygen levels in the atmosphere or some other cause remains unknown.[15] The increase in atmospheric oxygen was brought on by another critical invention of life, photosynthesis, in which chlorophyll molecules absorb sunlight and use its energy to convert water and carbon dioxide into glucose and molecular oxygen. The basic function is to convert light energy to chemical energy (stored in the glucose); oxygen is only a waste product, indeed initially a poisonous one that life had to overcome. Overcome it life did, with gusto; the products of photosynthesis allowed animals to grow larger, and today most complex life requires oxygen. On Earth most plants appear green because photosynthesis evolved to absorb the Sun's blue and red light, while reflecting green, although there are many examples of other pigments that absorb different wavelengths on Earth, such as red and purple. On a planet around a hotter star, plants might absorb blue light, resulting in red plants. Plants on planets around dim M dwarf stars (most of the stars in our Galaxy) might absorb all the light and appear black. All of this becomes very practical in the search for biosignatures of exoplanets.[16]

The chances for complex life developing in the universe remain very much open to debate. At the turn of the millennium, paleontologist Peter Ward and astronomer Donald Brownlee contended in a stimulating volume that while microbial life may be more abundant in the universe than we have thought, complex life may be rare. In *Rare Earth: Why Complex Life Is Uncommon in the Universe*, they argued that the conditions necessary for the evolution of complex life are so demanding that it would be extremely rare, certainly at the level of intelligence.[17] More recently researchers have proposed complex life is difficult to develop due to energy considerations. Whereas the single-celled bacterium could generate all the energy it needed within its cell membrane, the much bigger and more complex nucleated cells required an "industrial revolution" in terms of energy production if they were to survive. This revolution, the researchers argue, came only when cells gained mitochondria, which they view as an extremely rare event.[18]

These arguments for the rarity of complex life in the universe, however, are far from airtight. In the early years of the twentieth century Alfred Russel Wallace, the cofounder with Darwin of the theory of natural selection, used a Ward–Brownlee type of argument to conclude the Earth must be the center of the universe. To put it mildly, we now know this not to be true, suggesting that the "just so" story may not be a good argument.[19] And the rarity of complex life from the industrial revolution argument for complex cell energy rests entirely on the assumption that cellular absorption of mitochondria is an extremely rare event, an assumption also open to question. Nevertheless, the 3 billion years it took complex life to develop on Earth is a significant data point. Complex life may indeed be rarer than microbial life, but this must be weighed against the 13.8 billion-year history of the universe, of which at least the past 8 billion years might have been suitable for the development of life once enough heavy elements had been generated. Indeed, Harvard astronomer Avi Loeb has argued that complex life was possible in the first billion years of the creation of the universe.[20] The case of the frequency of complex life in the Galaxy thus remains wide open.

But if it does arise, what assumptions lay hidden about the nature of complex life? The evolution of complex life remains very much part of the astrobiology research program, undertaken by NASA's MIT Astrobiology team and other research groups around the world. The history of complex life on Earth raises a set of questions distinct from the issues of the definition of life and alien biochemistry associated with the origin of microbial life. Because the development of complex life during the Cambrian is associated with the creation of numerous body types, the "morphological" question arises of how these types were affected by environmental conditions on Earth, and how conditions on other planets would affect body types. The question of necessity and contingency in the evolution of complex life now comes to the fore, just as it does for the origin of life, but at a different level.

The underlying assumption behind all morphology and contingency arguments is that wherever life originates in the universe, Darwinian evolution by natural selection operates. Although this has given rise to many strange and wonderful morphologies on Earth based on diverse environmental conditions, it is also the case that where natural selection operates, convergent evolution often takes place. For example, the eye has been reinvented many times independently, as have wings in insects, birds, and bats (Figure 4.2). Fish and marine mammals such as dolphins have evolved streamlined shapes for their water environment. Evolutionary paleobiologist Simon Conway Morris has argued that convergent evolution will result in aliens very similar to us in morphology. On the other hand, evolutionists such as George Gaylord Simpson and Stephen Jay Gould have stressed contingency in the history of life, the latter precisely in the context of the Cambrian explosion of body types. Gould argues that rerunning the tape of life would result in very different lifeforms. Whether contingency or convergent evolution dominates the history of life in any given setting is still unknown, but their joint action will determine just how strange alien life will be.[21]

Such overall principles have given rise to many scenarios in both science and science fiction. The power of convergent evolution in an extraterrestrial context was emphasized 30 years ago in Gene Bylinsky's delightful book *Life in Darwin's Universe: Evolution and the Cosmos*. There Bylinsky argued that complex organisms beyond Earth would in general look like terrestrial life, because only a limited number of engineering solutions are possible when it comes to the necessities for life – similar to the argument in Conway Morris's book *Life's Solution*. Features ranging from a head well placed for sensory organs to bipedalism evolved repeatedly on Earth and will likely do so on other planets in response to the organism's needs as molded by natural selection. On the other hand, because they would have been shaped by their own unique planetary environments, organisms would be different from us in the particulars, just as there is great diversity of life on Earth, including the different requirements of land and water organisms. More generally, gravity imposes size limitations on life; from the cell to the whale is a large range indeed, but the food system of the whale (and the dinosaur on land) must strain to feed such a large structure, even as the heart struggles to sustain its blood flow. Life on a low-gravity planet might be free to soar upward in both the plant and animal kingdoms, while life on a high-gravity planet would be correspondingly stifled.[22]

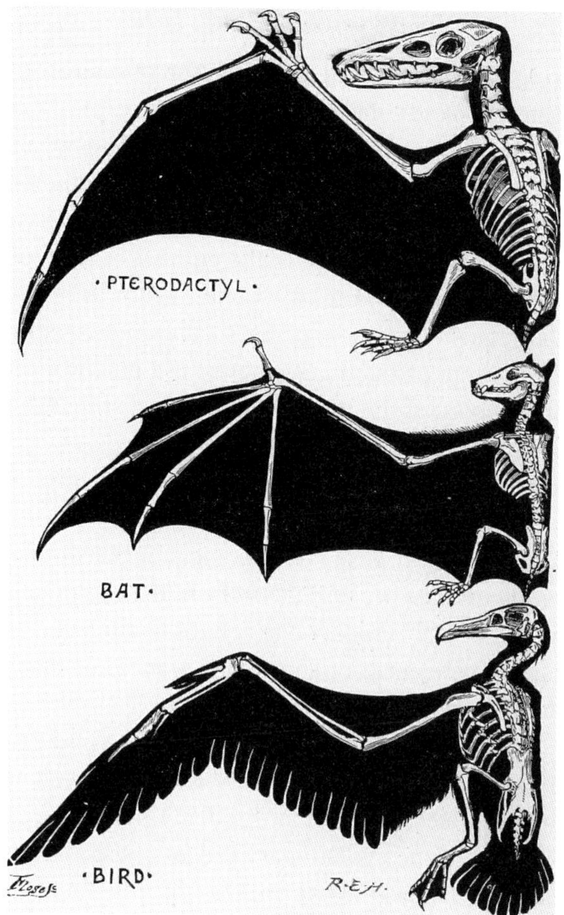

Figure 4.2 Functional convergent evolution of vertebrate wings as mechanisms for flight in pterosaurs, bats, and birds. From evolutionary biologist George John Romanes, *Darwin and after Darwin* (1892). Convergent evolution may argue for similar morphologies in alien life.

These themes as played out in science fiction are too numerous to mention, but they have been seriously and delightfully addressed in books such as Jack Cohen and Ian Stewart's *Evolving the Alien*, Terrence Dickinson's *Extraterrestrials: A Field Guide for Earthlings*, and Susan and Robert Jenkins's *The Biology of Star Trek*. There are even books describing how to create aliens and alien societies in fiction, based on scientific principles. And the best of science fiction itself represents a body of thoughtful knowledge that should not be underestimated. That is why we repeatedly turn to science fiction throughout this book in our efforts to transcend biological anthropocentrism.[23]

Intelligent Life: What Is It?

All the questions regarding the origin and evolution of microbial and complex life are multiplied when it comes to the evolution and nature of intelligence. Our usual concept of intelligence in the context of SETI is dominated by the Drake paradigm, whose central icon is the Drake Equation (Figure 4.3), including the crucial term representing the fraction of planets on which intelligence arises.[24] Because the Equation goes on to winnow this down to the number of technologically communicative civilizations, we normally think of intelligence as an entity that can conceive and build the requisite technology for such communication. This, however, is a very restricted concept of intelligence, even if a practical one in terms of our ability to detect it. In short, at the outset we need to distinguish intelligence from technologically communicative intelligence.

What, then, do we know about intelligence? The difficulties with that slippery concept are immediately revealed by posing the simple question "when did intelligence arise on Earth?" Even if posed to biological or paleontological experts, the answers would wildly diverge because the story of the evolution of intelligence on Earth depends entirely on how broadly one defines intelligence. Is it hominid intelligence, mammalian intelligence, or some earlier ancestor that gets the prize for the first "real" intelligence? Surely the only reasonable answer is that there was no one day, year, or even period in the history of life when intelligence was lacking, and then suddenly appeared. It is not a binary choice – you have it or you don't. It follows there was no one lifeform that suddenly possessed the quality we call intelligence. It was part of the evolutionary process, just as physical morphology was part of that process; indeed the two are inextricably intertwined.

That said, in tackling the question "what is intelligence?" scholars most often start with the evolution of human intelligence, even while recognizing that it may

$$N = R_* \times f_p \times n_e \times f_l \times f_i \times f_c \times L$$

Astronomical Biological Cultural

N = The number of technological civilizations in the galaxy.
R_* = The rate of formation of stars suitable for the development of intelligent life.
f_p = The fraction of those stars with planetary systems.
n_e = The number of planets in each planetary system with an environment suitable for life.
f_l = The fraction of suitable planets on which life actually appears.
f_i = The fraction of life-bearing planets on which intelligent life emerges.
f_c = The fraction of planets with intelligent life that develop technological civilizations.
L = The lifetime of a technological civilization.

Figure 4.3 The Drake Equation, a heuristic for estimating the number of technologically communicative civilizations in the Milky Way Galaxy, assumes intelligence with a certain kind of technology. The Equation incorporates astronomical, biological, and cultural factors, many of which are unknown, especially toward the right of the Equation addressing cultural evolution.

be only one manifestation. Carl Sagan's Pulitzer Prize-winning book, *The Dragons of Eden: Speculations on the Evolution of Intelligence*, attempted to synthesize what was known about human intelligence on Earth as of the mid-1970s. The book featured the triune brain consisting of the primal reptilian R-complex, the mammalian limbic system, and the primate neocortex, a model based on comparative neuroanatomy and behavior and explicitly grounded in evolution. Sagan argued that to a first approximation the aggressive behavior, territoriality, ritualistic, and social hierarchy aspects of our lives are influenced by the primal R-complex; the altruistic, emotional, and religious aspects are localized in the limbic system shared with non-primate mammals; and reason is associated with the neocortex. In other words, Sagan's view totally incorporated the idea that human intelligence was a product of a very long evolution. There is no doubt that he is correct in the general evolutionary sense, but the details of the triune brain have been largely discredited and are no longer accepted by most neuroscientists.[25]

Coming from an entirely different viewpoint, computer architect Jeff Hawkins reaches a similar conclusion based on evolution. In his book *On Intelligence*, he takes a broad view of intelligence, admitting that the simple nervous system of multicellular creatures by one definition constitutes intelligence, but arguing that "the story of real intelligence" began with our reptilian forbears some 300 million years ago. After all, these forbears had sophisticated senses and a primitive brain that allowed them to carry out complex behaviors. As with Sagan, he noted humans still have that reptilian brain, but have added a large cortex, known as the neocortex. The reptilian brain in us still carries out many physiological functions, but the neocortex evolved only a few tens of millions of years ago with mammals. And only 2 million years ago did the neocortex expand dramatically with our human ancestors, giving us abstract reason. Although memory capacities are strewn throughout the brain at various levels, the neocortex is an important part of the memory system, along with other components such as the hippocampus. For Hawkins, memory is the key to intelligence, because it allows prediction, and prediction, not behavior, is the key to intelligence.[26]

As we saw in Chapter 3, archaeologist Steven Mithen offers yet another approach to the evolution of human intelligence grounded in paleoanthropology, in which a variety of specialized intelligences, which he terms social, technical, and natural history intelligences, were integrated over millions of years, giving rise to a "cognitive fluidity" that allowed for language and a more generalized intelligence leading to tools, religion, art, and the modern mind less than 100,000 years ago.[27] Scientists study human intelligence from a multitude of other points of view, including evolutionary psychology, mapping the mind, climate and brain-cooling, the computational metaphor, and the memory and pattern analysis employed in the neocortex of the human brain as understood by neurobiology. From the point of view of neurobiology, there is no doubt that neurons are the empirical markers of information processing and eventually intelligence, and that they are one way of tracking the evolution of the mind using a concrete metric. None of these approaches has as yet proven definitive, but all have in common the Darwinian concept that intelligence is a product of cognitive evolution brought about by natural selection acting on the environment. It may be that some combination of these approaches will eventually result in a general theory of intelligence, but we are nowhere near that desired goal.

While most research has focused on the evolution of intelligence in hominids on Earth over the past 5 to 7 million years, for astrobiology, the evolution of intelligence in a more general sense is of primary importance. Here broader definitions of intelligence apply, such as "the ability to respond flexibly and successfully to one's environment and to learn from experience," or, put another way, "the ability to respond with an appropriate behavior in a given context."[28] Such broad definitions clearly apply to a wide range of animals. Accordingly research has been undertaken especially on dolphins and other mammals and some of the results have been sporadically summarized and applied in an extraterrestrial context at astrobiology meetings over the past several decades. Research on dolphin communication, brain encephalization (a normalized measure of brain size in a variety of animals evolved over the last 500 million years), convergent brain evolution on Earth, and progressive trends in the history of life have all been occasionally represented at these meetings. By some definitions, there is a good deal of consensus that the higher primates and some cetaceans such as dolphins and whales share some of our reasoning abilities.[29]

These studies raise the burning question of whether we can find general principles of intelligence analogous to attempts to find general principles of life. Recently researchers have sought to find such principles of intelligence inferable from the evolution of life on Earth in a broader sense than just hominids, and to apply them in the context of astrobiology. Indiana University evolutionary biologist Marc Bogonovich argues that the evolution of intelligence needs to be understood in the context of events spanning at least the past 540 million years of life on Earth. He outlines 10 hypotheses related to both human intelligence and "broader sense intelligence" in animals, and calls for a research program to determine what might be relevant to the evolution of a more generalized intelligence in the astrobiological context. Other researchers go even further. Employing the broad perspective encouraged by astrobiological thinking, neuroscientist and animal behaviorist Lori Marino frames intelligence in terms of vertebrates, invertebrates, and social insects. Among vertebrates Marino emphasizes that it is important to keep in mind that all vertebrate brains are evolutionary variations on a common theme laid down very early in vertebrate evolution. Tool-making and use, once thought to separate humans from animals, are now known to occur in many other species, including primates, elephants, birds, dolphins, whales, sea otters, and octopuses. Culture, self-awareness, and communication are also traits other animal species share with humans, although in diverse manifestations. She concludes that intelligence in this sense is a ubiquitous phenomenon in the animal kingdom, lending credence to the idea of great diversity if extraterrestrial intelligence exists. It might not be the intelligence that SETI programs seek, but such simple intelligence might give rise to more complex intelligence in many forms given the proper ecological niche, extinction events, and environmental circumstances. It may not be intelligence that can build radio telescopes, but forms of life that do other things we have not yet recognized.[30]

In order to illustrate this, let's take the case of octopuses, that cephalopod branch of the tree of life that has also evolved what many consider an exotic form of intelligence. In his book *Other Minds: The Octopus, the Sea, and the Deep Origins of Consciousness*, philosopher of science Peter Godfrey-Smith gives us a firsthand glimpse at the life of the octopus based on his own research and scuba diving adventures. These animals (Figure 4.4) have large brains, but very different nervous systems from mammals. The majority of their neurons (500 million in some species compared to 100 billion for humans) are not centralized in their brains, but distributed in their eight arms. Their "intelligence" is based not on what they think about but on what they can do, a suite of activities including navigating mazes, unscrewing jars to obtain food, and adapting to new and unusual circumstances. Although this form of intelligence is very difficult to compare to human or insect intelligence, Godfrey-Smith finds octopuses curious and flexible, adventurous and opportunistic. Unlike insects, they seem to have little social sense, mostly living as hermits. They therefore have no "culture" that we associate with intelligence, though Godfrey-Smith has also documented gregarious octopuses in the area he dubs "Octopolis," 50 feet below the surface off the coast of east Australia. He believes octopoid intelligence is the closest we will come to meeting an alien – until it actually happens. Where might the evolutionary path lead for octopoids on another world? Although it is difficult to imagine octopuses developing culture and technology in their watery world, nature does not preclude their evolution onto land in the same way amphibians once did on Earth. Indeed, author Michael Chorost has richly imagined land-dwelling octopoids in his book *How to Talk to Aliens*.[31]

This broader view of intelligence and its evolution greatly expands the research possibilities for astrobiology. Even invertebrates, representing 95 percent of all living animals on Earth, shed light on the evolution of complex nervous systems, and thus the evolution of intelligence. Writing about the latest research in this area as part of the astrobiology project, David Gold asserts, "If we think about intelligence as an organism's ability to adaptively respond to problems, there are clearly varying

Figure 4.4 The octopus may be the closest thing to an alien intelligence on Earth. Octopuses have evolved from the cephalopod tree of life rather than the mammalian. They have large brains, but their 500 million neurons are distributed throughout their eight arms. Credit: Ernst Haeckel, *Kunstformen der Nature* (Art Forms in Nature), 1904, Plate 54.

degrees of intelligence throughout the invertebrate world." Even if one does not accept this broad view of intelligence, the evolution of complex nervous systems is clearly applicable to astrobiology. Broadening even further, neurobiologist Maggie

Wray argues that while we usually think of intelligence in the form of individuals, "alien intelligence could alternatively take the form of a distributed, or collective, intelligence in which the cognitive abilities of cooperative *groups* of organisms far surpass the intellectual capabilities of any single individual." She illustrates this with colonies of social insects, which have complex communication systems, collective problem-solving and decision-making, and social learning. Again, Michael Chorost has delightfully imagined this possibility with "Bugworld" in his book *How to Talk to Aliens*.[32]

Intelligence researcher Denise Herzing has gone a step further with an approach to profiling organisms according to five criteria, including not just the standard encephalization quotient but also communication signal complexity, individual complexity, social complexity, and interspecies interaction. Dubbed COMPLEX (Complexity of Markers for Profiling Life in Exobiology), the effort attempts to profile a range from microbes to machines across these five categories, specifically with the goal of helping us assess other species discovered beyond Earth.[33]

Given all these possibilities in the landscape of intelligence, the question remains: will intelligence in any form actually develop, or will the "rare Earth" scenario prevail? That depends on your point of view in a contentious argument that centers once again on the idea of evolutionary convergence, just as we saw with the eye. One of Sagan's motivations for writing *Dragons of Eden* was to gain hints or insights into extraterrestrial intelligence; he concluded:

> once life has started in a relatively benign environment and billions of years of evolutionary time are available, the expectation of many of us is that intelligent beings would develop. The evolutionary path would, of course, be different from that taken on Earth … But there should be many functionally equivalent pathways to a similar end result. The entire evolutionary record on our planet, particularly the record contained in fossil endocasts, illustrates a progressive tendency toward intelligence.[34]

Alas, that conclusion embodies many questionable assumptions: the fact that bigger brains needed more time to evolve doesn't necessarily indicate a progressive tendency toward intelligence. Evolutionists such as George Gaylord Simpson and Theodosius Dobzhansky had already argued just the opposite of Sagan's claim, and Harvard evolutionist Ernst Mayr also differed strongly with Sagan, arguing that intelligence (by his definition) had emerged only once on Earth.[35]

Outspoken Harvard evolutionist Stephen Jay Gould agreed with the nonprevalence of *humanoid* intelligence, arguing in his book on the fossils of the Cambrian explosion that if we "wind back the tape of life to the early days of the Burgess Shale; let it play again from an identical starting point, and the chance becomes vanishingly small that anything like human intelligence would grace the replay." That, of course, depends on what we mean by "human intelligence," which is simply one variation on a theme that has played out over tens, if not hundreds of millions of years. By contrast Conway Morris, the king of convergence in all things biological, has argued that evolutionary convergence applies not only to morphology, but also to intelligence, if only the proper conditions are present. He is, however, skeptical that the proper conditions often obtain, summarizing his position in the subtitle of his 2003 book *Life's Solution: Inevitable Humans in a Lonely Universe*. In this he reached the same conclusion as had Ward and Brownlee, who argued that complex life and thus intelligence in the universe will be rare, not from a lack of convergence, but because so many factors must come together in order for it to exist. The question cannot be decided with the knowledge at hand. Given the number of extrasolar planets now known, and that a significant fraction are Earths or super-Earths in terms of mass, what is obviously needed is more knowledge of conditions on those planets, conditions on which natural selection might act in producing not only life but also intelligence.[36]

One thing does seem likely, however: evolutionary convergence notwithstanding, the specific sequence of events seen on Earth will not reappear on another planet, and natural selection will accordingly affect both morphology and forms of intelligence. Loren Eiseley's famously poetic dictum "In the nature of life and in the principles of evolution we have had our answer. Of men elsewhere, and beyond, there will be none forever" may be true in a specific sense. But the jury is still out in the most general sense of whether intelligence is a trait selected for competitive advantage that would lead to its widespread existence in the cosmos, and whether such intelligence, once formed, would be similar to ours or vastly different in its physical foundation, internal workings, sensory output, and ability to communicate. It is perhaps the greatest unanswered question in the universe.[37]

Postbiological Intelligence: Taking Cultural Evolution Seriously

Finally we broach the fourth type of life in Table 4.1: postbiological intelligence, a product of cultural in addition to biological evolution, the final two crucial parameters of the Drake Equation. The key realization here is that cultural evolution is not often taken seriously enough in imagining what extraterrestrial intelligence might be like. It may well give rise not only to technology and war, literature and painting, mosques and cathedrals (Figure 4.5) but also to changes in human nature, including postbiological intelligence. This is especially true if we recognize the necessity of thinking in what I call Stapledonian timescales, on the order of millions or billions of years, when we are contemplating the nature of intelligence in the universe today. One result of such thinking is artificial intelligence (AI), which does not yet exist on Earth in any sophisticated form that would be judged better than human intelligence. But the past few years have indeed seen tremendous progress in AI technology in specific domains. The question remains whether artificial

Figure 4.5 Cosmic evolution's factors of astronomical, biological and cultural evolution, represented sequentially by a globular cluster of stars, atoms viewed along the axis of DNA molecule, and the stained glass window from the south transept of the Notre Dame Cathedral in Paris. All three forms of evolution are governed by underlying scientific principles that lead to increasingly complex structures. Taking cultural evolution into account leads to the possibility that the universe may be postbiological. From Dick and Lupisella (2009). Credit: NASA, adapted from Chaisson (2009).

general intelligence is possible. If it is, because the laws of physics inherently limit a biological brain from increasing its intelligence even over long timescales, long-lived technical civilizations, for whom the improvement of intelligence may be a driving force of cultural evolution, will likely seek to increase intelligence through other means.[38]

I have formalized this concept in what I call the Intelligence Principle: "*The maintenance, improvement, and perpetuation of knowledge and intelligence is the central driving force of cultural evolution, and that to the extent intelligence can be improved, it will be improved.*" Failure to do so may cause cultural evolution to cease to exist in the presence of competing forces. It is true that artificial intelligence is only one possible outcome of cultural evolution (biotechnology and genetic engineering are others), but in my view it is the most likely outcome for a variety of reasons I have argued elsewhere. It could take the form of self-enhancement, where biologicals become cyborgs eventually approaching AI. Or it could be that AI creations not descended from biological life supplant biologicals, giving rise to something very different from biological intelligence. In either case, given the long timescales in the universe, on the order of billions of years since life has been possible after the creation of heavy elements and rocky planets, cultural evolution may thus have given rise to artificial intelligence. The idea has become increasingly accepted over the past two decades. Among others who consider postbiologicals as likely are physicist Paul Davies, astronomers Martin Rees and Seth Shostak, and philosopher Susan Schneider.[39]

Famously, the transition from biological to postbiological intelligence is an event that some AI experts such as Hans Moravec and Ray Kurzweil see as happening in a few generations on Earth. Whether this is a good or bad thing is in the eye of the beholder, especially if it results in leaving us biologicals behind. This is the fear of Stephen Hawking, who warned the BBC in 2014 that "the development of full artificial intelligence could spell the end of the human race." Concern is high enough that in England, for example, the mission of the new Leverhulme Centre for the Future of Intelligence is to prepare society for the benefits and dangers of machine intelligence. In the United States, Carnegie Mellon University, among others, has inaugurated a research program on the ethics of artificial intelligence. And in 2016, the White House held a series of workshops and released a detailed report on the consequences of AI. Much of the latter naturally deals with practical issues of AI at a lower level, but it will undoubtedly lead to further study of the concerns that Hawking expressed.

Whether or not sophisticated conscious and self-aware AI is developed on Earth in the next few generations, my point is that given the long timescales in the universe, it might have already happened long ago elsewhere in the universe, perhaps yielding a postbiological universe rather than the biological universe we usually envision. While this may seem like science fiction, the concept arises and is given force by taking cultural evolution no less seriously than physical and biological evolution in the course of cosmic evolution. Of course, reasonable people may argue about the direction of cultural evolution, but it is hard to think of anything more basic than improving intelligence, which affects everything else.[40]

The postbiological scenario does have one shortcoming: it presupposes that artificial intelligence is possible in principle. While many in the computer field take this for granted, philosophers such as John Searle have long argued that "strong AI," the idea that the mind is just a computer that could be reconstructed at will, is demonstrably false. His famous "Chinese room" experiment, in which Chinese symbols are passed into a room and "translated" by a person manipulating symbols without any knowledge of their meaning, illustrates that a computer can manipulate formal symbols without necessarily knowing the meaning of what it is doing. In other words, it has syntax but no semantics. And if the mind is not just a computer, the implication is that we can never build a computer equal to human intelligence, much less exceeding it, since there is no way to get from formal symbols to meaning, which is what many take to be the core of intelligence. And if this is true "in principle," presumably extraterrestrial intelligence would not be able to construct artificial intelligence either.[41]

Searle's arguments are certainly not airtight even in a purely terrestrial context, and ambitious artificial intelligence experts are banking on them not being true. If strong AI is possible and such advanced intelligence does exist, we are justified in asking "what are its minimum characteristics and what is its nature?" Pursuing this question soon leads us down the rabbit hole of consciousness, cognition, mind, and brain, all concepts related to intelligence but with their own definitional problems. Surely one of the minimum qualities for advanced AI is consciousness, a concept that remains a subject of great controversy, despite books with optimistic titles such as philosopher Daniel Dennett's *Consciousness Explained*. I once posed the question of the nature of intelligence to Searle. "Forget intelligence," he vigorously replied, "the origin and nature of consciousness is the real problem!"[42] Some ascribe lower consciousness to animals, and identify the evolution of a higher-order consciousness with higher primates. That too is a subject of great debate. One philosopher quipped that "a newborn is barely conscious, if that; a philosopher is fully conscious – too much so sometimes."[43] The related concept of cognition generally refers to the mental processing of sensory input that goes on inside the brain, and is also ascribed to both animals and humans. All of this in turn relates to the broader "mind-body problem," one of the long-standing issues in philosophy, dating back at least to Cartesian dualism, which most of the general public in the Western world still accepts.

What is certain is that the concepts of mind, consciousness, and cognition are increasingly amenable to research by neuroscientists, paleobiologists, anthropologists, psychologists, and philosophers of mind, studying both animals and humans. But taken as a group, even experts are far from consensus on any of these issues. We cannot even answer the question "is intelligence possible without consciousness?" Philosopher David Chalmers has argued it is logically possible; indeed these are the "zombies" of the human imagination, or at least the imaginary "philosophical zombies" that some philosophers like to discuss to illuminate problems about consciousness such as physicalism and dualism when it comes to the mind-body problem. Basically, the zombie problem asks how we can ever know whether something is physically alive but mentally empty.[44] Science fiction writer Peter Watts places zombies in an extraterrestrial context in his novel *Blindsight*, a remarkable

First Contact narrative set in the late twenty-first century where humans explore a spaceship inhabited by weird starfish-shaped creatures that are intelligent but not conscious. "Imagine you have intellect but no insight, agendas but no *awareness*," Watts has his character Siri muse in attempting to understand the alien "scramblers" his crew has discovered on the spaceship *Rorschach* in the Oort cloud at the edge of the solar system. "Your circuitry hums with strategies for survival and persistence, flexible, intelligent, even technological – but no other circuitry monitors it. You can think of anything, yet are conscious of nothing." Whether astrobiology needs to worry about such creatures remains to be seen, but philosopher Susan Schneider has argued that advanced aliens need not have consciousness.[45]

Zombies aside, and widespread public acceptance of a dichotomy between mind and body notwithstanding, today almost without exception experts in philosophy, psychology, cognitive science, neurobiology, and artificial intelligence accept some version of Darwinian materialism (or physicalism, as philosophers like to call it these days to avoid issues associated with the term "materialism"). This is what Francis Crick termed the "astonishing hypothesis," whereby consciousness and intelligence are seen to arise from the workings of the brain, itself a product of natural selection, even if the mechanism is still under dispute.[46] That view, at least, makes the problem scientifically tractable, because Earth can be seen as a laboratory for intelligence. But there are prominent objectors, ranging from A. R. Wallace, Darwin's cofounder of natural selection a century ago, to philosopher Thomas Nagel today. In his book *Mind and Cosmos: Why the Materialist Neo-Darwinian Conception of Nature Is Almost Certainly False*, Nagel, like Wallace, still sees no path whereby the mind could originate by natural selection. But here we are. Though Nagel denies it, his conclusion seems to point to a religious solution, unless some non-supernatural superintelligence played a role. All this by way of saying that we do not understand consciousness in humans or other animals, much less whether postbiologicals would be conscious.

This is not the place to elaborate on these fascinating problems, but we can ask "If it does exist, what would postbiological intelligence be like? What would it be *doing*?" Nick Bostrom has given us some idea in his book *Superintelligence*, where he defines superintelligence as "any intellect that greatly exceeds the cognitive performance of humans in virtually all domains." He argues that there are many paths to superintelligence, including artificial intelligence as we usually think of it with a silicon substrate (or carbon nanotubes or graphene), but also whole brain emulation (scanning and uploading a brain), biomedical or genetic enhancement to biological cognition, and human–machine interfaces. In his view AI may well be the first to cross the finish line of superintelligence on Earth. Any of these superintelligences could be of several types, based on speed, collective performance (hive minds like the Borg on Star Trek), and quality of thought. In all cases, "An artificial intelligence need not much resemble a human mind. AIs could be – indeed it is likely that most will be – extremely alien." He argues that the goals of superintelligence might be self-preservation, goal content integrity (maintaining its goals through any future developments), cognitive enhancement, technological perfection, and resource acquisition. A great deal more could be said about these goals and their potential dangers and benefits to humanity, but enough has been said to give an idea of the possibilities. Cognitive enhancement and technological perfection support the central idea of our Intelligence Principle, and resource acquisition leaves the imagination to wonder about just what this might mean.[47]

While Bostrom's speculations are largely terrestrially based, philosopher Susan Schneider, a former student of Searle also inspired by Bostrom, has taken up the challenge of AI in the context of extraterrestrial intelligence. She does so in the context of the philosophy of mind, particularly ideas of consciousness with regard to AI. She relates the possibility of extraterrestrial AI to "the hard problem of consciousness," which focuses not on the "easier" problem of information processing in the brain, but rather asks why there is a subjective "felt quality" to our conscious experience such as the human response to music or the rich hues of a sunset. While AI might solve problems, would it *feel* a certain way? Could such a felt quality exist in AI with its silicon rather than carbon substrate, even if it were conscious? The even harder problem is whether AI can be conscious. Not according to Searle's "biological naturalism," which claims that mental phenomena are based on biological processes in the brain, which require a carbon rather than silicon substrate. So Schneider argues against the biological naturalism embodied in Searle's Chinese room argument. She emphasizes that while the person in the Chinese room might not understand meaning and semantics, the larger system as a whole does understand it. Perhaps a complex silicon-based superintelligence *could* understand: just because brains cause consciousness doesn't mean there is no other path to consciousness. She concludes not only that most advanced forms of superintelligence in the universe might be postbiological artificial intelligence, but also that they might well have conscious experiences, especially if they are biologically inspired, that is, based on reverse engineering an alien biological brain. Alas, because of what is known as "the problem of other minds," we may never be quite certain whether AI has consciousness. Schneider has also explored these questions in relation to Spike Jonze's popular film *Her*.[48]

Does the argument against strong AI "in principle" apply to minds beyond Earth? Must semantics (meaning) be the core of all intelligence? Might a million- or billion-year-old intelligence have found a way around this? The viability of the postbiological intelligence argument depends on the answer no to the first and second questions, and yes to the third. More generally, the question is whether intelligence needs to be equated with various forms of computation. The answer may well be no.

All of this has practical implications. If AI is possible in principle and has been carried out in practice, a postbiological universe would likely affect the nature of any signal transmitted as well as methods of communication, important considerations for SETI. Jeffrey Hawkins's model of intelligence gives us a hint of how different such an intelligence could be

> There is no reason that an intelligent machine should look, act, or sense like a human. What makes it intelligent is that it can understand and interact with its world via a hierarchical memory model and can think about its world in a way analogous to how you and I think about our world … its thoughts and actions might be completely different from anything a human does, yet it will still be intelligent. Intelligence is measured by the predictive ability of a hierarchical memory, not by humanlike behavior.[49]

Nor does the lifetime of a technological civilization, the famous parameter L in the Drake Equation, need to be very long for the postbiological universe to be realized. When one considers the accelerating pace of cultural evolution as we enter the third millennium of our era, radical change of the sort foreseen by Moravec and Kurzweil does not seem so far-fetched.

We thus come to a startling conclusion, codified in Table 4.2. Based on what experts see happening on Earth, L need not be 5 billion, 1 billion, or a few million years. It is possible that a postbiological universe would occur if L exceeds a few hundred or a few thousand years, where L is defined as a technological civilization that has entered the electronic computer age, which on Earth was almost simultaneous with the usual definition of L as a radio communicative civilization. If L is less than a few hundred years, less than the time it takes for a technological civilization to conceive, design, construct, and launch its intelligent machines, we do not live in a postbiological universe. If L is between 100 and 1,000 years, a transition zone may result populated by human/machine symbiosis, sometimes referred to as "cyborgs," genetically engineered humans, or some other form. But if L is greater than 1,000 years, we almost certainly will have made the transition to a postbiological universe. And the question arises: would postbiologicals be likely to undertake electromagnetic SETI, or some other form of seeking intelligence in the universe? Since they and/or their descendants would be essentially immortal, would they not simply travel among the stars?

As an extrapolation of a contemporary terrestrial theme through analogy, the greatest weakness of the postbiological scenario may well be that it is not bold enough. As Arthur C. Clarke would say, it is perhaps a failure of imagination. Other means may exist to advance intelligence rather than through artificial intelligence as we conceive it, including some of the ways Bostrom envisions. The new fields of artificial or synthetic life, where chemists attempt to synthesize molecules that might support alternative genetic systems or metabolisms, might also shed light on the evolution of life and intelligence beyond Earth. Moreover, the entire

Table 4.2 *Lifetime of a Technological Civilization (L) and Effects on SETI**

L (Years)	Stage of Cultural Evolution	Effect on SETI
<100	Biological	- Civilizations Scarce but Comparable Level - Electromagnetic SETI Possible
100–1,000	Machine/Biology Hybrid (Cyborg)	Hybrid Techniques
>1,000	Postbiological - Advanced Artificial Intelligence (AI) - Brain Emulation (ems)	- Direct Electromagnetic SETI Unlikely? - Not Confined to Planetary Surfaces - Interstellar Travel Likely?

* Elaborated from Dick (2003). Note that civilizations at a level comparable to humans are unlikely, given the timescales of the universe.

postbiological scenario presupposes that, when all is said and done, science and technology progress over long timescales, an Enlightenment idea that is itself problematic, even in the terrestrial context. Still, the postbiological universe remains a logical possibility, and in my view a highly probable one, even if it embodies the idea of "progress."[50]

The Landscape of Life and Intelligence

In this section I have tried to indicate the general principles underlying the possible kinds of life in the universe, the landscapes of life and intelligence far beyond our normal anthropocentric views of terrestrial life. Table 4.3 summarizes these general principles, namely biochemistry, morphology, and cognitive capacity, all of which may operate through evolution according to physical conditions much broader than we now realize, just as the extremophiles on Earth have expanded our view of the possibilities of life. We have found that in considering even the most basic life beyond Earth we need to take into account the scaffolding, genetic material, and solvents available at any given location. For complex life, we need to consider the body patterns in terms of size, shape, and mass such as originated in the Cambrian explosion, all closely associated with issues of chance, necessity, and evolutionary convergence. For intelligence, we need to consider scientific issues about what constitutes intelligence, as well as age-old philosophical issues of mind, body, and consciousness, and again issues of convergence. And I have argued for the necessity to think even further out of the box of intelligence by emphasizing that the universe may be postbiological. Table 4.3 also lists one other form of intelligence I mentioned only in passing earlier in connection with social insects, a collective or distributed form often known as hive, or swarm, intelligence.

Table 4.3 *Principles for Possible Types of Life beyond Earth*

BIOCHEMISTRY/ GENETICS/SOLVENTS	**MORPHOLOGY**
Carbon/DNA/Water	Body Patterns:
Carbon/RNA/Water	Size
Carbon/Protein/Water	Shape
Silicon/?/Nitrogen	Mass
Ammonia/?/Methane	
SPACE-TIME FILTERS	**COGNITIVE CAPACITY**
Geometry	Sensory Apparatus
Mathematics	Consciousness
Sense of Time	Cognition
	Biological Intelligence
	Postbiological Intelligence
	Collective “Hive” or “Swarm” Intelligence

Some of these possibilities have already been imagined in science fiction. The largely anthropocentric aliens of *Star Trek* and *Star Wars* are well known, but they are only the tip of the iceberg. Lesser known are Fred Hoyle’s 90-million-mile-diameter sentient cloud of interstellar hydrogen, Stanislaw Lem’s sentient ocean, Hal Clement’s Meslkinites laboring on a planet where gravity varies from 3 to 700 times its strength on Earth, and Arthur C. Clarke’s winged Overlords. Postbiological machines of varying capacities roam the universe in any number of science fiction novels, sometimes bringing good (as in Michael McCollum’s *Lifeprobe*), sometimes destruction (as in Fred Saberhagan’s *Berserkers*). Swarm intelligences range from Olaf Stapledon’s *Last and First Men* (1933) and *Star Maker* (1937), to Michael Chorost’s Bugworld in *How to Talk to Aliens*. We should conclude by emphasizing that none of these and numerous other depictions of the alien in science or in imagination means that we can actually know what it is like to be an extraterrestrial. In fact, if we agree with Thomas Nagel’s conclusion that we cannot answer the question “what is it like to be a bat?, we certainly cannot know what it is like to be an extraterrestrial, perhaps even if we discover them.[51]

Our fourth principle in Table 4.3, space-time filters, is a recognition of the possibility of the unknowability of the alien. The perceptions of space and time by conscious/cognitive/intelligent beings may be very different from ours. Surely the fundamental basis of life’s interaction with the external world will depend on how any given lifeform sees that world in terms of geometry and mathematics (if it sees it at all in those terms), and how it employs its perhaps very different sense of time. This is difficult for humans to visualize, but as I indicated at the opening of this chapter, a valiant and delightful attempt at the spatial dimension was made 130 years ago when British schoolmaster Edwin A. Abbott penned *Flatland: A Romance in Many Dimensions* (1884). There a creature named "A Square" visited Pointland, Lineland, Flatland, and Spaceland, representing zero, one, two, and three dimensions, respectively. The creatures in each dimension thought it ridiculous that any other dimensions could exist.

Similarly, fundamental questions about time arise in the contemplation of aliens. We are not talking about "mechanical time," which in the case of Earth involves timekeeping based on planetary rotation being divided into 24 hours, 1,440 minutes, or 86,400 seconds. There is an interesting history about the representation of Earth's rotation through the painstaking construction of mechanical clocks ranging from pendulum clocks to chronometers for longitude and navigation and the intricacies of atomic clocks, with all the associated problems of leap seconds and so on to keep those timekeepers in sync with astronomical reality. If aliens keep time also based on planetary rotation (though surely not with our hours, minutes, and seconds and perhaps not even with their equivalent in some other number system), that would be an interesting story too. But I am talking here about the perception of time, sometimes called "subjective time" or psychological time, related to how any given nervous system processes sensory information. We know that on Earth animals have very different perceptions of time, and that this perception is related to body mass: the smaller the animal, the faster its metabolic rate and the slower time passes. A mouse has a very different perception of time than an elephant.[52] The same may be true with aliens, or their body clocks may operate according to some other variable. This difference in time perception in the terrestrial context is the subject of much research, and has occasionally been the subject of science fiction. Most notable for one extreme case is physicist Robert Forward's imaginative novel *Dragon's Egg* (1980), where the sesame-seed-sized intelligent cheela live out their lives on a neutron star with 67 billion times the surface gravity of Earth. Their biology utilizes nuclear rather than chemical reactions, so they think and develop a million times faster; one minute for humans is many years for a cheela. Their civilization therefore develops very rapidly as the humans watch with astonishment. In another example of a different variety, the aliens in Kurt Vonnegut's *Sirens of Titan* do not distinguish among past, present, and future, just as the heptapods in the movie *Arrival* do not (see later in this chapter). Other examples could be multiplied. I return to this "space-time filter" idea in the next chapter in the context of universal knowledge, for it is a problem not only of transcending anthropocentrism, but also of philosophy, psychology, and cognitive science. In fact, space-time filters may affect our search for extraterrestrial intelligence and may even account for the Fermi Paradox of why we do not see extraterrestrials everywhere.

Culture and Civilization

Culture and civilization arise in the astrobiological context because if we are going to deal with intelligent life in any of the forms of contact in Table 2.2, we assume culture and civilization will be involved. That is the assumption when we discuss the Drake Equation's fraction of intelligent species that produce civilizations, and the longevity of those civilizations; when we speculate about the evolution, behavior, and possible technological signatures of civilizations; and when we address the effects of encounters among them, as we did in Chapter 3. Obvious as culture and civilization may seem to humans embedded in our terrestrial history, the extension of these anthropocentric concepts beyond Earth is still the wildest of assumptions, a leap of faith warranted mostly by analogy rather than fact. Even given the existence of intelligence, the very idea of cultures or civilizations beyond Earth, and especially that their evolutionary trajectories might be at all similar to ours, are preconceptions that need to be examined as we attempt to transcend anthropocentrism.

Despite this precaution, if we are going to discuss extraterrestrial cultures and civilizations based on what we know rather than on pure conjecture, we need to start with their terrestrial counterparts, constantly keeping in mind the promise and pitfalls of analogy we have emphasized throughout this volume. As anthropologists John Traphagan and Kathryn Denning have both independently emphasized, SETI researchers who use terrestrial analogs with abandon in the case of extraterrestrial cultures and civilizations are treading on particularly thin ice. In order to appreciate the force of this point, we need to understand a few of the basic controversies that swirl around these concepts, without becoming too embroiled in the scholastic quagmire of definition.[53]

A Cultural Universe?

In the most basic sense, the terms "culture" and "civilization" are notoriously vague and often used interchangeably, both in astrobiology and in the general literature of the humanities and social sciences. The terminology often depends on the scholarly tradition using the term: more than 40 years ago, the American Anthropological Association, for example, published a book titled *Cultures beyond the Earth* (1975), while the term "extraterrestrial civilizations" is more likely to come from SETI scientists or from historians, when they address such a grandiose theme at all.[54] Both concepts, however, are useful for astrobiological purposes. Both were broached in an astrobiological context in 1975, when NASA was forming its SETI program and Nobelist Joshua Lederberg led a two-day "Workshop on Cultural Evolution," focused more specifically on "evolution of intelligent species and technological civilizations." Among the conclusions of the group – which included several scholars in the social sciences – was that "our new knowledge has changed the attitude of many specialists about the generality of cultural evolution from one of skepticism to a belief that it is a natural consequence of evolution under many environmental circumstances, given enough time." Here cultural evolution is seen – logically, it seems to me – as underpinning the idea of civilization, since it is difficult to see how a civilization can exist without the evolution of culture. Civilization does not spring from the proverbial head of Minerva. The concepts of culture, cultural evolution, and civilization have often been found in subsequent SETI literature, though not in any nuanced form, so it is important to examine their conceptual complexity.[55]

As with life and intelligence, the idea of culture is difficult to define; nevertheless, anthropologists and others have spent a great deal of time attempting to do just that in the terrestrial context. More than 60 years ago, two anthropologists collapsed 164 distinct definitions of culture into one: "Culture is a product; is historical; includes ideas, pattern, and values; is selective; is learned; is based upon symbols; and is an abstraction from behavior and the products of behavior." This is indeed a definition that seems designed by committee. Anthropologist Clifford Geertz defined culture more intuitively as "an historically transmitted pattern of meanings embedded in symbolic forms by means of which men [translate people these days] communicate, perpetuate and develop their knowledge about and attitudes toward life." Today, some anthropologists such as John Traphagan locate culture in individual brains and think of it as an "indeterminate social variable," where the behavior of any collective of individuals approaches a fixed value or at least a degree of consistency of behavior within the group, a consistency that can be determined only probabilistically, and thus indeterminately. Whatever concept one adopts, if, as Harvard biologist E. O. Wilson logically asserts, each society creates culture and is created by it, the idea of "culture" is a moving target, evolving in space and time on Earth – and presumably anywhere else in the universe.[56]

It follows, then, that just as evolution is of key importance in the physical and biological realms of astrobiology, so too is cultural evolution, and all are an integral part of cosmic evolution. Studies of culture in the cosmic context have been hampered by our understanding of cultural evolution on Earth, which can only be described as rudimentary by comparison with our knowledge of astronomical and biological evolution. Cultural evolution is a realm in which anthropologists and other social scientists dominate rather than the paleobiologists, neurobiologists, and evolutionary biologists central to the life and intelligence debate. They have in general judged as too simplistic Spencerian models of cultural evolution, based on the nineteenth-century work of Herbert Spencer, which view society as evolving through well-defined stages from simple to complex. Darwinian models have proliferated in recent decades but have been highly controversial. Attempts to Darwinize culture include the research programs of sociobiology, gene–culture coevolution, universal Darwinism, and memetics. All these models have considerable problems, and some social scientists still resist evolutionary models of culture altogether.[57]

A more robust knowledge of the idea of culture, and the mechanisms of cultural evolution, is nevertheless central to the SETI endeavor, arguably more central than SETI practitioners realize. For example, as I argued in the previous section, one indication of the power and relevance of cultural evolution to astrobiology is the distinct possibility that it has generated postbiological intelligence, including artificial intelligence, over timescales much shorter than biological evolution. Humans were not much different from us biologically 10,000 years ago, but they were certainly different culturally. This realization could change the scope and strategy of SETI, and needs to be kept in mind considering the large timescales we are dealing with in cosmic evolution. But we need not posit postbiologicals to realize the importance of cultural evolution to any attempt at discovering or communicating with cultures of any type beyond the Earth.

Civilizations in Space and Time

What, then, of civilizations? It seems axiomatic that cultural evolution need not always lead to what we call civilization – at least that is the situation on Earth. Without falling into the nineteenth-century trap of formulating evolutionary models of human social organizations on a linear scale ranging from savagery to high civilization, it is nonetheless true that scholars commonly refer to most of the hundreds of pre-Columbian native American Indian groups as cultures, while other social groups such as the Mayans, Aztecs, and Inca are seen as civilizations. A common demarcation point is a system of writing, but if that is the main criterion, the Incan system of quipu, an accounting device based on ropes and knots, might not qualify them as a civilization. What, then, do we mean when we say "civilization"? Where culture is the domain of the anthropologist and sociologist, in the elaboration of the idea of civilization they are joined by historians, although only a few are bold enough to tackle such an all-encompassing concept. Surprisingly, the term dates only from the first decade of the eighteenth century, and its modern sense was only established in the second half of the eighteenth century in the wake of the French Revolution.[58] Civilization has become one of the master concepts of modern times, the "unit of history" of Arnold Toynbee's massive 12-volume *A Study of History* (1934–1961), the central theme of Kenneth Clark's widely acclaimed *Civilisation* book and television series (1969), and the central point of contention in Samuel Huntington's *The Clash of Civilizations and the Remaking of the World Order* (1996). From whatever angle, many consider civilization the ultimate achievement of humanity, even if it is often characterized by war rather than the finer things of life we often associate with the word.

Because civilization is usually discussed in the SETI context in connection with the Drake Equation, which is interested only in detectable technological civilizations, it is important to think more deeply about what actually is meant by the term. It seems to me that "civilization" can be taken in at least two senses: individual civilizations, each of which has risen and fallen on a particular timescale, or the global planetary civilization existing at any given period. In the first sense, civilizations on Earth are generally considered to have arisen in the late Neolithic period, after the agricultural revolution about 10,000 BC, becoming more mature around 3,000 BC with the Sumerian and Egyptian civilizations, followed by the ancient Indian, Minoan/Mycenaean, and Chinese civilizations in the Old World, and the Mayan, Aztec, and Incan in the New World. SETI practitioners and theoreticians, however, almost always consider civilization in the second, more global sense. This is not surprising, since SETI practitioners are not experts on the rise and fall and complexities of civilizations. But when communicating with an extraterrestrial civilization, it is entirely possible we will be dealing with civilizations in the first sense, each having its own geopolitical challenges as we do on Earth, fantasies about progress and planetary-scale government notwithstanding. There is, logically speaking, a third option: on many or all planets cultures and civilizations do not exist because social organization does not occur at this level or in this way – the case on Earth for most of the history of humanity. The latter option would seem to be a dead end for SETI in the electromagnetic sense, and in any case there is good reason to believe social organization must occur for reasons of culture evolution. This third case would be ripe for astronaut anthropologists.[59]

It is in the sense of non-global individual states that historian Samuel Huntington considers civilizations "the broadest cultural entities," and offers six "central propositions concerning the nature, identity, and dynamics of civilizations": 1) civilization can be considered in the singular and the plural, as I noted earlier; 2) civilization is a cultural entity, "a culture writ large"; 3) the components of civilizations cannot be understood without reference to the overall civilization, and while the latter have no clear-cut boundaries, they are meaningful entities; 4) civilizations are mortal but also very long-lived, "they evolve, adapt, and are the most enduring of human associations"; 5) civilizations are primarily cultural rather than political, not principally identified by their governmental components; and 6) scholars generally agree on the identification of major civilizations.[60]

Scholars have challenged most of Huntington's points; on the latter, for example, Toynbee had designated 21 (later 23) civilizations over the history of the world (Table 4.4), while Huntington speaks of the world's "seven or eight civilizations" remaining today: the Sinic (Chinese), Japanese, Hindu, Islamic, Orthodox,

Western (Christian), Latin American, and "possibly" African. It is true that many of Toynbee's civilizations are now extinct, such as the Mesopotamian, Minoan, and Mayan, but even in Huntington's accounting of those remaining today, some would add, subtract, regroup, or rename many. For example, is "Western Christian" a really meaningful entity, and does it include South America and the broader "global South," where Christianity is in many ways now most vibrant? Should Russian and Eastern Orthodox peoples really be considered separate from the rest of Europe?[61] And so on. Beyond such groupings, are civilizations really united by causal bonds, especially given internal conflict? One might posit a hierarchy of bonds, which become increasingly loose (and meaningful) as one goes down the hierarchy. "We gain nothing by lumping cultures into broader conglomerates," one critic wrote, "and [to the question of the 'clash of civilizations'] we can be seriously misled if we assume that difference inevitably means hostility. Life, and politics, are not so simple." Given such problems, some scholars deny any structural coherence in groups beyond the nation states familiar to all of us, of which there are some 193 in the United Nations (up from 55 when it was born in 1945), virtually all of the 196 nation states generally considered to exist on planet Earth.

Table 4.4 *Terrestrial Civilizations According to Toynbee (1954)**

Primary Civilizations	**Secondary Civilizations**	**Tertiary Civilizations**
Egyptiac		
Andean		
Mayan	Yucatec	
	Mexic	
Sumeric	Babylonian	
	Hittite	
Indus Culture	Indic	Hindu
Minoan	Syriac	Iranic
	Hellenic	Arabic
		Western Christian
		Orthodox Christian
		Orthodox Christian in Russia
Shang Culture	Sinic	Far Eastern Chinese
		Far Eastern Korea and Japan

* Toynbee (1954) posited 23 civilizations that have existed throughout Earth's history, not including what he termed "abortive civilizations" such as the Scandinavian, and "arrested civilizations" such as the Polynesian and Eskimo. Toynbee defined primary civilizations as those derived directly from "primitive" societies, secondary civilizations as those derived from primary, and tertiary as those derived from secondary through churches "constructed by their internal proletariats." Based on table IV, volume 7 of Toynbee's *A Study of History*. The delineation of such civilizations is controversial even on Earth, but may be crucial in dealing with any alien civilizations.

Historian Jack Matlock Jr. invokes a colorful analogy for terrestrial civilizations: "A civilization by definition is infinitely more complex than, say, a garden. Nevertheless, describing it is in principle no different." Each garden is unique, yet some will have common characteristics not shared by others, in terms of geometry, soil, color, and so on. "Gardens, like civilizations, can be described, analyzed and interpreted," he concludes. "But one thing is certain. It would be absurd to speak of a 'clash of gardens.' It is equally absurd to speak of a 'clash of civilizations.'"[62] Some would say this is mere semantics, that we can fall back on "cultures" or "nation states" as the meaningful unit. Or, going in the other direction, considering "civilization" in the second sense as a global planetary civilization, that would surely be an objective entity rather than a mental convenience. The question, then, is do global planetary civilizations really exist?

Global civilization is the assumption of not only SETI practitioners around the world but also SETI theoreticians. This is nowhere more evident than in the Soviet Union, perhaps because of its recent experimentation with different forms of society. There, at the height of the Cold War in 1964, astronomer Nikolai Kardashev ranked global extraterrestrial civilizations by their energy consumption, one way of measuring their technological advancement. In his view, three types of civilizations could be envisioned in this sense: Type I harnessing all the energy of its planet, Type II harnessing the energy of its star, and Type III harnessing the energy of its galaxy. This corresponds to 10^{16}, 10^{26}, and 10^{37} watts, respectively. Others have since expanded this ranking to Types IV and V, harnessing the power, respectively, of all galaxies and the entire universe. Human civilization does not yet attain Type I status, which provides a certain perspective from the point of view of energy consumption. While some scientists such as Michio Kaku have elaborated on this view, historian of technology George Basalla has pointed to the weakness in this energy–civilization equation, by comparing it to those in the early twentieth century who said that because humans then used 100 times more coal than their ancestors, they were 100 times more civilized. Does our increased use of energy in the twenty-first century make us more civilized? Basalla also made the telling point that just because the Soviet Union and the United States controlled more energy in the form of atomic bombs than most other nations combined, that was no good measure of civilization.[63] This is one typology of civilizations ranked in a particular way by energy consumption.

Other typologies, based on other characteristics, are possible. The important point is that by whatever characterization we choose, these concepts about the nature of civilizations will likely have real impact if we make contact. And conversely, we will of course be interested in the nature of extraterrestrial civilizations in order to illuminate our own, perhaps leading to a truly universal theory of civilizations. In fact, one approach to applying our knowledge of terrestrial cultures to astrobiology is through the idea of universals. In the same way that we looked for universals in life and intelligence, there may be cultural universals. This is exactly the question that anthropologist Michael Ashkenazi asked in his brief but important essay "On the Applicability of Human Cultural Models to Extraterrestrial Intelligent Civilizations." He suggested all human cultures share at least five "meta-concerns": nurturance, in the sense both of raising young and of perpetuation of the species and culture; organization of groups to achieve larger purposes than possible with the individual alone; allocation of resources and power throughout the society; the presence of worldviews that drive the society in certain directions; and technical strengths and interests of the society. To the extent that this is a successful research program illuminating human cultures and civilizations, it might cautiously be applied in the extraterrestrial context. As Ashkenazi concludes, "We can assume that a civilization will be a complex collective of individuals linked together by some common general ideas, with some degree of shared institutions, and the need for a mechanism to ensure continuity."[64]

What is inarguable is that astrobiology vastly expands the possible cultural and civilizational landscape. In the cosmic context, our terrestrial ideas of culture will surely be broadened if we discover cosmic cultures and civilizations, in which case the natural history of culture and its theoretical underpinnings will be taken to a new level. In this way, cosmos and culture are inextricably intertwined, and will be even more so in the future. As John Traphagan points out with the eye of an anthropologist, our speculations about intelligence, culture, and civilizations are products of our own terrestrial experience. Different civilizations, whether global or non-global, may well have developed very different worldviews when it comes to these concepts, as well as to the ideas of human nature and progress. At the very least we should make use of all worldviews when considering extraterrestrial cultures and civilizations.[65]

Technology and Communication

A bedrock assumption of most SETI programs is that all civilizations produce technology, either in the form of detectable leakage or intentional electromagnetic signals, or as a result of megastructures built around their parent stars, structures that produce infrared radiation. The search for more alien artifacts (SETA) in a more general sense also assumes technology. But do intelligence and culture necessarily result in technology, and if so, what kind of technology? That question is of utmost importance in any operational search for intelligence or technology beyond Earth. SETI pioneer Jill Tarter often makes the point that SETI is really the Search for Extraterrestrial Technology (SETT): "SETI is a misnomer because there is no known way to detect intelligence directly across interstellar distances. Even on Earth we argue about exactly what constitutes intelligence, and we have no reliable way of measuring it at a distance (either spatial or temporal). In the case of extraterrestrial intelligence, the best we can do is to search for some manifestation of another technology." Such technologies could take many forms, but it is not surprising that for operational purposes astronomers can search only for what they might detect with available instrumentation. In other words, their very research programs are shaped not only by assumptions about extraterrestrial technology but also by the limits of our own. On both ends, then, there is a basic question: must extraterrestrial civilizations produce technology? If so, what is the range of technologies they might produce? And do we ourselves have the proper technology to detect whatever extraterrestrial intelligence might produce? These questions bear strongly on whether we will ever be able to communicate.[66]

Does Technology Converge?

There is no doubt that technology is one of the products of culture on Earth. But it is not necessarily true that every culture must have technology, unless one defines culture by its tool-making capacity rather than by its capacity to use language or some other measure. Even less is it true that every culture must have radio communicative technology. Nevertheless, such technology is one of the options, and the practical question is how the evolution of culture affects methods of communication with extraterrestrial intelligence, and ultimately our chances of successful communication. Here we stand to learn much from work in the history and philosophy of technology, which probes not only the evolution of technology on Earth but also the foundational concepts of technology and its relation to technology and culture. Among the issues philosophers of technology discuss is technological determinism, meaning the extent to which culture shapes technology or technology shapes culture.[67]

This is exactly the point historian of science George Basalla takes up in his book *Civilized Life in the Universe*, where he argues that SETI practitioners have a mistaken view of the nature of technology. In particular, "they assume that technology moves progressively toward goals predetermined by the universal laws of science," and that this results in contact via interstellar travel or communication. Their mistake, he asserts, is that they do not consider the influence of culture on the evolution of technology. Although his book often takes what seems to me to be an unwarranted polemical stance against SETI, Basalla in this case has a valid point based on his primary area of expertise. Technology on Earth, he argues, could have developed in many different directions, and although technology is constrained by science, science is not universal when considered in the context of the universe, and neither is technology. It is true that the development of technology on Earth is well known. But "historians have no proof that technology follows this or any other predictable sequence of stages. If a technological tradition begins with stone hand tools, it need not end in electronic communications. The history of technology is filled with technological paths never followed." Moreover, "once a particular technology is developed, and social, cultural and economic commitments made to it, then other technological possibilities are closed. The opening of a door to one technological solution closes off outlets to its alternatives." In short, Basalla sees technology as a many-branched tree, with some branches developed, and others unexplored or abandoned.

It is true that as time has gone on, technology use among terrestrial cultures has tended to converge. But on another planet, technology may converge on another solution. This is not esoteric reasoning; from the point of view of real science, astronomer Martin Harwit has recently argued, reasonably it seems to me, that our astronomical instruments shape our research, our ideas of what can and should be observed, and our very conceptions of the universe. If we are to be "in search of the true universe," we need to take a step back and realize that the tools we use to observe it are only a small subset of the possible. Had World War II not given rise to infrared technology transferred to astronomy, observational priorities might have gone in another direction, and we might by now have solved the mysteries of dark matter and dark energy, and thereby have a different view of the universe. Again, as in biology and history, so in technology. And as seen in the next chapter, perhaps also in science and mathematics. Technological determinism, in the sense of converging on the same set of technologies, may not be the way of the universe. To put it another way, Conway Morris's argument for convergence in the case of life and intelligence may have met its limit with technology.[68]

Can Humans Communicate with Aliens?

These sobering considerations of technology also bear on the possibilities of communication, one of the intriguing possible outcomes in the Search for Extraterrestrial Intelligence, and the most basic goal of those who favor the more active version of SETI known as Messaging Extraterrestrial Intelligence (METI). In either case, the ultimate goal is not only a search, but also a communications exercise, in effect what the Russians from an early date termed Communication with Extraterrestrial Intelligence (CETI). If such a search is successful, it may be that we receive only a "dial tone," a signal recognizable as artificial but not carrying information. Or it may be that we receive a message that might be deciphered, from a civilization long dead. But the most spectacular scenario is receiving a message that requires (or at least makes possible) an attempt at communication, the case I discussed in the previous chapter in connection with the Mayan decipherment analogy. Both SETI and METI require us to think in novel ways about the methods and substance of such communication.[69]

Whether searching or messaging, once again we need to look at fundamental assumptions, this time about communication. The vast panoply and cacophony of communication that surrounds us in the animal world is remarkable enough even in our terrestrial experience. If communication is seen only as the transfer of information, this is accomplished even by bacteria. At a higher level, animal communication on Earth is accomplished by many means, including visual (gestures, color changes, bioluminescence), auditory (vocalizing), and olfactory (pheromones). The familiar birdsong, whale and dolphin calls, chimpanzee vocalizations, and other established means of auditory communication on Earth are only the tip of the iceberg in a symphony of animal communication that includes even more exotic means such as color change. On Earth at the human level, auditory and written language is the key to communication among fellow humans, though visual gestures and body language are also important. The bottom line is that life on Earth has evolved numerous methods for communication at a variety of levels, and that on other planets these and other means may well extend to intelligent beings. We would expect no less as natural selection works its will beyond Earth.

It is also important to emphasize that interspecies communication is more problematic than intraspecies communication. The earliest attempts at human–animal interspecies communication involved dolphins, still the subject of robust study. Since then, humans have attempted to communicate with a variety of other animals, including chimps and parrots. Some researchers have concluded communication among animals is a window into their cognition and consciousness. Even so, what are the chances for real interspecies communication? Ludwig Wittgenstein gave a pessimistic opinion with his famous statement, "If a lion could talk, we could not understand him." This is commonly taken to mean there is such a radical incommensurability between animal and human minds that the idea of communication is impossible. On the other hand, here the idea of convergence may once again come into play. Some suggest "there may be a convergence or continuity in the communication and cognitive abilities in animals from different evolutionary paths." Simon Conway Morris cites this idea favorably as fitting in with his work on biological convergence, stretching convergence one more step. In this case, perhaps we could understand a talking lion. But a great deal more research needs to be done before such far-reaching conclusions can be accepted. And even so, would it apply to beings evolved under different planetary conditions?[70]

In all cases, the question is how do these studies translate to the extraterrestrial realm? There the questions run even deeper, for problems of communication, including language, depend on the possible domains of mental structure. I have suggested that at the topmost level of generality we may envision three cases when comparing terrestrial and extraterrestrial minds (Figure 4.6). In case 1, our mental structures and modes of perceiving and thinking may overlap entirely, in which case dialogue may be relatively "easy." In case 2, they may overlap only partially, yielding some common basis for dialogue. In case 3, there is no overlap at all, in which case there is no dialogue at all. But all hope is not lost; for in case 4, a "dialogue chain" of partially overlapping mental structures may eventually enable dialogue. These Venn diagrams of "mental structures" at the physiological level, or "modes of thinking" at the output level, embody many problems and require more research in diverse areas before any conclusions can be reached about their effect on communication. They are also related to the idea of space-time filters in Table 4.3, and ultimately to the philosophical problem of objective knowledge discussed in the next chapter.[71]

Many science fiction scenarios might be used to illustrate these problems. Let us focus for a moment on one, made popular in the recent cerebral movie *Arrival*, based on American writer Ted Chiang's "Story of Your Life." Here a race of alien cephalopods, known as heptapods for their seven-legged morphology, arrive on Earth and initiate first contact – an example of a direct terrestrial Type 1 encounter in Tables 2.2 and 2.3. Communication is difficult, and centers around understanding their written language, which is expressed when the alien emits from its

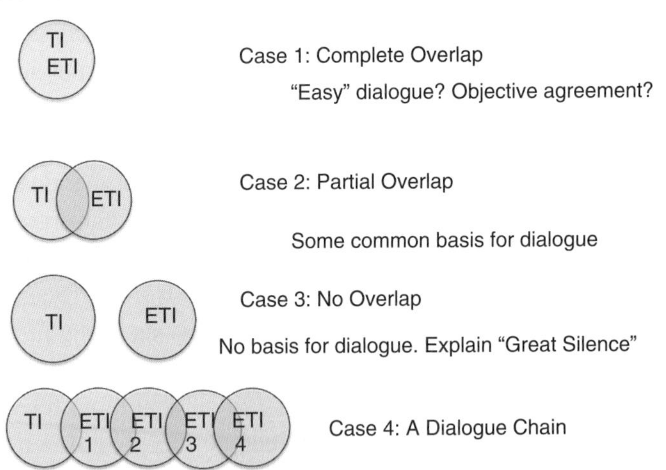

Figure 4.6 Possible scenarios in the relationship between extraterrestrial intelligence and terrestrial intelligence. The Venn diagrams may be taken as representing mental structure or modes of perceiving and thinking, and will affect our abilities to communicate. Credit: author.

starfish-like hands an inky substance that takes shape as complex circular symbols ("semagrams") in the foggy air surrounding the creature on the other side of a glass-enclosed wall on its spaceship. It is a form of nonlinear orthography (Heptapod B), which reflects the heptapod's nonlinear thinking, based on the alien's radial symmetry, as well as its understanding of mathematics and Fermat's principle of least time. This is an expression of the so-called Sapir–Whorf hypothesis, a real-life theory in linguistics that states that language determines thought and that differences in language structure affect how different cultures see the world. This is an idea, though it must be said one not widely accepted today, based on real experience with cultures on Earth. In the case of the heptapods, this nonlinear thinking affects the way they perceive time, and means they know the future – an ability used to good effect in the film. The story is thus narrated as case 2 in our Venn diagrams, a partial overlap of mental structures and modes of thinking and perception that results in communication between humans and heptapods. Whether that would actually be the case for creatures with a nonlinear perception of time is far from certain. My biggest criticism of the film is that it is very unlikely the semagrams would be deciphered in the few days depicted in the movie for dramatic effect; rather, as Philip Morrison has suggested, this would be the work of years, decades, or centuries. Meaningful communication might take even longer.[72]

The Problem of Universals

This provocative scenario brings us to the practical problem of how we communicate with extraterrestrials. Two strategies suggest themselves, located at the opposite ends of a spectrum of possibilities. The first is assuming that certain universals would be applicable in any planetary context. Usually science and mathematics are seen as the universals to be communicated, but music, evolution, and other aspects might also be seen as universal. By contrast to this strategy of universals, we could assume that communication is "embedded," situated in the creatures themselves and very much dependent on their environment. Swedish historian David Dunér has emphasized that our cognitive and communication skills are adapted to our environment, rather than being universal, an example of "embodied" thought theory prominent in cognitive science. As Dunér has pointed out, a theory of situated cognition holds that the environment has an active role in driving cognitive processes, producing what some researchers call "the extended mind": "Aliens, developed in and adapted to an entirely different physical and cultural environment would probably have very different modes of thinking. Thinking in the universe could be very different from what we are used to in our westernised, anthropocentric and earthbound human culture." The nonlinear thinking characteristic of heptapods in *Arrival* resonates with this idea of embodied cognition, and with the idea that space-time filters may influence thought, as suggested in the fourth box in Table 4.3. It must be said, however, that the idea that a human studying the heptapod language would be able to acquire nonlinear thinking and thus see the future like the heptapods, is a stretch, even though the idea was used to good effect in *Arrival*.[73]

What we choose to communicate with extraterrestrials very much depends on which of these scenarios, or some combination of them, we accept. In the past half-century, since the construction of radio telescopes, the pioneering paper by Cocconi and Morrison, and Frank Drake's Project Ozma, communicating with extraterrestrial intelligence has almost always meant sending or receiving signals by electromagnetic waves. Such signals can be in the form of digital pictographs such as those sent from Arecibo in 1975, or in the form of various symbolic systems, including language, both transmitted as mathematics. These attempts thus clearly fall in the "universals" camp, specifically "mathematics as universal." Indeed two of the most important inventors of such attempts are mathematicians: Hans Freudenthal, who in 1960 created a *lingua cosmica* he called Lincos, and Carl DeVito, who invented a numeric language to build a vocabulary of symbols. This is also the case with engineer Brian McConnell, who at the turn of the millennium devised a method for sending messages in the form of symbols, images, or abstract languages using only binary numbers. He was explicit in his assumptions: an understanding of "electronics, digital computing, math, and Boolean arithmetic." In such a system an understanding of electromagnetic radiation is assumed to be the lowest common denominator.[74]

The "universals" approach is also used by two researchers with an aesthetic bent, Guillermo Lemarchand and Jon Lomberg, the latter well known for his artistic collaborations with Carl Sagan, including the "Sounds of Earth" sent on the *Voyager* record. They ask the right questions, namely "Do we perceive the real world or just models of it in our brain? Is there anything in physics and mathematics that we consider truly universal? How about art, ethics, or culture? How alien can the alien be?" Their search for universal cognitive maps leads them to consider four types of cognitive universals that move beyond the laws of nature: physical-technological, aesthetic, ethical, and spiritual. By "cognitive maps," they invoke the paradigm or worldview epistemology of Thomas Kuhn, whereby intelligence "makes representations of its environment in its own processing system," in our case, the human brain. They reach no definitive conclusions, but call for a research program among physical, social, and humanist scholars that would investigate such cognitive maps.[75]

Similarly, reaching beyond the usual scientific and mathematical approaches, Doug Vakoch suggests ways of communicating our altruism, our music, even our aspirations, arguing that if the principles of science are universal, extraterrestrials will already know about science and mathematics, and probably much more if they are more advanced. Extraterrestrial historians and anthropologists, however, will be interested in our level of science and mathematics, and Vakoch also recommends communicating our understanding of cosmic evolution in interstellar messages. Since we cannot be sure what characteristics humans and extraterrestrials have in common, he argues, logically it seems to me, that it is far better to take a variety of approaches to message construction, each with distinct sets of assumptions. One of these approaches may allow us to build a conceptual framework for mutual understanding. Vakoch's work has inspired others to enter the growing field of extraterrestrial communications and semiotics. The subject is very much tied to the question of whether human knowledge is universal, which I treat in detail in Chapter 5.[76]

It is not easy to choose between the strategies suggested by the ideas of universals on the one hand and embodied cognition on the other, indicating that it is prudent to employ both of them. There is reason to be optimistic that one or both strategies will work. From the point of view of artificial intelligence, Marvin Minsky has concluded that communication will be possible because we will think in similar ways (my case 1 or case 2), because "all intelligent problem-solvers are subject to the same ultimate constraints – limitations on space, time and materials." His arguments, however, apply "only to those stages of mental evolution in which beings are still concerned with surviving, communicating, and expanding their control of the physical world. Beyond that, we may be unable to sympathize with what they come to regard as important," perhaps leading to my cases 3 and 4. In fact, one of the solutions to the Fermi Paradox is that terrestrial and extraterrestrial minds are not commensurable.[77]

The stakes are high. Should communication with extraterrestrial intelligence be possible, the effort holds open the possibility of solving long-standing problems in philosophy, including the epistemological problem of objective knowledge. As I discuss in the next chapter, by comparing knowledge derived from many independently evolved mental structures, what remains would surely constitute objective knowledge on a higher level than fathomable by terrestrial standards alone. And we could then begin to choose whether the concept of universals or embodied mind theory, or something in between represents the true nature of intelligence in the universe. As Darko Suvin suggests, the fifth-century BC Greek philosopher Protagoras may have been wrong. Man may not be the measure of all things in the universe.

"Given what we are, can we think about the world in different ways than we now do?," asked philosopher Joseph Pitt in an insightful essay on the limits of science fiction. His answer was clearly no, to the extent that he concluded, "science fiction that tries to describe alternative worlds in believable ways is doomed to fail for philosophical reasons."[78] His point is well taken. Clearly, we think about the world differently than the ancient Greeks. And clearly, we can extend today's familiar concepts to some extent, as we have in this chapter. But despite the search for universals, the use of science and imagination, and the possible existence of convergence in the evolution of life, intelligence, culture, civilization, technology, and communication, in the end we must admit we cannot completely transcend anthropocentrism in nontrivial ways. The only way to do that is to find life itself. Before we do, however, a great deal of work can be done at least attempting to transcend anthropocentrism, especially here on Earth with real data from other species, realizing our limitations all the while.

5

Is Human Knowledge Universal?

> We must never forget that human knowledge is *human* knowledge. Vanity and ignorance alone support the claim that human reason has a privileged status. Because we are the product of a long, directionless, evolutionary process, we are forced to accept that there is something essentially contingent about our most profound claims.
>
> *Michael Ruse, 1998*[1]

> Is there any scientific evidence that … what is proved in human mathematics is an objective universal truth, true of this physical universe or any possible universe, regardless of the existence of any beings? The answer is no. There is no such evidence!
>
> *George Lakoff and Rafael Nunez, 2000*[2]

> Is there a general biology? Do the same laws and biological principles that are used to explain life on Earth apply elsewhere?
>
> *Baruch S. Blumberg, 2011*[3]

Across the street from the US Capitol stands the Library of Congress, one of the great cultural centers of the world. Within its marbled halls more than 155 million items find their place, not only books and manuscripts but also music and maps, photographs and artwork, films and sound recordings. As I headed for my office in the Jefferson Building every day while this book was in the making, I walked past magnificent columns and statues (Figure 5.1), as well as walls adorned with the wisdom of the past: "knowledge is power"; "the inquiry, knowledge and belief of truth is the sovereign good of human nature"; and, from Confucius, "Give instruction unto those who cannot procure it for themselves." The main reading room is a marvel to behold, the eight statues beneath its massive cupola representing eight categories of civilized life and thought: philosophy, art, history, commerce, religion, science, law, and poetry. Beneath them 16 statues represent scholars and their seminal achievement throughout history. The staircase landing leading to the observer's gallery is dominated by a marble mosaic of Minerva, the Protector of Civilization. The message is not subtle: this *is* civilization. As a whole, the late nineteenth-century Jefferson Building symbolizes, according to the Library, "an optimistic era when Americans thought it possible to create a universal collection of knowledge in all fields of endeavor." Embodying as it does a significant fraction of the intellectual output of the world over the past several thousand years, it is a cathedral of knowledge in every sense of the word, almost worshipful in its attitude and bearing. But as I settle into my office overlooking the Supreme Court in one direction and the Capitol in the other – architectural wonders representative of one of the great civilizations in the world descending from classical Greece and ancient Rome – I contemplate a different set of questions: Just how universal is this knowledge? In the landscape of intelligence in the universe, is our knowledge *their* knowledge? Would aliens understand it, admire it, or laugh at it? Would it occupy one tiny corner in the Galactic Library?[4]

The contemplation of extraterrestrial intelligence almost immediately raises these questions, a set of issues that extends well beyond those discussed in the previous chapter, involving not just our concepts of life and intelligence, culture and civilization, technology and communication but the validity and universality of human knowledge itself. These questions go to the heart of what it means to be human – and what it means to be alien. Humans have evolved in a very specific set

Figure 5.1 The holdings of the Library of Congress in Washington, DC, and other libraries of the world, represent all of human knowledge. But is this knowledge universal? Photo by Steven J. Dick.

of environmental conditions, affecting not only physical form but also brain development, sensory apparatus, perceptions, and mental capacities, all resulting in specific human strengths and limitations. Aliens would presumably also have evolved according to their environment, assuming that evolution by natural selection is a universal phenomenon, as seems very likely. Given this fact, the crucial question is how much of our knowledge – especially our seemingly "objective" natural science and mathematics – is contingent on the history of evolution on planet Earth, and how much is universal? Even more problematic, would our approaches to the social sciences and humanities – the "human sciences" – be recognizable to aliens? These questions inform two practical matters, in addition to the possibility of communication discussed in the previous chapter: How much of our knowledge would actually be useful in assessing the alien in the event of a discovery? And how might our knowledge be universalized if we find life beyond Earth, in the sense of placing our possibly parochial ideas in a much broader and perhaps more objective context?

In this chapter I argue that much of human knowledge may not be universal, and therefore that universalizing knowledge in the wake of the discovery of alien life would constitute one of the major transformations in our thinking. Whether we are seeking to universalize knowledge or trying to determine how limited our knowledge inherently is, answers to such fundamental questions would not only affect the search for life in the universe, but would also shed light on us as humans. Even if extraterrestrial intelligence is never discovered, one of the great benefits of astrobiology is that it stimulates us to step back and analyze our usual assumptions in a generalized way. As the ultimate thought experiment, examining our assumptions about knowledge can lead to new insights. Only in the event of an actual discovery, however, would we have the potential to universalize knowledge in a cosmic context, resolving one of the great questions of philosophy, the problem of objective knowledge. As a bonus, such a discovery might also simultaneously resolve the philosophical stance of what is known as "relativism," the postmodern idea that knowledge has no absolute truth in itself, but only in relation to culture or historical context.

On Human and Nonhuman Understanding

In asking whether human knowledge in general is universal, we can draw on research from at least three fields: philosophy of knowledge, cognitive science, and evolutionary biology. The philosophy of knowledge, a field given the intimidating but harmless moniker "epistemology," normally proceeds with a discussion of the relative merits of the schools of rationalism and empiricism, realism and idealism, and their variations and elaborations by philosophers over time. Cognitive science brings to bear disciplines such as neuroscience, animal cognition, and artificial intelligence in an attempt to understand the nature of the human mind. And evolutionary biology, along with everything else it encompasses, sheds light on the evolution of the mind. A great deal has been written in all of these areas in the terrestrial context; for us, these fields are relevant as the musings of one terrestrial species, but just the beginning of the problem of knowledge in a cosmic context.

Philosophy of Knowledge: Is the Universe Real?

The epistemological quest for the nature, sources, and validity of human knowledge has a long and venerable history that still forms a major part of modern philosophy. Today robust disciplines such as the philosophy of mathematics, the philosophy of the sciences, and the philosophy of technology seek to uncover and analyze the fundamental assumptions in each of their fields related to our knowledge of the world around us. So do the philosophy of history, religion, anthropology, and sociology in the realm of the humanities and social sciences. But from its earliest beginnings, philosophy has also questioned the foundations and acceptance of knowledge itself, what is certain and what is probable, what is knowledge and what is belief.

The scientific successes of Johannes Kepler, Galileo Galilei, Isaac Newton, and their colleagues in Europe during the seventeenth-century scientific revolution constitute a well-known and laudable episode in human history. It was these successes in understanding Nature that inspired in the late seventeenth and eighteenth centuries a quest to assess just how valid the rapidly amassing knowledge was. Among the first to undertake this quest was French philosopher René Descartes, who developed what we now call a "rationalist" philosophy whereby true knowledge comes solely from reason, the "innate ideas" of the human mind, independent of sense experience. He began a long line of rationalists that includes Baruch Spinoza, Gottfried Leibniz, and Immanuel Kant.[5]

Rationalism was clearly not the whole story with the new science, as indicated by the importance of observation and experiment. Even Newton, that "silent, thinking, sober lad" and ornament of planet Earth, had recourse to data all around him as he formulated his principles of gravitation and optics. With this in mind, especially Newtonian mechanics, British philosophers John Locke, George Berkeley, and David Hume explored the empirical basis of science. Locke's *Essay Concerning Human Understanding* (1689) argued that human knowledge and understanding is based on our experiences through "sensation," combined with our reasoning abilities. He spent the entire first book of his essay attacking rationalist notions that much of human knowledge was based on innate ideas independent of our experience. Only mildly skeptical compared to what was to come, Locke's work resulted in a philosophy of *realism*, arguing that the world of material objects exists independently of the human mind. In its time Locke's insistence on empiricism as the foundation of knowledge rather than divine revelation or other methods was so radical that his ideas were initially banned at Oxford University.[6]

Both Berkeley and Hume found Locke's empiricism wanting, but in different ways. In his *Principles of Human Knowledge* (1713) and other works, Berkeley, an Irish Anglican minister, argued against Locke's realism, and espoused a philosophy of *idealism*, in which reality is fundamentally mental rather than material. In his view, no matter exists in the world: our experiences are misleading in the sense that they encourage us to perceive a physical universe as something external to our minds when actually it is nothing other than the experiential content of our minds, which is produced, ordered, maintained, and underwritten by an omnipresent divine mind. Berkeley came to this view at least in part to save religious ideas, since God was not material and could not be observed. Hume, on the other hand, was an empiricist like Locke, arguing that all knowledge should be justified through experience. But in his *Treatise on Human Nature* (1739), Hume famously was unwilling to follow Locke all the way. Rather, he criticized Locke because of what is now known as the problem of induction, the idea that our experiences today do not guarantee our experiences tomorrow. Seeing the Sun rise day after day does not guarantee it will rise tomorrow. Seeing that all swans are white constitutes a generalization about them only until one black swan is observed. Along with his contemporary Adam Smith and many others, Hume became a major figure in the Scottish Enlightenment, and is still known today not only for his criticisms of empiricism even while preferring it to rationalism but also for his program of naturalism, the idea that humans and their world can be explained in natural terms. In a 50-year period from 1689 to 1739, Locke, Berkeley, and Hume laid out the set of problems that initiated the study of the foundations of human knowledge, an endeavor that continues today.[7]

Today empiricism, through its observational and experimental components, is widely recognized as the basic epistemology of science, even if it is also recognized that observation can sometimes be problematic. The reasoning central to the rationalist school is of course important, but the human mind cannot gain much valid knowledge without the sense perception of the empirical school, with all its faults. In short, we can surely agree that human knowledge depends on both perception and reason. Indeed, some argue that the separation of the rationalist and empirical schools is misleading, since both schools embody parts of the other. It is more a matter of predominance: rationalism stresses reason, empiricism stresses experience. It is in these rather less confrontational terms that rationalism is now distinguished from empiricism. The crucial question in the present context is, if empiricism conditioned by reason is an essential part of acquiring knowledge, is the knowledge thus acquired from sense perception really objective knowledge of the external world? Is what is sensed the "real thing"? Would aliens sense the same thing? If not, how would their knowledge differ from ours? In short, is the universe as we think of it real, or is our knowledge of it compromised by our mental filters, which might well be different among aliens?

It is here that, 40 years after Hume's *Treatise on Human Nature*, German philosopher Immanuel Kant (Figure 5.2) provided a shocking answer as he expanded previous philosophic inquiry to embrace the question of what we could know with certainty given our limited minds. Awakened from his "dogmatic slumbers" by having read Hume around 1770, in his critical philosophy espoused in the *Critique of Pure Reason* (1781), Kant attempted to determine what is contingent and what is

Figure 5.2 German philosopher Immanuel Kant (1724–1804) questioned what is contingent and what is necessary in human knowledge, an issue relevant to the extraterrestrial life debate today. Credit: Grafissimo/DigitalVision Vectors/Getty Images.

necessary in human knowledge (Table 5.1). He did so by distinguishing a rational a priori category of knowledge as necessary truths (all white houses are white), as well as an empirical a posteriori component of knowledge based on experience and the senses (the house is white). But Kant's original contribution was to argue that in addition to these two components, our cognitive apparatus yields another type of truth, what he calls "synthetic a priori" judgments about the world, depicted as knowledge type 2 in Table 5.1. Whereas a priori truths by definition contained the predicate in the subject (a triangle has three sides, or an intelligent man is intelligent), synthetic a priori truths contained more than the predicate in the subject (the angles of a triangle add up to 180 degrees, or 7 plus 5 equals 12). In Kant's view, not only are mathematical judgments synthetic a priori, so also are geometrical judgments, such as "A straight line is the shortest distance between two points."

Table 5.1 *Types of Knowledge According to Kant*

	ANALYTIC STATEMENTS MUST BE TRUE PREDICATE CONTAINED IN SUBJECT DERIVED FROM LOGIC	**SYNTHETIC STATEMENTS** MAY BE TRUE OR FALSE PREDICATE NOT CONTAINED IN SUBJECT DERIVED FROM EXPERIENCE
A PRIORI KNOWLEDGE - Necessary, pure thought/innate ideas in the mind. - Gained through understanding without empiricism. - Conditions what is apprehended through the senses. - Yields real knowledge only in conjunction with the senses	**Knowledge Type 1** RATIONALISM Definitions such as: a triangle has three sides Tautologies such as: all white houses are houses; all white houses are white Concepts such as: space and time, Nature's laws All bodies are extended	**Knowledge Type 2** [Does not rest on empirical evidence] Mathematical judgments (7 + 5 = 12) Pure geometry judgments - 180 degrees in a triangle - causality, space, and time Metaphysical judgments - God - Free will - Immortality
A POSTERIORI KNOWLEDGE Contingent, empirical, derived from experience and phenomena via senses ("sensibilities")	**Knowledge Type 4** [nonexistent by definition]	**Knowledge Type 3** EMPIRICISM All sensory experience All empirical statements such as "the house is white" Objective for us but not for all beings

What Kant called "synthesis" is required to get from the subject to the predicate, so synthetic a priori truths "are truths about the independent nature of reality and cannot be deduced simply from the proposition."[8]

Why is this so shocking? Kant's delineation of this new category of knowledge bears directly on the question of whether human knowledge is universal in the broadest "transcendent" sense above and beyond humans. As one of his recent biographers has pointed out, since synthetic a priori judgments (like all a priori judgments) depend on our cognitive apparatus, these claims are about reality "only as it is experienced by beings such as we are … these claims cannot be claims about the world as it is independent of our conceptual apparatus."[9] In other words, in Kant's view, our experiential knowledge must first pass through a set of a priori "forms," or filters (acknowledged in my Table 4.3), which he calls sensibility, understanding, and reasoning. Space and time are the forms of sensibility; without them we cannot view the world of our experience. They are objective for us, but not for all beings; they are "empirically real" but "transcendentally ideal." As Kant's biographer puts it, "if we were constituted differently, we might be able to 'see' (or intuit) things as they are in themselves and not just as they appear to us. As finite beings, we cannot experience anything without our senses. Space and time are necessary conditions of any experience for us. As such, they provide us with *a priori* knowledge of the world of our experience."[10] For Kant, geometry and arithmetic deal with synthetic a priori judgments: they are based on our experiential senses filtered by our a priori categories of space and time. And since even our empirical judgments (knowledge type 3) must pass through our space-time filters, perception is reality, but only *our reality*, not the absolute reality. That absolute reality is transcendent, forever beyond our reach. Even if we could compare our knowledge with extraterrestrial knowledge, Kant might claim that we still had not traversed the boundary between the empirical and the transcendent (the noumenal divide), but that we and the aliens had been subsumed under some broader classification within the realm of the phenomenal divide. I myself, however, would tend toward a neo-Kantian position that such a comparison of human and alien minds might approach absolute reality.

With this scheme of knowledge, Kant is often seen as having revolutionized our understanding of reality in the same way Copernicus revolutionized our cosmological worldview. Indeed, that is how Kant saw himself in the second edition to the *Critique of Pure Reason* (1787): "We here propose to do just what Copernicus did in attempting to explain the celestial movements." In other words, as Copernicus placed the Sun in the center of the universe rather than the Earth, Kant placed the human mind at the center of reality *as we perceive it*, turning around the relation between subject and object. Synthetic a priori knowledge implies that rather than being a passive recorder of experience of the outside world as the empiricists claimed, "the mind is active, playing a part in shaping the world of experience and constituting the objects of knowledge. The mind imposes categories, such as cause and effect, and ideas of space and time, upon the incoming sense-data of experience to help us interpret the world and generate knowledge." The result is that we can only know the phenomena of the world as determined by the structure of our minds, a structure that makes us see objects in space and time. There may be reality beyond the phenomena, which Kant called the "thing in itself," the noumenal world, but the human mind could never know this reality.[11]

For SETI researchers and social scientists who take seriously Kant's epistemology or more general claims about the problematic role of the mind in relation to reality, this is disturbing news indeed. Surely alien minds (maybe even terrestrial animal minds?) might have an entirely different structure than ours, and thus an entirely different set of categories, resulting in a different view of the universe. Their synthetic a priori judgments might well mediate between rationalism and empiricism in a different way than ours. This is bad news for attempts to communicate, and perhaps bad news even for any attempts to search for alien signals. Aliens viewing the world in a fundamentally different way than humans may have no use for science, much less radio astronomy. And even if they did, our ability to communicate would surely be at serious risk. In short, the Kantian view of knowledge is another solution to the Fermi Paradox. Where are they? Why do we not see them? Because their mental structures are different from ours; their categories of thought are different from ours; their very ways of understanding the world are different from ours. And the ways in which we might see them could manifest in very different ways.[12]

Not only might alien senses differ from ours, their a priori reasoning filters might also. It is perhaps relevant in this respect that among Kant's earliest work as a young man in Prussia was a full-scale treatise on the universe and extraterrestrial life. His *Universal Natural History and Theory of the Heavens* (*Allgemeine Naturgeschichte und Theorie des Himmels*, 1755) was primarily a work of cosmogony, the material origin of the universe based on Newtonian principles, with no role assigned for God or teleology. In it he held that the universe was infinite in both space and time, that numerous planetary systems had been and continued to be formed, and that intelligent life existed on many of these planets.[13] Although not widely known in Kant's time, the *Universal Natural History* is seen today as a serious precursor to modern cosmogony. The important point for our purposes is that Kant may have developed his critical philosophy on epistemology in part because he contemplated the possibilities of knowledge among extraterrestrial intelligent beings, which he saw as progressing through "all infinity of time and space with degrees, growing into infinity, of perfection of the ability of thinking, and bring[ing] themselves gradually closer to the goal of the highest excellence, namely, to divinity." There is no direct, smoking-gun evidence that such views were a precipitating event for Kant's philosophy, but Kant affirmed his "strong belief" in extraterrestrials in the *Critique of Pure Reason*, and continued to advocate extraterrestrials in his later writings. At the very least this belief in aliens may have helped to frame Kant's ideas as he developed his epistemology. And while Kant himself may not have claimed that knowledge of "the thing itself" in the "transcendent ideal" may be the knowledge gained by extraterrestrials pooling their knowledge among themselves or with us, a neo-Kantian stance could certainly make that claim. If we are ever able to compare categories of cognition with aliens, and if we find there are strong overlaps with aliens, then it is likely that Kant was wrong about the noumenon being absolutely distinct from the content of our synthetic apperception.[14]

We can now begin to see what all of the fuss is about when it comes to Kant's types of knowledge, both in the past and in the future. Many of Kant's predecessors and contemporaries placed concepts such as God, free will, and immortality in the a priori category, deduced from pure reason and therefore necessary by definition. By removing them from that category and insisting they were synthetic a priori concepts, Kant placed all three concepts in doubt, even if that was not his primary motivation. Moreover, by insisting on such a category of knowledge, Kant placed into doubt the validity of human knowledge in these metaphysical areas and many others, including mathematics and geometry. In doing so, he also placed into doubt the universality of human knowledge in an extraterrestrial context. Knowledge, Kant argued, is not derived from the unconceptualized experience of the empiricists, nor from the unalloyed reason of the rationalists: "sense experience is needed to provide the content of knowledge and concepts give it its form. In this picture the judgments that Kant calls synthetic a priori hold centre stage because they determine the manner in which sensory experience and concepts are conjoined."[15]

So is the universe real? While few philosophers today would use the terminology or accept the details of Kant's epistemology, almost all would endorse his conclusion that our knowledge of the external world is compromised by the human filters through which our perceptions must pass. As philosopher Derek Malone-France says, Kant's general and foundational insight that conscious experience (cognition) is "always necessarily at least partially self-constructive, rather than being a pure, unadulterated reflection of an external objective reality," is almost universally accepted, in one form or another. For example, while some philosophers still argue about the analytic-synthetic distinction, whether or not we use those terms the result is the same in the eyes of most philosophers: different cognitive structures would give rise to different experiences of reality. Kant held that the "categories" of thought are universal among humans, but might not be universal when compared with other types of intelligence.

At the same time, there may be some hope for common ground: some philosophers believe there may be meta-universal categories such as "temporal successiveness" that occur among all intelligences. In this case, "other forms of intelligent life elsewhere in the universe might understand the universe and interact with it in ways that are quite different from us, but the range of that difference is not unlimited."[16] Suffice it to say that contact with extraterrestrial intelligence would solve many age-old philosophical problems, or at least raise them to a new level, if indeed mutual communication and understanding were possible at all. We might find out at last if the universe is real. We would at last learn if "cognitive relativism" is a valid philosophical stance. And, not least, we might at last find out if there is such a thing as "truth."[17]

Mind and Body: From Embodied Mind to Astrocognition

More than two centuries have passed since Kant formulated his transcendental idealism, the concept that the mind can only perceive appearances, not the thing itself. Kant's view of knowledge transformed philosophy and is still very much discussed today. But what more can we say today about the universality of human knowledge based on more modern research? Here we must turn to the flourishing discipline of cognitive science, the study of the mind, in principle including not just the human mind, but also animal minds, alien minds, and artificial minds. The field encompasses broad research programs in cognitive neuroscience (brain structure and functioning), evolutionary psychology (evolution of the mind), embodied cognition (the environment in which cognition is embedded), and animal cognition (both animal brains and behavior), among others. Cognitive science also overlaps with the field of philosophy of mind, which studies the nature of the mind, including the classical mind-body problem.[18]

As we saw in the previous chapter with the work of Carl Sagan, Jeff Hawkins, and Steven Mithen, brain structure and evolution are important for concepts of intelligence. But intelligence is not the same as knowledge, and the ideas of "embodied cognition" and animal cognition relate more directly to knowledge in the epistemological sense of "how we know things" that is the main subject of this chapter. For some researchers, the classical mind-body problem has now been recast as the new mind-body problem: how does the body shape the mind, whether human or animal, since both body and mind are related by evolution? (Figure 5.3). I broached this also in the previous chapter in connection with the ability to communicate with extraterrestrial intelligence, for if mind depends on body, communications might be very difficult. Not only that, but if the body shapes the mind, it also shapes thought, and extraterrestrials will have knowledge of the external world very different from us – just as Kant suspected on different grounds.

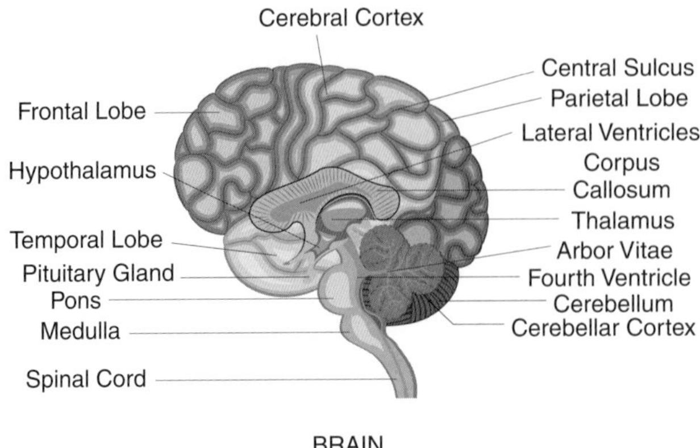

Figure 5.3 The amazing human brain weighs only three pounds, but is the most complex biological object on Earth. It is a product of the evolutionary contingencies of its history, as alien minds would also be. How this affects thought and communication is an open question. Credit: Stock_shoppe/iStock/Getty Images Plus/Getty Images.

But what exactly does "body shapes mind" mean? To say that embodied cognition means that cognition "arises from bodily interactions with the world" seems noncontroversial. But consider the implications of this further elaboration of embodied cognition: "cognition depends on the kinds of experiences that come from having a body with particular perceptual and motor capacities that are inseparably linked and that together form the matrix within which memory, emotion, language, and all other aspects of life are meshed." In sharp contrast with the idea that the mind is a calculating device, or a device to manipulate symbols, this implies that extraterrestrial minds are determined by their bodies, in particular their sensory and motor capabilities.[19]

Philosopher George Lakoff, one of the most outspoken proponents of this "embodied mind theory," believes this implies "the mind isn't what we thought it was. Philosophy wasn't even close in its speculations" about a disembodied mind, which was an a priori declaration rather than an empirical discovery. Rather, in the new philosophy of embodied cognition, thought is largely unconscious, embodied, and metaphorical. Thought in the new view is all about neural circuitry rather than manipulating abstract symbols as in a computer; as such it is something physical and subject to science. This new view of the mind changes everything massively, Lakoff argues, and in a shocking way. We cannot know our own mind, as Descartes thought with his rationalism, because it is not directly accessible to us in the sense of conscious introspection since most of the mind is unconscious. In his view, the old dichotomies such as rationalism versus empiricism and realism versus idealism go out the window. Philosophical ideas like time, causation, and morality are shaped by the body. There is no transcendent universal reasoning and no transcendent universal morality. All of this, Lakoff argues, is the result of taking seriously the most recent research in cognitive science. This research includes fields such as cognitive linguistics, neuroscience, cognitive anthropology, and cognitive and developmental psychology. This is not the place to review this vast literature, and such a vast literature does not mean embodied mind theory has been proven true. But in my view it is intriguing and approaching convincing, and has profound implications for whether human knowledge is universal.[20]

Embodied mind theory is also compatible with results in the field of animal cognition, since animal minds are embodied no less than our own. The new field of cognitive ethology has arisen to study this problem. Some researchers have concluded animals are conscious, thinking beings, and that they are thinking about different things in a different way than we do. Others criticize this research as too anthropomorphic and unscientific in its methods. In short, the study of a subset of the amazing diversity of animals on Earth is a possible example of the problems and reactions we will encounter if and when we have to deal with extraterrestrials.

When extended to extraterrestrials, David Dunér has coined the term "astro-cognition" to study the origin and evolution of cognitive abilities in extraterrestrial environments, as well as a variety of other issues related to space. In the context of the evolution of cognition, the basic premise is that cognitive abilities are adapted to "the physical and biological environment of their celestial body in order to understand and interpret, interact and deal with, and orientate itself in the particular physical and biological environment, in relation to its specific conditions." Dunér, Mathias Osvath, and others have suggested a research program for astrocognition, including the mechanisms behind similar behavior in animals, the ways in which intelligent behavior is produced, the preconditions for intelligence, and the terrestrial ecological factors that have produced intelligent behavior across animal groups. Embodied cognition and its extension to astrocognition are also in keeping with the similar idea of "situated cognition" that I discussed in connection with communication in the previous chapter, giving rise to the four cases in Figure 4.6 of complete, partial, and no overlap in mental structures and modes of perceiving. Movies such as *Arrival* give concrete meaning to the relevance of these ideas in the event of discovery. Indeed, an argument can be made that the search for intelligence beyond Earth is meaningless unless we understand the underlying cognitive basis of intelligence. Whatever term is used, whether "embodied mind theory," "cognitive ethology," or "astrocognition," this is a field wide open for future development based on real data.[21]

Evolutionary Biology and Epistemology

Lakoff's embodied realism and Dunér's astrocognition programs would seem to make sense if the mind and brain are the products of evolutionary biology – a point that Darwin believed but the cofounder of the idea of natural selection, Alfred R. Wallace, did not. So what do evolutionary biologists have to say on the subject today? Or more accurately, what do philosophers of evolutionary biology, who are much more likely to take up these kinds of problems, have to say? Do the principles and actions of Darwinian evolution shed any more light on the nature and limits of human knowledge other than embodied mind theory?

Some philosophers might argue that epistemology should not use or depend on empirical knowledge at all. Naturalistic epistemologists, on the other hand, argue that not only should the latest results of science be taken into account, but those results may also be central to the problem of knowledge. In this category would fall those in the field known as evolutionary epistemology, which deals with the evolution of animal and human cognition, in particular with the generation and maintenance of our senses and cognitive mechanisms. Whereas embodied mind theory states that cognition depends on the body in which it is embedded, evolutionary epistemology now claims that since the body is a product of evolution and cognition is related to the body, cognition will also depend on the environment in which it evolves. In the words of philosopher of biology Michael Ruse it means "taking Darwin seriously," not only for biology, but also for all of its by-products, just as we took cultural evolution seriously in our discussion of the postbiological universe in Chapter 4. In this view, all the tools of evolutionary biology not only can but must be applied to cognition, and epistemology itself is a product of evolution. Epistemology without evolution is seen as bankrupt. The evolution of cognitive processes during the long history of life on Earth can now be studied, rather than being tied to a narrower approach such as linguistics or even broader approaches such as cultural evolution. Just as some see evolutionary biology as the starting point for an entire suite of cultural evolution possibilities such as evolutionary psychology – the commonsense but controversial idea that our psychology and behavior have evolved over time according to natural selection – so also for epistemology.[22]

But exactly how should we take Darwinian evolution into account when it comes to epistemology? Ruse argues that traditional approaches to evolutionary epistemology do not work because they focus too much on knowledge itself, and not enough on the humans producing the knowledge. He argues that Darwinian evolution by natural selection, on the other hand, does provide a possible mechanism for knowledge and understanding in the form of "epigenetic rules," innate dispositions or capacities in humans based on genetics. Taking science as the highest representative of human knowledge, Ruse contends that these innate dispositions play a key role: "The nature and development of science is constrained and informed by the biologically channeled modes of thinking imposed on us by evolution – a consequence of the reproductive struggles faced by humans today, and even more a consequence of those struggles faced by humans in the past." In particular, he argues that the methodology of science grows out of these ongoing struggles in the form of these Darwinian-selected rules. A simple example is that a proto-human who sees two tigers entering a cave and only one reemerge, would be at a selective advantage if he decided not to go into the cave because he deduced 1 + 1 = 2 (and therefore 2 – 1 = 1), as compared to his less mathematically inclined cousin who did go into the cave because he thought 2 – 1 = 0. Through this and many other examples, Ruse concludes that "science is a human interpretation of the physical universe."[23]

Although very different from Kant's philosophy, it seems to me that embodied cognition and evolutionary epistemology come to much the same conclusion in terms of knowledge and extraterrestrials. But not exactly the same conclusion. As Ruse points out, although both the Kantian and the Darwinian emphasize that the mind does not passively receive knowledge but actively shapes it, unlike knowledge, evolution does not progress or have an end goal in the opinion of most biologists. Rather, Ruse finds a closer analogy between Darwinian epistemology and David Hume's philosophy, since both see reasoning as closely intertwined with the preservation and propagation of the species.[24]

The implications of evolutionary epistemology for astrobiology are clear: as an evolutionary product of its environment, not only the mind but also its ways of knowing will be different depending on the long history of life under any given planetary conditions. Whether Kant or Hume is the closer analogy, the conclusion is the same for extraterrestrials: if all minds are embodied and Darwinian factors are at work in epistemology, then it is at least possible that our knowledge is not their knowledge, our science is not their science, our mathematics is not their mathematics. If, 330 years after John Locke's *Essay on Human Understanding* we were to write an *Essay on Nonhuman Understanding*, the result would be very different and much more nuanced than his conclusion that empiricism combined with reason is the optimal way to knowledge. From Hume to Kant, from embodied mind theory to evolutionary epistemology, our ideas about knowledge have undergone a wholesale revolution in ways that radically affect how we might interact with aliens, and they with us.

Taking one more step back, Ruse, though holding that the mind is a product of evolution by natural selection, still finds our ability to peer into the mysteries of nature mysterious in itself. He believes there is a relativism in epistemology no less than ethics: "I think that is something that the Darwinian has to accept as given … there is no guarantee that tools forged by natural selection to help us survive and reproduce are going to be able to see and understand everything."[25] Ultimate reality, he says, may be beyond the ken even of Darwinian explanation, at least given what we know at this time and on this planet. In this sense, the long-running "science wars" theme known as relativism, the idea that mathematics, science, and ethics are culture-dependent, may be more relevant in the extraterrestrial context than on Earth. While postmodernism in its most extreme form makes little sense to me in a terrestrial context, its proponents may well have been ahead of their time when it comes to extraterrestrial intelligence.

We can see why an Earth-bound and locally invented system like theology would not be the same among extraterrestrials as among Earthlings – diverse as theologies are even here on Earth. But the philosophy of knowledge, cognitive science, and evolutionary biology all suggest one conclusion: the different cognitive structures bound to exist among extraterrestrials leave open the real possibility that human knowledge, including even science and mathematics – might not be the same as theirs.[26] At the very least, the differences might be enough to affect SETI programs in terms of communications. We cannot know for sure, but studies with "intelligent" animals on Earth may shed some light on this conclusion, which is indeed shocking to those of us who have spent careers in science and science studies. So let us not stop there. Since science and mathematics are considered the most objective part of human knowledge, and since their lack of universality would be a bitter pill to swallow for their practitioners, in the next section I more closely explore the possibility – at once disturbing and exhilarating – that even they are not universal.

Universal Science and Mathematics?

Before examining the possibilities of universal science we need to ask what "science" means in the extraterrestrial context. We are in luck, for in a pioneering but little-known essay, prolific philosopher of science Nicholas Rescher has addressed just this question of "extraterrestrial science." He points out that we cannot equate intelligence and social organization with science; after all, modern science did not begin until the seventeenth-century scientific revolution, despite roots in ancient Greece and substantial progress during the Middle Ages, which were not "dark" by any means. Intelligence, including animal intelligence, can exist without science or the more general systematic investigation of nature, as it has for most of Earth's history, and still does by some of the definitions in the previous chapter. So perhaps complex intelligence could exist in outer space without science, though it is questionable whether such a situation is sustainable in the long term. Moreover, even if alien science does exist, it does not mean aliens have reached the same conclusions as we have. After all, the history of science shows that our scientific conclusions are quite different from those of the seventeenth century, and are constantly changing, even while approaching (most scientists believe and hope) the "truth" about the universe, Kant notwithstanding. Rescher concludes it is not similarity of content or conclusions, but equivalent functions, that should define alien science. The fundamental issue is whether aliens embrace the goals of science, which are, in his estimation, "description, explanation, prediction and control over nature."[27]

In addressing this question, Rescher emphasizes the potential diversity of science in a universal context: the *mathematics* used for alien science might be very different; the *orientation* of alien science might be different in the sense that any particular extraterrestrial intelligence might concentrate on social science rather than natural science (bad for SETI), and their natural science might be different too; and the alien *conceptualization* of science might be different, embracing radically different cognitive points of view just as our conceptual apparatus (think relativity and quantum mechanics) is radically different from a century ago. This formulation provides a good framework for our discussion of universal science and mathematics.

Is 2 + 2 = 4 for Aliens?

Let us begin with mathematics, already broached in the previous chapter in connection with optimal ways to communicate with aliens. There we saw scholars often argue that mathematics is the universal language. Even though many systems developed in the fascinating history of mathematics on Earth, at a basic counting and arithmetic level for everyone, and at a more sophisticated level for the cognoscenti, mathematics is certainly a universal language among cultures on Earth. But is it necessarily so throughout the universe? Naturally, there are different views on this subject, with ancient roots still reflected in the modern philosophy of mathematics. Mathematical Platonists argue that natural numbers (1, 2, 3, and so on) really exist, and that they are discovered in the same way as scientists discover other objects. If this is so, aliens should discover them too. Formalists, on the other hand, argue that numbers are the creation of the human mind, in which case there is no reason to believe aliens would ever discover them. (We might, however, some day *explain* it to them, in the hope they would understand and find it a logical system, if a provincial one.) In his fascinating book *Is God a Mathematician?*, astrophysicist Mario Livio lays out the long history of this debate.[28]

The formalist position leaves us with the problem that Princeton physicist and mathematician Eugene Wigner enunciated more than 50 years ago in his pioneering essay "the unreasonable effectiveness of mathematics in the natural sciences." Galileo 400 years ago marveled that the laws of nature are written in the language of mathematics. But why? More to the point, why should our minds, shaped by natural selection, be able to fathom laws of nature in mathematical terms that we invented? Wigner pointed out that Newton's laws of planetary motion (the universal law of gravitation), which Newton verified at an accuracy of 4 percent, has now proved accurate beyond all reasonable expectations, at less than 10,000th of a percent in 1960. It is verified with even more accuracy now. He went on to point out that the same is true in elementary quantum mechanics, the mathematics of complex spectra, and in quantum electrodynamics. So the question is, why does this invented mathematics work so well in our analysis of the universe? Wigner dubbed the uncanny accuracy of our mathematical abilities to match nature's reality "the empirical law of epistemology," but for him, it remained a mystery, "a wonderful gift which we neither understand nor deserve." More than once, he referred to it as a "miracle" that we have come to take for granted because it works so well. The statement resonates with Albert Einstein's more general remark that the most incomprehensible thing about the world is that it is comprehensible.[29]

Responses to Wigner came from scholars in many fields, including physicist Max Tegmark and mathematician Ivor Grattan-Guinness. Tegmark argued that mathematics is so effective because the world is not only described by mathematics but *is* mathematics, and it is therefore not surprising it is so effective since we are gradually uncovering its structure – Platonism with a vengeance. Grattan-Guinness argued that since science motivates much mathematics and both use theories and analogies, the close correspondence is no surprise at all.[30] Such explanations have not satisfied everyone, and the problem is still a burning question in the philosophy of mathematics. As the reader will have noticed, the embodied mind theories espoused by George Lakoff and others from cognitive science largely adopt the formalist view, and argue that mathematics works so well because the human brain evolved in response to the physical world around it. The concept of number arises because we count concrete objects, and mathematics works because the brain evolved to understand its surroundings. Mathematics is a construction of the brain, not something to be discovered. There are also positions in between the Platonist and the formalist. Mathematician Carl DeVito thinks natural numbers do exist independently of us, while the rest of mathematics is invented. Thus, for Plato and DeVito, 2 + 2 does equal 4 for the alien, but for the formalist this may not be true, since the alien may not have invented the concept of number, or if it has, may not have invented the same numbers.[31]

Unlike the physical sciences, there is nothing "observed" to prove whether mathematics is universal – unless, as Tegmark says, in observing the universe we are observing mathematical structures. Certainly we can say that alien counting systems would not be the same, as witnessed by the terrestrial history of mathematics, which in its development took many forms in many cultures before eventually agreeing on the common system now used. Nor would alien mathematical tools likely be the same, whether geometry, trigonometry, or calculus, much less the way they came about on Earth, as in the Newton–Leibniz controversy over the calculus. Livio argues that the question "is mathematics a discovery or an invention?" is ill posed because it is actually a combination of both: "Mathematics is partly created and partly discovered. Humans commonly invent mathematical concepts and discover the relations among those concepts."[32] Thus, he sees the axioms of Euclidean geometry as inventions, but the theorems as discoveries, since theorems link the different concepts. Whether this is convincing, we can now see how the problem of invention versus discovery in mathematics is crucial to how SETI and METI programs should proceed.

The Orientation of Alien Science

Rescher's second point in addressing the universality of science is that the orientation of science might be different. Little can be said prior to discovery about whether extraterrestrials might have developed social science to the exclusion of natural science – except that if they have no natural science, it is hard to see how SETI would ever work, since they would not have developed radio astronomy. Rescher argues logically that the physical, biological, and social constitution of the aliens will determine the agenda of scientific questions in terms of what they see as interesting, important, relevant, and significant: "the cognitive posture of the inquirers must be expected to play a crucial role in shaping and determining the course of scientific inquiry itself."[33] So even if aliens have natural science, intelligent porpoises might only develop hydrodynamics; they might not know about electromagnetic waves if their environment does not have loadstones or electrical storms.

But we can address whether their natural science might be *different*, an issue that brings us squarely back to the ubiquitous question of universality. We believe the physical sciences are universal, since all our physics and chemistry going back to Newton's physics, the Periodic Table, and the Standard Model of elementary particles indicate their laws work anywhere in the universe. That was hard-won knowledge. A legendary falling apple was one thing, but Newton's achievement was a theory of "universal gravitation," which he demonstrated by the motions of celestial bodies (Figure 5.4). The discovery of the chemical elements over a long period of time was a triumph of hard work. But to create the Periodic Table of Elements and to prove through spectroscopy the uniformity of these elements throughout the universe was another order of achievement (Figure 5.5). And particle physics continues to reveal important facts about cosmology, since cosmic rays are the universe's version of the same particle physics as occurs on Earth.

Figure 5.4 Isaac Newton (1642–1627), best known for his theory of universal gravitation, symbolic of human claims to a universal physics. Otherworldly Newtons would undoubtedly take a different route to scientific conclusions due to the contingencies of their history. But would they come to the same conclusions? Original engraving by unknown artist, courtesy AIP Emilio Segrè Visual Archives, *Physics Today* Collection.

In contrast to the universality of the physical sciences, biology remains an open question in this respect, and one of the main questions astrobiology hopes to address. Unlike physics and chemistry, with their remote observations using spectroscopy, life is much more difficult to detect remotely. Indeed, the earliest searches for life were undertaken by sending spacecraft such as *Viking* to make observations in situ on the surface of Mars, or from orbit around Mars, Jupiter, and Saturn. Only recently has the method of biosignatures held out hope of remote detection of life in planetary atmospheres or surfaces. Even then, the detection of life in this manner is likely to be ambiguous, and the proof of universal biological principles far from immediate.

There is, however, another approach, based not on observation but on principles – the universal principles of evolution by natural selection. In the absence of other evidence, it seems eminently reasonable to believe natural selection operates throughout the universe, whether carbon and DNA based, or silicon based with some other genetic system. Indeed, I believe most scientists would argue this is the case, and many have actively done so. Almost 40 years ago, in his pathbreaking

1 H 1.0079																	2 He 4.0026
3 Li 6.941	4 Be 9.0122											5 B 10.811	6 C 12.011	7 N 14.007	8 O 15.999	9 F 18.998	10 Ne 20.1797
11 Na 22.989	12 Mg 24.305											13 Al 26.981	14 Si 28.085	15 P 30.974	16 S 32.066	17 Cl 35.453	18 Ar 39.948
19 K 39.098	20 Ca 40.078	21 Sc 44.955	22 Ti 47.867	23 V 50.9415	24 Cr 51.9961	25 Mn 54.938	26 Fe 55.845	27 Co 58.933	28 Ni 58.6934	29 Cu 63.546	30 Zn 65.38	31 Ga 69.723	32 Ge 72.63	33 As 74.921	34 Se 78.971	35 Br 79.904	36 Kr 83.798
37 Rb 85.467	38 Sr 87.62	39 Y 88.9058	40 Zr 91.224	41 Nb 92.9063	42 Mo 95.95	43 Tc (98)	44 Ru 101.07	45 Rh 102.90	46 Pd 106.42	47 Ag 107.8682	48 Cd 112.414	49 In 114.818	50 Sn 118.710	51 Sb 121.760	52 Te 127.60	53 I 126.90	54 Xe 131.293
55 Cs 132.905	56 Ba 137.327	57–71*	72 Hf 178.49	73 Ta 180.94	74 W 183.84	75 Re 186.207	76 Os 190.23	77 Ir 192.217	78 Pt 195.084	79 Au 196.96	80 Hg 200.59	81 Tl 204.38	82 Pb 207.2	83 Bi 208.98	84 Po (209)	85 At (210)	86 Rn (222)
87 Fr (223)	88 Ra (226)	89–103**	104 Rf (267)	105 Db (268)	106 Sg (271)	107 Bh (272)	108 Hs (270)	109 Mt (276)	110 Ds (281)	111 Rg (280)	112 Cn (285)	113 Uut (284)	114 Fl (289)	115 Uup (288)	116 Lv (293)	117 Uus (294)	118 Uuo (294)

*	57 La 138.90	58 Ce 140.116	59 Pr 140.90	60 Nd 144.242	61 Pm (145)	62 Sm 150.36	63 Eu 151.964	64 Gd 157.25	65 Tb 158.92	66 Dy 162.500	67 Ho 164.93	68 Er 167.259	69 Tm 168.93	70 Yb 173.054	71 Lu 174.9668
**	89 Ac (227)	90 Th 232.0377	90 Pa 231.03	92 U 238.02	93 Np (237)	94 Pu (244)	95 Am (243)	96 Cm (247)	97 Bk (247)	98 Cf (251)	99 Es (252)	100 Fm (257)	101 Md (258)	102 No (259)	103 Lr (262)

Figure 5.5 Periodic Table of Elements. Throughout the visible universe, the same elements occur as on Earth. The numeration and alphabetic systems of extraterrestrials would be different, but the underlying atomic weights would be the same. Would an alien mind order the elements in the same way? Credit: Juliedeshaies/iStock/Getty Images Plus/Getty Images.

book *The Selfish Gene*, evolutionary biologist Richard Dawkins argued that any life we find beyond Earth would surely be strange and unearthly. But, he asked:

> Is there anything that must be true of all life, wherever it is found, and whatever the basis of its chemistry? If forms of life exist whose chemistry is based on silicon rather than carbon, or ammonia rather than water … if a form of life is found that is not based on chemistry at all but on electronic reverberating circuits, will there still be any general principle that is true of all life?

His unequivocal answer was spoken like a true Darwinian: "If I had to bet, I would put my money on one fundamental principle. This is the law that all life evolves by the differential survival of replicating entities."[34] In other words, evolution by natural selection. Dawkins went on to argue for the "meme" as the analog to the gene for cultural evolution, and more generally for the idea of universal Darwinism in a number of fields, an idea I say more about in the next section in the context of the human sciences. But the important point here is that evolution by natural selection provides a principle for the elaboration of universal biology, in the absence of observational evidence beyond the Earth.

Universal biology as a robust research program is just beginning to evolve. Perhaps its most important impetus comes from astrobiology, and it has been an explicit goal of NASA's astrobiology program. Its 2008 roadmap stated that "we must move beyond the circumstances of our own particular origins in order to develop a broader discipline, 'Universal Biology.'" In doing so the authors of the roadmap proposed to use the universal laws of physics and chemistry to understand biological processes such as polymer formation, self-organization, energy utilization, information transfer, and Darwinian evolution. Accordingly, the University of Illinois at Urbana-Champaign, one of the research teams of the NASA Astrobiology Institute, harbors the Institute for Universal Biology at its Institute for Genomic Biology, as part of its research on the origin and evolution of life in the universe. It uses the fields of microbiology, geobiology, computational chemistry, and genomics to find general principles for living organisms. "It is important to develop the field of universal biology, because we may never find traces of life on other planets. But if we understand that life is generic, maybe even an expected outcome of the laws of physics, then we'll know for sure that we are not alone," wrote the institute's director, Nigel Goldenfeld.[35]

Work on universal biology is an up-and-coming field, and not only from biologists but also from philosophers. The latter have in general worried about whether it is possible in principle to devise a universal biology, to find those long-sought general principles, when we know of only one type of life. This is the famous "N = 1" problem, made real because all life on Earth is believed to have evolved from a last universal common ancestor, and all subsequent forms of life are based on DNA. Philosopher Carol Cleland relates this to the problem of defining life that we broached in Chapter 2. There we noted that she argued life could not be defined in the absence of a general systems theory, which we do not and cannot have with a sample of one. Nevertheless, philosophers are tackling the problem along with scientists. Taking the lead from Richard Dawkins, philosopher Carlos Mariscal argues that a biological claim can be judged universal only if it includes principles of evolution, but no contingent facts of life on Earth. He concludes that biological generalizations are possible, and that we might in fact learn something about the nature of life beyond Earth by studying life on Earth. This conclusion is consistent with those who use terrestrial life as analogs to life beyond Earth, but is carried out at a much deeper level of general principles.[36]

"Perhaps reluctantly," Rescher concludes, "we must face the fact that on a cosmic scale the 'hard' physical sciences have something of the same cultural relativity that one encounters with the materials of the 'softer' social sciences on a terrestrial basis."[37] The same may be true of the natural sciences in general, including biology, despite a universal principle of evolution by natural selection. A bitter pill indeed for some to swallow.

The Conceptualization of Alien Science

Rescher's third area for framing the problem of universal science is the conceptualization of science. Today we do not conceptualize science in the same way as did Aristotle in ancient Greece; his ideas of natural motion and natural place are far from our present-day physics of universal gravitation, and even further removed from Einstein's ideas of space-time and general relativity (Figure 5.6). Tomorrow our descendants may look back on our era in the same way. How much more different would alien science be, even if the aliens were at the same level of development as we are – a highly unlikely proposition in a universe so old. Moreover, alien science would surely depend on aliens' circumstances; again a porpoise may have a highly developed hydrodynamics, but no astronomy at all. "It is only reasonable to presume," Rescher concludes, "that the conceptual character of the 'science' of an alien civilization is so radically different from ours as to orient their thought about 'the nature of things' in altogether different directions." This does not necessarily mean that we could not explain our science to aliens and see if we come to the same conclusions. That would be an interesting conversation!

These theoretical musings of Rescher as philosopher are given more concrete form by the study of the history of science. Many examples could be cited, but let's look at just one that illuminates different conceptions of science in terrestrial history. In a remarkable collaboration between historian of Greek science Gerald E. R. Lloyd and historian of Chinese science Nathan Sivin, the two historians compare the development of scientific concepts and approaches in ancient Greece and China during the period 400 BC to 200 AD. They conclude that the sciences as

Figure 5.6 Albert Einstein and Niels Bohr. Einstein's work on space-time and relativity, and Bohr's work on atomic theory, are symbolic of how our basic scientific concepts radically evolve over time. Photograph by Paul Ehrenfest, courtesy AIP Emilio Segrè Visual Archives, Ehrenfest Collection.

developed in the two civilizations were very different. "In many cases, the central preoccupations of the inquirers, the way they construed the issues to be investigated, differed. So did the fundamental concepts they used to articulate much of their work." The Greeks focused on nature and the elements such as earth, air, fire, and water, while the Chinese focused on the concepts known as *tao, ch'i, yin-yang*, and the five phases (if these Chinese concepts are unfamiliar to you I have proven my point). The dominant Greek approach was the search for foundations and causation; the dominant Chinese approach was to explore correspondences, resonances, and interconnections. The Greek intellectuals saw themselves as masters of truth, while the Chinese investigators' goal was to advise and guide rulers. Unlike the Greeks, they produced "a synthesis in which heaven, earth, society, and the human body all interacted to form a single resonant universe." As in concepts and goals, so also in applications: both cultures meticulously observed the heavens, but only among the Chinese was the regulation of the calendar a matter of state concern.[38] To say that the two conceptions and goals of science eventually converged is true to a certain extent. We have no separate Chinese and Greek laws of universal gravitation. But the pathways toward modern science were not a steady and continuous march, nor were its conceptual underpinnings an unbroken record of uncovering the truth. Nor is there any guarantee our current science is the final word – Einstein taught us that with his conceptual revolution of relativity and space-time.

Moreover, whether one is a rationalist or an empiricist, whether ascribing to Kant's epistemology or not, it is important to recognize that even science is not free of metaphysics. Philosopher E. A Burtt demonstrated a century ago the metaphysical foundations of modern physical science. More recently philosopher Iris Fry asked the question "is science metaphysically neutral?" and answered that no, it is not. This does not mean that all things are possible in science. To the contrary, she makes a strong argument for evolutionary naturalism, against the arguments of those who say it is just an assumption, by pointing out that "it is supported by vast amount of data from a wide range of scientific fields and is currently so strongly established as to be considered by the majority of scientists as not just another option … up for grabs along with … competing world views."[39] Yes, science has its own metaphysical assumptions, but in interaction with empiricism we can have some assurances it is a much better option than theistic guidance of evolution, or other competing worldviews that have no empirical foundation. That does not mean that evolution as it now stands is the final word, or that theism cannot be useful in other ways.

Given all this, is it plausible to conclude with Rescher that "natural science – broadly construed as inquiry into the ways of nature – is something that is in principle almost infinitely plastic" in the extraterrestrial context? For Rescher, the development of natural science "will trace out a historical course closely geared to the specific capacities, interests, environment, and opportunities of the creatures that develop it. We are deeply mistaken if we think of it as a process that must follow a route roughly parallel to ours and issue in a comparable product. It would be grossly unimaginative to think that either the journey or the destination must be the same – or even substantially similar."[40] This conclusion, while cogently argued, is far from certain on several levels. After all, the intellectual descendants of the ancient Greeks and Chinese did eventually converge on the science we accept today, even if it is not the final word. And Kant would likely have been ambiguous about such a conclusion, since he was far from being a complete skeptic about human knowledge, even if it may not be universal. For him, empirical science was the best we could do on our world. But the man who wrote the *Universal Natural History* would likely have been a fan of SETI, for by combining the knowledge of different alien civilizations gleaned from a successful SETI program, there would be at least some chance of approaching "the thing itself." At the same time, a certain plasticity in science among many worlds – especially considering our likely very different levels of development – does not make any easier the job of SETI or the job of determining its impact.

Are the "Human Sciences" Universal?

Given our doubts about the universality of science, mathematics, and human knowledge in general, could the human sciences possibly be universal? The question is in some sense dead on arrival. They are, after all, the *human* sciences, and unlike physics, no one has claimed them to be universal. Human behavior is too unpredictable, and grouping humans together, whether in small groups or large cultures, only makes things worse. This is, however, an important question because the human sciences – which include the social sciences of anthropology, political science, economics, psychology, and sociology, as well as the humanities such as history and archaeology – might be helpful, or even necessary, to discover the alien in the first place. And these disciplines become immediately relevant when and if we find intelligent life. Put another way, how much of our knowledge about human social origins, evolution, and behavior would actually be useful in assessing and interacting with the alien in the event of a discovery? Conversely, given the distinct possibility that the human sciences do not apply to aliens, once we find life, we will want to know in what ways our social knowledge is parochial and might be universalized, or at least further generalized. What might seem esoteric academic questions quickly become very practical in the event of discovery. It behooves us to tackle the problem before discovery occurs.

The Contingency of the Human Sciences

There are two good reasons not to consider the human sciences universal. First, the human sciences are of very recent origin, at least as a systematic area of study. True, even Socrates and his compatriots speculated on concepts still relevant to social sciences and humanities today, but "know thyself" was hardly a well-defined field. Even in the sixteenth and seventeenth centuries, as the scientific revolution in the natural sciences was hitting its stride in Europe, the social sciences were in embryonic form with broad speculations on matters such as the dignity of Man, ethical principles, and Cartesian dualism. In some ways, this is surprising, since the social sciences would seem more directly relevant to humans than the natural sciences. But they are also notoriously more complicated. In any case, the eighteenth century finally saw a blossoming of these and other areas, with many of the ideas emanating from the same philosophers discussed earlier, including John Locke and Gottfried Leibniz. Others such as Jean-Jacques Rousseau opined on human nature, while foundational treatises, such as Adam Smith's *Wealth of Nations* (1776) and Hume's *Treatise on Human Nature* (1739) were written as part of the Scottish Enlightenment. But only in the nineteenth century did the social sciences hit their stride, with Auguste Comte and Karl Marx writing on society, Darwin and Wallace writing on human evolution, and the formation of academic disciplines in psychology, sociology, and anthropology. All of these disciplines were further developed in the twentieth century, but there is still no widely accepted or workable theory of society, culture, or cultural evolution even on Earth, much less any laws of human behavior or cultural evolution.[41]

The history of the human sciences, young and subjective as they are, thus strongly hints that the discipline as it now stands, with all its myriad but embryonic components, cannot hope to embrace the possibilities of extraterrestrial culture, perhaps not even those at our stage of development. Indeed, French philosopher Michel Foucault, among many others, has critically questioned the objectivity of the modern human sciences, even as practiced in the terrestrial context. Foucault's main point, made in books such as *Les mots et les choses* [*The Order of Things: An Archaeology of the Human Sciences*], is that the human sciences "purport to offer universal scientific truths about human nature that are, in fact, often mere expressions of ethical and political commitments of a particular society. Foucault's 'critical philosophy' undermines such claims by exhibiting how they are just the outcome of contingent historical forces, and are not scientifically grounded truths."[42] Put another way, Foucault and his followers pose a critique of historical reason, in the same way Kant posed a critique of pure reason. I should point out that I am no fan of Foucault's postmodernist philosophy in a terrestrial context, but as hinted earlier, it may well make sense in an extraterrestrial context.

Second, quite apart from the youth of the human sciences, it would be very surprising if the questions that have arisen with regard to individuals and social systems on Earth adequately represent what must be a vastly larger set of individual behaviors and social systems if intelligent life is common beyond Earth. In the same way that we almost certainly currently lack a universal biology in the face of all the possibilities discussed earlier and in the previous chapter, we almost certainly lack universal social sciences in the face of all the social possibilities. And although it is a common belief that we have a universal mathematics and physics that applies throughout the universe, we have seen in the previous section that even this can be seriously brought into question. Thus, solely on the basis of the small sample size, it seems almost certain that the human sciences are not universal.

Third, when we examine our own knowledge from a historical point of view, our understanding of ourselves as embodied in the human sciences changes. Foucault is again relevant here. In his book *The Order of Things*, he gave a historical account of what knowledge meant from the Renaissance to present. In his work he used a historiographic technique he called "archaeology": "The premise of the archaeological method is that systems of thought and knowledge (epistemes or discursive formations, in Foucault's terminology) are governed by rules, beyond those of grammar and logic, that operate beneath the consciousness of individual subjects and define a system of conceptual possibilities that determines the boundaries of thought in a given domain and period." He found two great discontinuities in the episteme of Western culture: around 1650 and 1800. He argues that the very way we think changed during these times.[43] This is true of the human sciences no less than the natural sciences. Whether one takes Foucault's approach, the paradigm approach of Thomas Kuhn, or some other analysis, the conclusion is the same: with their rapidly changing knowledge of human behavior and culture, the human sciences have the least claim to being universal. In the same way we had to wrestle with the contingency of biological and social forms in the previous chapter, the human sciences must be judged contingent, a product of our own space and time.

Aspirations for Universality

The contingency of the human sciences does not mean, however, that they cannot aspire to universality, even given their current parochial nature. Its youth, incompleteness, and changing nature notwithstanding, the philosophy of the social sciences has already posed a series of fundamental questions that have considerable relevance in a more universal framework. What philosophical assumptions underpin the social sciences and humanities? How are the goals and methods of the human sciences different from those in the natural sciences? Are there universal social laws even in the terrestrial realm that might be extrapolated to the extraterrestrial realm? What might such laws look like?[44]

These questions are all interconnected, but unfortunately a survey of the social sciences gives the disconcerting impression that there is little agreement on almost anything. At the broadest level, the practitioners of the human sciences argue whether their field should even model itself on the natural sciences, aspiring to formulate social laws and perhaps even make predictions, or whether their goal is rather to interpret the facts with the goal of improving society. Indeed, this "naturalist versus interpretive" argument has been called "perhaps the most central and enduring issue in the philosophy of social science," affecting all other issues. Many in the social sciences would deny that universality is or should be an aspirational goal. Critics of this school of naturalism point to the lack of any apparent social laws, the complexity of society compared to moving bodies in physics, and the difficulty of controlled experiments. Both sides at least make their archetypal assumption clear: on one side, the assumption is that the social sciences can be modeled after the natural sciences, and on the other side, they are in a category apart.

In my view, an empirical approach for the social sciences, combined with prudent use of analogy, would seem to be the surest path to having any hope of studying alien behavior and cultures before their discovery. Just as biologists tend to extrapolate from what they know about biology on Earth to astrobiology, so social scientists can cautiously extrapolate from what they know about social systems on Earth to astropsychology or astrosociology. Beyond that, the search for universal aspects of biology may find its counterpart in the search for universal aspects of social science, and both may help discover life beyond Earth, or cope with life when it is discovered. This search can be aided by an evolutionary approach, which resonates with the field today known as Big History, the idea that in the scheme of cosmic evolution there is continuity between physical, biological, and cultural evolution. Or as historian of the human sciences Roger Smith puts it, at least for some approaches to the human sciences, "evolutionary theory appeared to establish continuity between nature and human nature and between natural science and social science." Not all would agree, but it is one way forward. It is our way forward.[45]

Adopting the naturalistic point of view has the added bonus that with such an approach we can use analogy from what the natural sciences have achieved to what universalists in the social sciences hope to achieve, even keeping in mind that humans and their cultures are more complicated than atoms and all their congregated forms. Thus, as we saw earlier in this chapter, aspirations for universality in the biological sciences have proceeded by appeals to the universality of evolution by natural selection, using empirical data about life on Earth but attempting to leave out what is contingent about that life. Might evolution again come to the rescue in the problem of universalizing the human sciences? Could we, for example, use some version of Darwinian evolution to better understand cultural evolution, especially considering the rising importance of both cooperative and nonrandom, directed forms of evolutionary dynamics? Might certain theories of sociocultural evolution, if true, provide the basis for a universal social science? As I hinted in the previous chapter in connection with attempts to transcend anthropocentrism, there has been no lack of attempts at forging evolutionary perspectives on human behavior. Among these perspectives are sociobiology, meme studies, gene–culture coevolution, evolutionary psychology, and behavioral ecology. As one philosopher pointed out, after four decades of research in these areas, there is no end in sight to the Darwinization of the social sciences.[46] Not only might such theories of cultural evolution have a direct impact on our understanding of the nature of intelligence in the universe, as we saw in connection with postbiological intelligence, they might also have a direct impact on our understanding of extraterrestrial cultures. If, for example, biology really does determine to some extent social characteristics and interactions, that would be a start in understanding aliens as individuals and in their societies.

The rub is in knowing, or guessing, what is contingent and what is necessary in cultural evolution. Thus we arrive at the same problem as in systematic attempts to find what is universal or contingent in biological evolution and other areas. As we saw in the previous chapter, anthropologists have attempted to do just that, in one case offering up as cultural universals the "meta-concerns" of nurturance, organization, allocation of resources and power, worldviews, and technical strengths. Another social scientist independently offered universals such as eating, drinking, and sleeping in biology; relationships such as dominance in sociology; universal grammatical rules in linguistics; universal activities and institutions such as tool-making and the family in anthropology; all the while holding out some hope for universals in psychology. Taking another step back to include animal species, another anthropologist has identified four "functional prerequisites" common to both human and animal behavior: dominance, kinship, specialization of tasks, and cooperation. The overlapping of some of these independently derived proposed universals would seem to lend them some credence. But as psychologist Albert Harrison warns, universalism is not necessarily the case in the social and behavioral sciences, and those who adhere to anthropology's idea of cultural relativism – which includes Foucault and the postmodernists – would argue that behavior cannot be understood outside its own place or time.[47]

Universalists would argue that commonalities with Earth would be necessarily found in cultures beyond Earth, and we should act accordingly. In my view, based on the continuities across cosmic evolution, this is likely a step in the right direction. But because of the breadth of the social sciences, the search for cultural universals is only one of many different approaches that might be tried. What might work for psychology might not work for sociology, and what might work for sociology might not work for history. There is as yet no consensus on whether evolutionary theory will be the final word on illuminating the human social sciences.

We are therefore a long way from determining whether any potential phases of sociocultural evolution would be universal.

The path from universal characteristics to universal laws is even more uncertain. At least limited universality in the physical sciences (tempered by our previous discussion) was achieved by using empirical data to devise universal laws, such as Newton's law of gravitation or Carnot's second law of thermodynamics. Whether universal social laws are possible or even desirable, they have been postulated both in fiction and in the real world, at least for parts of the human sciences. Most famously, in his popular Foundation series, Isaac Asimov introduced the science of psychohistory, which used psychology, sociology, history, and statistics to predict the behavior of large groups. In the novel, Asimov defined psychohistory as "that branch of mathematics which deals with the reactions of human conglomerates to fixed social and economic stimuli," where the size of the population as determined by "Seldon's First Theorem," had to be sufficiently large for valid statistical treatment. Moreover, the population itself had to be "unaware of psychohistoric analysis in order that its reactions be truly random." The "Seldon functions" underlying this endeavor were based on social and economic forces, and Asimov uses the idea to affect the future history of the Galactic Empire, reducing its coming Dark Age from 30,000 years to a mere 1,000. While this idea may seem far out, Asimov liked to point out that because of statistics such predictions were easier for large groups than for individuals, just as statistically predicting the behavior of groups of atoms in a gas was easier than determining the fate of a single atom. The 2008 Nobel Laureate in Economics, Paul Krugman, confessed that his economic methods were grounded in Asimov's idea of psychohistory, and held out hope that human destiny might one day be shaped by a more mathematical social science. As of now, economics is only able to make minor course corrections. Such methods, of course, raise all kinds of questions about determinism and free will.[48]

In the real world, much has been written about why determining laws of social action is so difficult, a failure often traced back to the philosophical problems of action, intentionality, and the mind-body problem.[49] Nonetheless, there are real, if peripheral, fields such as psychohistory, cliodynamics and social macrodynamics that aspire to determine the law-like behavior of groups of people. Whether these or some other variants can be developed in the near or far future may determine if universal social law is possible, or whether it will remain in the realm of fiction. Even if the latter, anthropologists are adept at comparing characteristics across cultures, and they will undoubtedly be in the forefront of any alien encounter. Conversely, aliens may already have developed their own version of psychohistory, which they may deploy at first contact.

Why Should We Care?

Why should we care if the human sciences are universal, or if they can be universalized? Because even short of universalizing the human sciences in the near future, they will be essential tools when alien life is discovered. Suppose a radio or laser signal of intelligent origin is received. The astronomers will rejoice in a job well done, but the discovery will be only the beginning of the process. And it will not be mainly the natural scientists but those in the human sciences who will do the job. Anthropologists, philosophers, historians, linguists, and cryptographers will convene, both to decipher the message and to interpret its meaning. Their relative numbers will depend on the discovery scenario. The human sciences will be stretched to their capacity and beyond, even as they become more universalized as the message is interpreted. As SETI pioneer Philip Morrison emphasized, the signal "is the object of intense socially required study for a long period of time. I regard it as much more like the enterprise of history of science than like the enterprise of reading an ordinary message … The data rate will for a long time exceed our ability to interpret it."[50] In a face-to-face meeting such as that depicted in *Arrival*, participation by those in the social sciences will be even more important – and urgent.

Given this likelihood of the centrality of the human sciences in the event of such a discovery, would it not be prudent to prepare ahead of time? In fact, it is already being done, though on a small scale and in no way systematically. Among the first to examine the utility of the human sciences in the extraterrestrial context were the participants in an American Anthropological Association symposium held in 1974 and published as *Cultures beyond the Earth*. The book's subtitle, *The Role of Anthropology in Outer Space*, is somewhat misleading for several reasons: only two of its eight authors were anthropologists (perhaps because of the stigma associated with the subject), it is a mixed volume including fictional stories as well as factual analysis, and it is not in any sense systematic. Nonetheless, it does include a brief but stimulating foreword by futurist Alvin Toffler and an afterword by anthropologist Sol Tax; it was sponsored as part of a "Cultural Futuristics" symposium; and, most important of all, it contains ideas that were at the time new and sophisticated. In his foreword, for example, Toffler pointed out that "what we think, imagine or dream about cultures beyond the earth not only reflects our own hidden fears and wishes, but alters them." He saw the book as important because "it forces us to disinter deeply buried premises about ourselves."[51] This is a straightforward but important point, one that we do not explicitly address often enough. Contemplating extraterrestrial cultures forces us to do that, raising, as Toffler said, "the critique of our cultural assumptions to a 'meta-level.'" Moreover, he argued, the cultures that anthropology traditionally studies are all human and less technologically advanced; analyses of such cultures leave vast areas of life unilluminated by contrast or comparison. Toffler went even further, asserting that extraterrestrial anthropology "calls into question the very idea of cultures based on a single epistemology, of single time tracks or merely human sensory modalities. It forces questions about intelligence and consciousness. It makes one wonder whether our assumptions about probability apply universally. In the course of all this, it also begins to give intellectual shape to the whole question of space exploration and its relationship to our world."[52] This statement, which resonates with what I have said earlier in this chapter, gives some indication of the unrealized potential of the human sciences in relation to the discovery of alien life.

It is one thing for a futurist to say such things. But in his afterword to the volume, Sol Tax, at that time a professor of anthropology at the University of Chicago, endorsed and elaborated on these ideas. Extraterrestrial anthropology, he said, removes itself from our planet to view "human nature" as a whole. It envisions the opportunity to study human behavior and the change or development of human cultures under extraterrestrial conditions; to test the applicability of anthropological knowledge to the design of extraterrestrial human communities; and to develop anthropological models for quite different species of sentient and intelligent beings by using, on a higher level, the comparative methods by which we have come to understand each earthly culture in contrast to others. Moreover, Tax noted, "Only when we have comparisons with species that are cultural in nonhuman ways – some of them maybe far more advanced than we – will we approach full understanding of the possibilities and limitations of human cultures." Nor was this a fruitless undertaking, because "even if we have no contact with nonhuman cultures in the immediate future, the models that we meanwhile make require that we sharpen the questions that we ask about human beings."[53] Studies of culture among animals are of course also relevant here, especially in the evolution of culture, but they inevitably fall in the more primitive direction. Contemplation of extraterrestrial cultures allows us to approach the problem from the direction of more advanced cultures, emphasizing that humans may not be on the upper end of a cultural spectrum that includes species from other planets.

Between Toffler and Tax in this volume were two anthropologists, Roger W. Wescott and Philip Singer. Wescott pointed out that anthropology brings both strengths and weaknesses to the problem of extraterrestrial intelligence. Among the strengths is the range of its inventory of cultures, primitive and literate, extant and extinct. Among the weaknesses is the fact that in his view, anthropology tends to study the primitive and prehistoric more than the modern cultures. SETI and space programs are the purview of modern industrialized countries, and anthropologists are less accustomed to operating within this context, much less with advanced extraterrestrial civilizations. In a broader sense, however, the tools of anthropology are applicable.

Since that pioneering volume in 1975, only a very few anthropologists and other social scientists have continued to take up the challenge of applying the human sciences in an extraterrestrial context. In the 1980s, Ben Finney, an anthropologist at the University of Hawaii, almost singlehandedly addressed some of the cultural issues, and even worked with the SETI community. Psychologist Albert Harrison has repeatedly emphasized the importance of the social sciences in connection with SETI. A recent overview of "social evolution" by anthropologist Kathryn Denning is particularly nuanced in discussing the problems and promise of the social sciences for SETI, while Debbora Battaglia has contributed substantially to this literature with her volume *E.T. Culture: Anthropology in Outerspaces*.[54]

By far the most diverse work has been done in part at the urging of psychologist Douglas Vakoch, who for many years held the enviable title of director of interstellar communications at the SETI Institute. It was he who organized three consecutive annual sessions at the American Anthropological Association from 2004 to 2006, the results of which were published in a NASA volume, *Archaeology, Anthropology and Interstellar Communication*. The topics in this volume range from archaeological analogs of interstellar message decipherment, to issues of culture and communication, and the vast anthropological database on culture contacts.[55]

Given these pioneering efforts, it is likely that more anthropologists will join the discussion, an outcome highly desired. The appearance of a cover story "Anthropology and the Search for Extraterrestrial Intelligence" in the journal *Anthropology Today* in 2006 indicates that the discipline may be ready to expand its boundaries. In the latter I argued that anthropology has already made sporadic contributions to SETI by lending its expertise on the evolution of technological civilizations, on culture contacts and diffusion, and on interstellar message deciphering and construction. All these areas merit further systematic development, and are an indication of what is to come when and if life is actually discovered beyond Earth.[56]

Nor is interest in the human sciences confined to anthropology. A new discipline dubbed "astrosociology" has arisen in the past few years that addresses the societal impact of space exploration, including extraterrestrial life. Astrosociology is defined as "the study of *astrosocial phenomena*, where astrosocial phenomena comprises a subset of all social, cultural, and behavioral phenomena … characterized by a relationship between human behavior and space phenomena." As James Pass, the sociologist who coined the term in 2004, puts it, "the astrosociological perspective brings the social sciences into the space age by fostering the creation and development of a field dedicated to the study of the impact of space exploration." The impact of astrobiology is an explicit part of this new field. As the founders of the field put it:

> Even without an announcement of success forthcoming in the near future, and even without consideration of the implications if such an announcement became a reality, the very attempt to seek out life in an organized manner merits the attention of astrosociologists from a number of disciplines, including sociology, psychology, anthropology, and history. If this is the case, astrosociology must investigate this behavior along with the implications of long-term failure and success. The social and cultural implications of this work make it too important to ignore. In fact, it is imperative that astrosociologists participate alongside their space-community counterparts to attain comprehensive knowledge; both for its own sake and for practical application should some type of reaction prove necessary.[57]

The transformation from the social sciences to the astrosocial sciences has already begun.

Enough has been said to show not only the utility but also the essential nature of the human sciences to any discovery of life beyond Earth, particularly when it comes to intelligent life. While the human sciences will of necessity need to be used in assessing alien cultures, and while some of its concepts can and have been expanded in the spirit of moving beyond anthropocentrism, the discovery of new cultures is likely to expand our current human sciences in ways we cannot now foresee. Beyond that, the goal will be to expand the human sciences beyond the human – in short, to universalize the humanities and social sciences, converting them to the astro-humanities and astro-social sciences, even if the goal of universal social laws is never reached.

6

How Can We Envision Impact?

> Notes of a busy life in distant worlds
> Beat like a far wave on my anxious ear.
>
> *Tennyson, Timbuctoo (1929)*

> There is no reason to conclude that the affairs of men are becoming more predictable. The opposite may well be true.
>
> *Nate Silver*[1]

> Modern science and technology have an apparently decisive influence on the fate of cultures, and even seem to produce fundamental upheavals in every dimension of cultural life.
>
> *Jean Ladrière*[2]

American statistician Nate Silver became famous for predicting the outcome of the 2012 US presidential election in all 50 states, the election in which Barack Obama won his second term. His subsequent best-selling book *The Signal and the Noise: Why So Many Predictions Fail – but Some Don't* demonstrates that empirical methods using real data can indeed make accurate predictions in fields ranging from baseball to elections. But most of all, the book demonstrates why predictions fail in many areas. Ironically, Silver himself failed spectacularly in the very area where he had achieved some measure of success before – the 2016 elections in which Donald Trump was elected president of the United States, to the astonishment of everyone and the chagrin of most. The attempt to "predict" the societal impact of discovering life beyond Earth is even more difficult: not only must any data be based on analogy and history rather than direct experience but also gauging impact is one more step removed, like trying to divine the effect Trump's election would have on the United States and the world in the years to come. Even were Trump a more predicable person, that would be a notoriously difficult task. Even with mountains of political data and armies of reporters and scholars focusing on the problem, human behavior is problematic and unpredictable. And yet, we keep trying, for very practical political, economic, and social reasons.[3]

In addition to transcending anthropocentrism and universalizing knowledge, a third critical issue when discussing the humanistic aspects of astrobiology is the very idea and meaning of societal impact, and how we can possibly approach it. Whether addressing science, technology, or some other domain of human life, determining impact is a hazardous undertaking, to say the least. This is true even when examining past impact, much less projecting its future. And it is true even when tackling local issues, much less global issues encompassing all of humanity. Among the critical questions here are what we mean when we say "societal impact," and in particular "societal impact of astrobiology," especially since society is notoriously diverse and large segments of society may not even have heard of astrobiology. Should we distinguish various phases of impact, at least in terms of short-, medium-, and long-term impact? Does it really make sense to speak about "humanity" as a whole being affected? Is there any general theory of societal impact that can help guide us on our way? Or is the known history of the societal impact of various aspects of science and technology a better guide?

These questions are daunting, but difficulties notwithstanding, we should not lose sight of the fact that science and technology impact studies are numerous, increasingly important, and frequently undertaken by governments wishing to ensure good stewardship of taxpayer funding. In the United States, the National Science Foundation has its Science, Technology and Society program, and the American Association for the Advancement of Science supports an active "Societal Impacts of Science and Engineering" section. Substantial studies have been undertaken on the societal impact of scientific endeavors such as the Human Genome Project, biotechnology, nanotechnology, and spaceflight, sometimes at considerable expense. The US Department of Energy and the National Institutes of Health devoted 3–5 percent of the $3 billion Human Genome Project budget to studying the ethical, legal, and social issues of their work, ranging from philosophical implications such as free will versus genetic determinism, to practical considerations such as the privacy and confidentiality of genetic information. The National Science Foundation funds a robust societal impact of nanotechnology program, and during my years as NASA chief historian, we initiated, in accordance with an obscure clause in the NASA charter, a program of studies on the societal impact of spaceflight. Even closer to astrobiology's core interests are NASA's planetary protection protocols, which are certainly studies of potential impact, and the policymaking that follows from them. In addition to these activities, which are relevant for both methodology and substance, science, technology, and society programs exist at many universities, reflecting the importance of public understanding of the broader issues in science and technology. Nor is interest limited to the United States. To cite only one example, the European Science Foundation and the European Space Policy Institute have recognized the importance of the humanities and social sciences to space exploration (and vice versa) through their series of published Studies in Space Policy, which embrace the impact of space exploration, including the discovery of life.[4]

Given this background, we should be under no illusion that millions of dollars are going to be spent on studying the implications of extraterrestrial life – not, that is, until it is discovered, in which case the floodgates may open as they did with the Human Genome Project, now in the form of a practical problem rather than a theoretical one. But the extraordinary importance of science in shaping culture – evident not only in the current activities just mentioned but also historically at least since the seventeenth-century scientific revolution – provides a firm foundation for proceeding, even while recognizing the problems and limitations. Beyond the history, discovery, and analogy approaches discussed in Part I, in this chapter I examine a possible framework for understanding societal impact, within which those approaches can be utilized. Within this framework I then examine how societal impacts of discovering life have been tackled in two very different ways: imagination in the form of science fiction, and deduction using an array of scholarly studies drawing on many disciplines. We shall find that although it is a difficult and multifaceted problem, studying the impact of discovering life beyond Earth is a definable and circumscribable endeavor, a necessary prerequisite to formulating policy for an event that may be essential to humanity's future.

A Framework for Societal Impact

To say that the impact of science on culture has been the subject of much study is not to say that a generally acceptable framework exists for such studies. In fact, such frameworks seem to be sorely lacking, perhaps stymied by the complexities involved and the recognition that such a broad array of problems should not be shoehorned into a single rigid framework. Nevertheless, as indicated earlier, societal impact studies are done every day in many fields. The question is, what approach should we take particular to our rather peculiar subject?[5]

The Anatomy of Impact

In the most basic sense, just as we have distinguished prediscovery, discovery, and postdiscovery stages in the discovery process, so we may now distinguish preimpact, impact, and postimpact phases in the impact process. In his book *The Challenge Presented to Cultures by Science and Technology*, based on a high-level meeting sponsored by the UN organization UNESCO, Belgian philosopher and scientist Jean Ladrière demonstrated that such phases are useful in understanding the impact of science on society. Implicit in his treatment is the simple but undeniable fact that any impact scenario consists of a past, present, and future. Before impact, a certain set of assumptions, ideas, and values (we would also say worldviews) are in place. These are then in flux for an extended period before the assimilation of new assumptions, ideas, and values is completed. During this period of flux, reinforcing and competing worldviews are also in play. In Ladrière's rather unwieldy terminology, the process of impact involves a "destructuration" phase during which the old order is torn down, and a "restructuration" phase during which the new order is built up. We will refer to these simply as destructuring and restructuring stages, each of which will have political, economic, and cultural components (Table 6.1).[6]

During the destructuring phase a new discovery of large enough magnitude affects all three components of society (political, economic, and cultural), and either challenges or confirms them. This is followed by growing acceptance of the new information and its implications for the future, which feeds into society's values and meanings. Finally, a new idea of the future is established, a future that may be rich in possibility and transformable by human activity, or alternatively, threatening and out of our control. In short, a critical mass of novelty is reached bursting with potential that begs to be reintegrated. This triggers the restructuring phase, in which there is a reintegration and reunification of culture and its subsystems based on the new meanings and values. Some aspects of culture may accept the new system of values and meanings, while others reject it. This is followed by criticism and resolution of conflicting views. And finally in the postimpact phase, there is restoration of a unity of vision – to the extent that such unity can ever exist in a multicultural world.

Table 6.1 *The Anatomy of Impact*

Impact Phase	Impact Status	Areas Affected
Preimpact	Established Assumptions, Ideas, and Values	
Impact	**Destructuring Phase (Disruption of Cultural Elements)** - challenge or confirm values - growing acceptance - new vision of future	Political Economic Cultural
	Restructuring Phase (Rebuilding Cultural Elements) - acceptance or rejection of new elements - criticism and move toward resolution - unity of vision restored	Proportional to magnitude of discovery
Postimpact	Acceptance of New Assumptions, Ideas, and Values	

This potentially valuable framework is obviously an idealized view, bound to be complicated by the messiness of human behavior, not to mention the problem of overlapping discoveries. Nonetheless, Jesuit astronomer William Stoeger has proposed that Ladrière's ideas may be useful as a way of understanding astronomy's impact on culture. He sees the impact of astronomy primarily in the area of culture rather than in the political or economic realms, in the sense that astronomy and cosmology usually impart a sense of our place in the universe rather than providing technical applications of economic value. Astronomy does indeed affect education, art, literature, and religion, not to mention generating wonder and inspiration. In short, "it presents us with new perspectives on ourselves, our world, our history, our destiny and our significance" – no small effect.[7] At the same time, Stoeger points out that over the long term, astronomy acts counter to our tendency to think that science and technology can help us to entirely control our future, since in the very long term, human destiny is bound up with the cosmic demise of the universe according to the laws of physics. For our purposes, however, we are dealing with decades, hundreds or thousands of years, not millions or billions of years in terms of impact.

What does this mean in practice? We can put this abstract schema in more concrete terms through specific examples. In the case of astronomy, the discovery of the Copernican worldview certainly illustrates the extended process of societal impact, where worldview is now seen acting as framework rather than analogy as in Chapter 2. First published in full form in 1543, the heliocentric theory challenged the traditional geocentric worldview (Figures 6.1 and 6.2). The ordered cosmos of Aristotle, as reflected to the public in literary forms such as Dante's *Divine Comedy*, had provided the framework for Christian thought and human life in general. The gradual dismantling (destructuring) of this framework is well documented, as is the restructuring of a worldview associated in science with the names of Galileo, Kepler, and Newton, and with Donne, Milton, and many others in the broader cultural realm. The growing acceptance of the new worldview and its assimilation into culture took centuries to complete, as indicated not only in far-reaching controversies such as the Galileo affair, in which the new model was seen to be at odds with the Christian worldview, but also in literature and the arts. As C. S. Lewis put it, the "discarded image" of the medieval geocentric universe affected the very frame within which the human mind operated. "In every age the human mind is deeply influenced by the accepted Model of the universe," he wrote in the 1960s as the search for life beyond Earth was gaining ground.

Similarly, a few years before his pathbreaking work on paradigms and the structure of scientific revolutions, Thomas Kuhn wrote, "Copernicanism required a transformation in man's view of his relation to God and the bases of his morality. Such a transformation could not be worked out overnight, and it was scarcely even

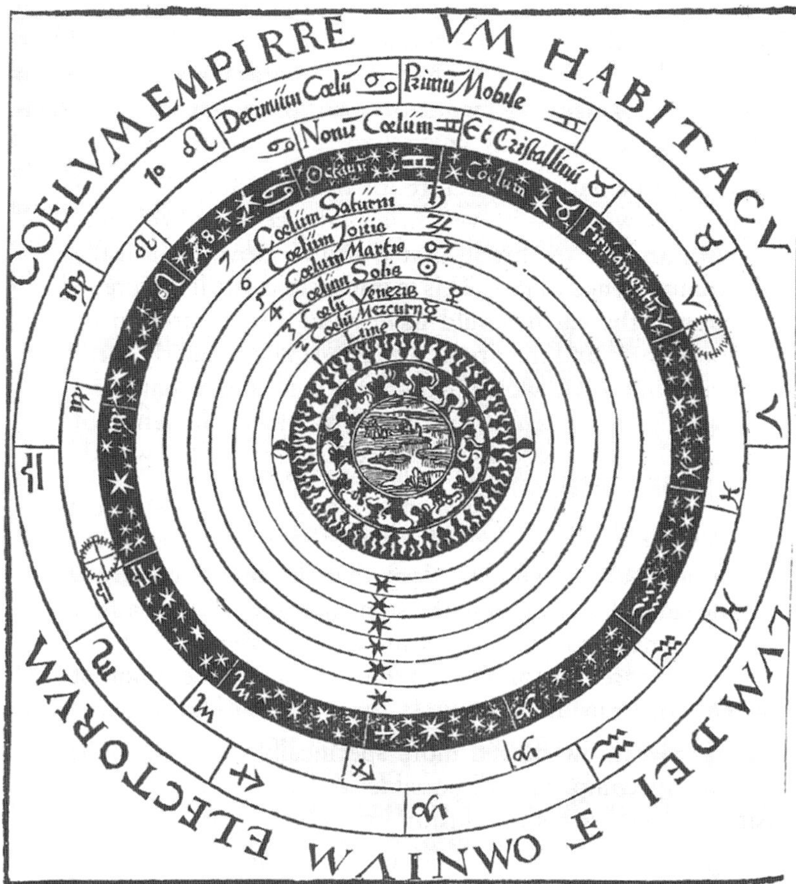

Figure 6.1 The medieval geocentric worldview from Peter Apianus, *Cosmographicus liber Petri Apiani mathematici studiose collectus* (1524). From Aristotle through the Middle Ages, this ordered cosmological worldview, the *kosmos*, provided the framework within which human life was carried out. In the Christian context, it was incorporated into such literary works as Dante's *Divine Comedy.*

begun while the evidence for Copernicanism remained as indecisive as it had been in the *De Revolutionibus*."[8] Sensing the tectonic shift in the making, in the cultural realm, John Donne wrote in 1611, "And new Philosophy calls all in doubt … The Sun is lost, and th' earth, and no mans wit can well direct him where to looke for it … 'tis all in peeces, all cohaereance gone; all just supply, and all Relation." Johannes Kepler pondered a cosmos full of worlds and asked "if their globes are nobler, we are not the noblest of rational creatures. Then how can all things be for man's sake? How can we be the masters of God's handiwork?" Kepler comforted himself with the perceived primacy of the Sun and its retinue of planets. But not everyone, as we know from history least of all the Catholic Church, was

Figure 6.2 The heliocentric worldview made the Earth a planet and the planets potential Earths. At left Copernicus points to his new worldview, while at right Galileo offers his telescope and Kepler wishes for wings so that he might visit the new world. The Sun says, "I give light, heat, and motion to all." This image is an elaboration of the title page from John Wilkins's *Discovery of a World in the Moone* (1638), and appeared as the title page to the combined edition of the *Discovery* and the *Discourse Concerning a New Planet* (1640). These books were published in the midst of the transition from the medieval to the modern worldview, a transition that may illuminate the kind of change that will occur if extraterrestrial life is actually discovered.

so sanguine about the prospects for humanity in the new universe. The discarded geocentric image represented much more than the loss of a physical model. It was the loss of human moral bearings.[9]

In short, in the wake of the Copernican theory, along the general outlines of the Ladrièrean schema, traditional values and scientific principles were challenged, and over a period of a century or more new values and scientific principles were gradually worked out and accepted, and a unified vision of the universe was restored. The implications of the new vision, particularly the theological implications, are still difficult for some groups to accept.

We should be quick to note that not every discovery has such a strong or global impact. In fact, Harlow Shapley's discovery around 1920 that our solar system is on the periphery of our Galaxy is an example of a silent revolution – at least in its early stages – most probably because of its more technical nature. Astronomers celebrated the discovery, the press routinely reported it, and the general population went about its business as usual despite humanity's slide from the center to the edge of the Galaxy. Richard Berenzden has noted this silent/noisy dichotomy in the reactions to certain worldviews. The proof of the eccentric position of the Earth in the Galaxy, he found, "caused almost no discussion whatsoever" in the press. But Edwin Hubble's announcement in 1929 that our galaxy was only one of millions (today billions) certainly did. Today, these revelations of the 1920s, along with new discoveries from the Hubble Space Telescope and other spacecraft, have worked their way into the public consciousness as the accepted cosmological worldview. At least for most people: Berenzden's cautionary note that "when scientific revolutions impinge upon metaphysics or social theory they are likely to become unusually polemical and possibly unacceptable," has also proven true among some religious elements even in the twenty-first century.[10]

The anatomy of impact, of course, applies to much broader areas than astronomy. And just how difficult such a shift might be is illustrated in the reaction to the worldview we know as Darwinism. One hundred fifty years after the Darwinian revolution, its implications remain very much controversial, especially among a large segment of the American public. As we saw in Chapter 3 in the context of worldviews as analogies, studies have shown how Darwin's theory had distinctive impacts over the short term and the long term and among scientists, theologians, and other segments of the population. The title of one of the studies, *Science, Ideology and World View*, is likely to express succinctly the general tenor of the debate in the aftermath of the discovery of extraterrestrial life. In short the restructuring of the new Darwinian worldview in Ladrièrean terms is still taking place, at different paces in different cultures.[11]

Examples could be multiplied, but we have said enough to illustrate the utility of the impact model. Stoeger went on to propose that although astronomy and cosmology play a role in both destructuring and restructuring, they most heavily play an integrative role in the latter, reunifying, stages of cultural evolution. In particular, they expand our horizons, open new possibilities, prod us toward new values, and enrich us with new information that we can use to determine our true origins and envision our real possible destinies – surely characteristics that will result from a discovery of life beyond Earth. More controversially (but not surprisingly given his Jesuit background), Stoeger suggests astronomy helps us appreciate the limitations of science, "leading us to see that there are ways of knowing beyond scientific ones and that the search for meaning must go far beyond science itself."[12] Both Ladrière and Stoeger point out that the last phase of restructuring culture can be negative or positive, negative (in Stoeger's view, at least) if the outcome is a radical scientism in which science is seen as the only way toward knowledge, positive if culture absorbs scientific knowledge by expanding theological, psychological, and philosophical interpretations. Whether we accept these terms and definitions, it would seem axiomatic that the goal of any discovery impact should be to have a positive effect on culture. Indeed, the raison d'être for studying societal impacts in a forward-looking way is to maximize the benefits of any discovery for society and to minimize its negative effects.

This seemingly simple schema of destructuring and restructuring, which may apply to impacts large and small over the long and short terms, provides a potentially powerful framework within which we can discuss the impact of discovering life beyond Earth under whatever scenario. It resonates with the rise and fall of worldviews I discuss in the next chapter, including stages in worldview development ranging from presentation based on observation to elaboration, opposition, exploration of implications outside the field, general acceptance, and final confirmation. It also parallels in some ways Thomas Kuhn's paradigms, though the latter are now so laden with baggage that it is perhaps best to avoid the complications of that comparison. Finally, the destructuring and restructuring process is arguably inherent at a very practical level in the long-term cycles of the global world order. In his book *World Order*, Henry Kissinger argues that the world is at its greatest peril when one world order is breaking down and another is forming, a situation in which we find ourselves today as the political framework established after World War II is breaking down. "Restraints disappear, and the field is open to the most expansive claims and the most implacable actors," he writes. "Chaos follows until a new system of order is established." Chaos can be good or bad; in one sense, it is creativity by another name. But it is most likely to have good outcomes only if we are prepared.[13]

In summary, astrobiology in its most expansive form – the search for exoplanets, biosignatures, and the past, present, and future of life ranging from microbes to intelligence – is an increasingly important part of astronomy, and one to which the Ladrièrean schema – fundamentally an outline of the rise and fall of world-views – bids fair to apply in accordance with astronomy, politics, and other areas of culture. We now examine more closely the potential of this general framework with regard to the two chief modes of reasoning relevant to determining the impact of discovering life: imagining impact and deducing impact.

Imagining Impact: Science Fiction and Alien Life

If we accept that the idea that a universe full of life is a kind of worldview or "bio-physical cosmology," then science fiction may be seen as one way of working out that worldview in popular culture.[14] Part of that process is studying the societal impacts if life is discovered under any of the various scenarios laid out in Chapter 2. As we have seen there in the context of the nature of discovery, a huge amount of fiction has been written on the subject of alien life encounters, constituting one of the earliest themes of the genre stretching back to the late nineteenth century and H. G. Wells's *War of the Worlds*. Some is clearly better than others, but my firm position is that science fiction should not be denigrated as a valid method for approaching the problem. As the organizers of a Royal Society meeting on the impact of astrobiology noted, "imagination … must not be underestimated as a valuable means to advance knowledge towards new frontiers, and is not at all an unscientific concept."[15] Science fiction, after all, has a long tradition of often being broadly right about the future; at the same time, it is almost never right about the details. And admittedly sometimes when the imagination is wrong, it is very wrong. There are failures of imagination, as in the failure of some to see the possibility of heavier-than-air flight, nuclear power, and desktop computers, even a few decades before they happened. We must also admit the possibility of failure of imagination in our particular case of the discovery of life beyond Earth. Shakespeare may well be right that there are more things in heaven and earth than are dreamt of in our philosophies.

Despite these caveats, the imaginative thought given this subject over the past century and more is a resource that should not be ignored. In Chapter 2 I looked at some classic science fiction scenarios from the point of view of discovery. I now examine some of those same classic stories with a view toward impact.

Direct Impacts: Face to Face with the Alien

As we argued in Chapter 2 (see Tables 2.1 and 2.2), it would seem likely that Encounter Type 1, direct contact with alien life in the vicinity of Earth, would have the greatest impact for the greatest number of people, followed by Types 2, 3, and 4, based solely on the distance of contact and (in the case of Type 4) its indirect nature even though in the vicinity of Earth. This is certainly borne out in those science fiction exemplars given in Table 2.3. In H. G. Wells's *War of the Worlds* (1898) the Type 1 impact was intense but brief: the invading Martians were in the end destroyed by the humble microbes of Earth, and, one presumes, life on Earth went on as usual after a period of rebuilding. Moreover, the action was localized to London, not a global impact, though it undoubtedly would have been global had the Martian invasion been successful. But if terrestrial organisms can be deadly to aliens, the reverse is also true. The potential impact of pathological alien organisms entering the Earth's biosphere is played out in Michael Crichton's *Andromeda Strain* (1969), where an alien organism arrives piggybacking on a military satellite that has returned to Earth. A government-sponsored team named Wildfire must cope with the resulting destruction and attempt to halt the spread of the microbe. It is notable that Wildfire in the novel was a preestablished protocol for just this scenario; in reality, while protocols are in place for biohazards and for back contamination, no Wildfire-team protocol exists today, except in a general sense of the Centers for Disease Control in the United States or the World Health Organization.

A half-century after *War of the Worlds,* the 20th Century Fox film *The Day the Earth Stood Still* envisioned quite a different impact when an alien spaceship lands in Washington, DC. The US Army surrounds it, but the alien Klaatu announces he has come in peace. As he attempts to unveil a gift for the president that would help him understand alien life, Klaatu is wounded by a soldier who thinks he has a weapon. The robot Gort disintegrates the army's weapons in a single flash of light and stands guard as Klaatu is taken to the hospital and basically heals himself. This is not a good start to the encounter, but it holds a lesson about the importance of early actions and misinterpreted motives. Klaatu has a message for the world, but suspicion and uncertainty reign. As Klaatu becomes more aware of the problems of Earth, he becomes more sympathetic to Earthlings. But he and Gort depart Earth with a final message that, having just joined the Nuclear Age and about to join the Space Age, they have a choice of joining other inhabited planets in peace, or of being destroyed if they extend human violence into space.[16] This, then, is a prototype of the powerful but essentially benevolent alien.

The powerful but benevolent and well-intentioned alien prototype was amplified two years later in Arthur C. Clarke's *Childhood's End* (1953), which depicts the Earth visited by alien Overlords possessing a bodily form resembling the devil, who attempt to incorporate humanity into the scheme of universal sentience. The Overlords impose world government and an end to war on Earth. Amid some resistance, the first contact theme is played out: the eradication of poverty and ignorance is set against negative aspects that include a decline in the creative arts and in science brought on by the Overlords' vastly superior knowledge. Yet the Overlords later reveal they are only the agents of an even more superior intelligence, and their ultimate goal is to usher the human race (by way of its children) into the Overmind, something the Overlords themselves cannot achieve. The concern here is with the ultimate destiny of humanity, which is intimately connected to extraterrestrials. The lesson here is that even though aliens may be benevolent and well intentioned, the impact on Earth could be overwhelming.

Clarke's *2001: A Space Odyssey* and its sequels features aliens who have also visited Earth, but in the distant past, and left behind an artifact in the form of a mysterious black monolith. The monolith turns out to be an active artifact, which has sparked human ingenuity and changed human history, as depicted in the opening scenes of the movie when the ape-man is inspired to use a bone as a tool, the beginning of technology, which, with an upward throw of the bone, quickly transfers to Space Age technology. The idea of the discovery of an active artifact in the vicinity of Earth, though now in modern times, is also the theme of David Brin's *Existence*. In this case, Earthlings must deal with an invitation to join the extraterrestrials, illustrating that an active artifact can have an impact almost equivalent to the arrival of the aliens themselves (which is why I have categorized it as Type 1).

How does the impact change if the contact is still direct but beyond Earth and its vicinity? The full force of Clarke's *2001* novel and movie comes only with such a direct contact scenario of Type 2 after a human expedition, aided by HAL the computer, is mounted to Jupiter, leading to the famous enigmatic encounter with aliens of unknown form. The result of that encounter is an implied end to an era in humanity, symbolized in the form of the Star Child. Such Type 2 encounters, where humans meet extraterrestrials in an off-Earth environment, have been played out in plots ranging from Ridley Scott's *Alien* series to Stanislaw Lem's more subtle treatment in his novels *Solaris* (1961), *His Master's Voice* (1968), and *Fiasco* (1986). In all these cases, the impact on the space travelers is certainly intense, but it is limited to those travelers, while the inhabitants of Earth remain safe. Whereas *Alien* depicts the extreme possibility of evil in the universe, all three of Lem's novels pose another possibility: the essential unknowability of the alien. In *Solaris*, for example, the planet Solaris is found to contain an ocean that is in some sense alive. Countless attempts to establish contact are frustrated. Although much is learned about the ocean, in the end scientists are convinced only that they are confronted by a protoplasmic ocean-brain enveloping the entire planet, idling its time in theoretical cogitation about the universe, a monologue "inevitably beyond our understanding." Similarly, *His Master's Voice* and *Fiasco* bring home the essential unknowability of the alien. In all these novels, because the action takes place far from Earth, the impact is less than a Type 1 contact.

The same holds true for contact with an alien artifact beyond Earth's vicinity. Clarke's *Rendezvous with Rama* – featuring an encounter with a very large alien artifact in the form of an alien spacecraft – is a prime example of a direct encounter with an active artifact beyond the Earth-Moon system. The object is first detected outside the orbit of Jupiter and believed to be an asteroid. Photography reveals it to be a 16 by 50 kilometer perfect cylinder, and an expedition is mounted to intercept and study the object. The explorers find apparently dead cities, factories, a sea, and other enigmatic structures. Eventually a crab-like being is found, determined to be robotic with some intelligence, but its builders, the Ramans themselves, remain enigmatic. The spacecraft is flung out of the solar system, essentially benevolent and mysterious. The spacecraft departs the solar system, its impact in the form of knowledge gained by the expedition's surveillance of the spacecraft.

Indirect Impacts: Coping with the Message

We can contrast these Earth-shaking events of direct contact on and beyond Earth with what is usually seen today as the more likely scenario, remote contact via some kind of electromagnetic signal. As noted in Chapter 2, this theme was particularly popular in science fiction in the wake of Frank Drake's Project Ozma in 1960, the first attempt to carry out a search for extraterrestrial intelligence by radio telescope. Logically there are several possible outcomes, the two chief ones being on the one hand detection of a "dial tone" with characteristics clearly indicating intelligence but either lacking a message or not decoded, and on the other hand the discovery of a message that is decoded and translated. Naturally, the latter is the mode most often portrayed in science fiction because of its potentially dramatic effect.

Within a year of Drake's Ozma, astronomer Fred Hoyle and TV producer John Elliot wrote *A for Andromeda*, one of the earliest depictions of the case where an alien message is actually decoded. It first appeared as a BBC TV production in 1961, before being novelized in 1962. The message, which is unusual in originating from the Andromeda Galaxy rather than within our Milky Way Galaxy, gives instructions on how to build an advanced computer. That computer in turn gives instructions for creating a living cell, which continues to grow and begins to affect the brainwaves of the people around it, killing one (actor Julie Christie in her first major performance). The computer gives new instructions for the creation of a human embryo, which makes the scientists suspicious, but they cannot control their curiosity. The creature grows to maturity and turns out to be a clone of Julie Christie's character, nicknamed Andromeda. Andromeda has unusual powers and is put to work by the military. But the creature is trying to take over humanity, and the story ends with the death of Andromeda. Here is a case of how indirect contact could become more direct, if humans decide to implement transmitted computer code. The difference with direct contacts is that because of the remote nature of the initial contact, humans have more time to make decisions – or not.

In 1968 Harrison Brown wrote *The Cassiopeia Affair*, another case where an alien message is received and deciphered, this time from Cassiopeia 3579, some 30 light-years away. Brown, a leading geochemist who worked on the Manhattan Project, became a faculty member at Caltech, as well as a member of the National Academy of Sciences, and was also a social and political activist. It is no surprise, then, that his novel concentrated on the social and political implications of receiving a message. After the decipherment of the message, Brown presents a realistic scenario of what might happen. The White House is informed, and practical considerations immediately arise: Should scientists immediately inform the public? Should the lead agency be NASA, the National Science Foundation, the United Nations, or some other organization? How should international and domestic political implications be addressed? In the novel a meeting of heads of state takes place, where a "Declaration of Rome" lays out the foundations for "a new world order to regulate contacts with other worlds, establish and maintain disarmament, and settle international disputes peaceably." It all fits rather well into the Ladrièrean scheme described earlier in this chapter.

James Gunn's classic *The Listeners* (1972) is another case where a message is decoded, with a twist. In 2027 astronomers detect a signal from the star Capella some 45 light-years distant, consisting of Earth's own transmissions received and beamed back to Earth. But then a message is found embedded in the signal, represented as a bird-like figure with wings. Scientists and theologians interpret this figure in different ways, and more information reveals that the Capellans are a dying race because their star has reached the stage of a red giant in its evolutionary history, a stage where the star expands and any life in the star system will be extinguished. A message is sent to Capella and 90 years later the reply is received, revealing an encyclopedia of Capellan knowledge, which gradually begins to change humanity. But it is also revealed that the Capellan civilization is long dead; the messages are being sent not by the Capellans but by an automatic transmitting device built by them in their final days – in other words, an active artifact. In addition to its plot line, the novel is notable for treating not only the institutional workings and frustrations of such a long-term search but also the possible implications of contact, with its portrayal of religious leaders who must deal with the existence of intelligence beyond Earth. The Capellans, by the way, lived on a moon orbiting a gas giant planet, a remarkable idea in the decade before the *Voyager* spacecraft even suspected an ocean with possible life on the Jovian moon Europa in our own solar system, an ocean later confirmed on this and other moons in our solar system.

I mentioned in Chapter 2 Fred Hoyle's offbeat novel *The Black Cloud* as an example of Type 4 indirect contact in the vicinity of Earth – although arguably it borders on direct contact due to the strange and extended nature of the intelligence. After the discovery of the cloud approaching Earth from a distance of 21 astronomical units, well within the solar system at about the distance of Uranus, astronomers in the United States approach the National Science Foundation, while British astronomers approach the Royal Astronomical Society. When the seriousness of the matter is realized, the heads of government become involved, and plans are made for the encounter. Special scientific centers to monitor the cloud are established in both the United States and Britain. Within days of the discovery of the Black Cloud astronomers realize based on its calculated density that it will block out the Sun's light entirely. In other words, the impact on Earth will be dramatic, since the Earth will be plunged into a prolonged period of darkness, perhaps lasting a month. Moreover, if the cloud (believed to be composed of hydrogen) engulfs the Earth, the effects on the atmosphere might be catastrophic. Just how dramatic the encounter will be the scientists cannot yet fathom, since they do not realize it is sentient. The cloud, which is about 90 million miles in diameter (the distance from the Earth to the Sun), unaccountably begins to slow down by firing off blobs of material. Its thin outer regions make contact with the Earth, and rather than engulfing the Earth, it settles down into a flat disk. Earthlings suffer through huge climatic changes brought about by the effects of the cloud on sunlight and the Earth's atmosphere. Because of its strange behavior, scientists finally realize the cloud is sentient. Hoyle depicts not only the societal impact in the form of climate change but also the scientific impact when the scientists learn to communicate with the creature. Planetary intelligence, the cloud reveals, is unusual because the gravitational force limits the size of its beings and the scope of their neurological activity. Detailed comparisons of neurological structure between an intelligent cloud and a human give a sense of just how unique – and inferior – humanity might be. Still, the novel is hopeful about communication and intention; despite the chaos and destruction the cloud brings to Earth, in the end it is convinced to retreat.

The Fate of Humanity

Ideas of contact and its implications have been repeated and elaborated many times in science fiction, but these classic exemplars are sufficient for our purposes of learning about impact scenarios, just as in Chapter 2 they informed our discovery scenarios. What do these novels tell us about the Ladrièrean destructuring and restructuring schema of societal impact? In Wells's *War of the Worlds* the destructuring phase takes the form of a physical attack that is overcome; aside from presumably cleaning up the destruction, we are not told what the longer-term psychological impact might have been, though one can imagine such impact would have taken place in the wake of such a traumatic event. In *Childhood's End*, in which humanity is ushered into a universal Overmind, both the destructuring (the end of war, world government, and so on) and the restructuring are in the hands of the extraterrestrials. The Ladrièrean idea of humanity's ability to control its future is decidedly compromised if not destroyed. Harrison Brown's vision of the foundations for "a new world order" fits rather neatly into the Ladrièrean scheme of restructuring following alien contact.

By contrast both Gunn's *Listeners* and Sagan's *Contact* deal with the shorter-term destructuration of the old world order. Sagan's *Contact* novel explores relatively short-term societal impact, depicting the immediate reaction by religious, military, and political groups. Readers are left to only imagine the longer-term impact, and in fact doubts are raised about the reality of the extraterrestrials. The point is that the Ladrièrean scheme of destructuring old worldviews, absorbing new information, and restructuring new worldviews with new assumptions, ideas, and values, works well as a framework for scenarios laid out in science fiction novels. And it could work well also as an overall framework for characterizing the impact of real events following the discovery of alien life.

Deducing Impact: Between Imagination and Prediction

We have seen that science fiction is very valuable in setting out scenarios, some of which may come true in broad outline, most of which will not, certainly not in terms of specifics. When we leave the realm of imagination and enter the real world, however, we must again issue our standard warning against prediction. Astronomers can predict the motions of the planets based on Newton's laws of motion; weather forecasters have improved hurricane landfall predictions within 100 miles; and even elections can occasionally be predicted with some degree of accuracy. But for the most part human behavior is notoriously unpredictable, even where large amounts of data are available. What hope, then, do we have of predicting the societal impact of discovering life beyond Earth, when we have no data available?

As we have repeatedly stressed, the simple answer is "none." Societal impact is the aggregate of many human behaviors, and lacking Asimovian psychohistory, accurate prediction of the impact of finding life beyond Earth is out of the question. But that does not mean we should throw up our hands. Whether in business, science, or other areas, it is the most human of activities to attempt to foresee what lies ahead for all kinds of practical reasons; not a few futurists make a good living out of it. More specifically, modern societies study impacts of various kinds all the time, and for a variety of reasons. Some studies, such as environmental impact statements, are prepared before a project is undertaken out of concern for what *might* happen. Others, such as the large programs studying the ethical, legal, and social issues of the Human Genome Project and nanotechnology, are ongoing because of what *is* happening, e.g., the availability of genetic information and the ability to manipulate it through such editing programs as CRISPR. Still others, including the impact of spaceflight, are historical studies undertaken because of what *has* happened. Sometimes it is a combination of these. Such studies are often controversial because large amounts of money are involved. More often they are controversial because they engage social issues of great import. The societal impact of astrobiology, particularly the impact of the discovery of life beyond Earth, largely falls in the category of what might happen. It is too important to ignore.

A Matter of Statistics?

I opened this chapter with Nate Silver's book *The Signal and the Noise*, which is all about how to make predictions, and why they often fail. That opening was framed as a cautionary note, but can we gain at least some insights from his conclusions? Our question is not whether we can pull an artificial signal from the electromagnetic noise of the universe. That is the problem of the astronomer. Rather, our question is, can we extract a signal about the impact of life beyond Earth from the cacophony of human behavior, and even more questionably, from alien behavior? The key to Silver's book is the idea of determining probabilities, particularly the method known as Bayesian statistics, developed by eighteenth-century English minister Thomas Bayes. And the key to Bayesian statistics is the idea of prior probabilities, the probability of a particular event happening in the absence of any evidence. This must then be combined with two other factors: the evidence and the reliability of the evidence. The mathematician Keith Devlin has given us a vivid example of how this works. Suppose you undergo a test for a rare cancer that has an incidence of 1 percent in the general population. The test is 79 percent reliable, with 21 percent false positives. If you test positive and conclude that there is a 79 percent chance you have cancer, you would be terribly wrong. You have not taken into account the prior probability that only 1 percent of the general population has this form of cancer. By an equation that Bayes worked out, your actual probability of having the cancer, taking into account this prior probability, is only 4.6 percent.[17]

The Bayesian method also applies to human behavior. Silver gives another vivid example of coming home from a business trip and finding a strange pair of underwear in your dresser drawer. What is the probability your partner is cheating on you? No matter what you may think, the answer hinges on the prior probability that about 4 percent of married couples cheat on the their spouses in any given year. The point is that Bayesian statistics can be not only counterintuitive but also very personal, and, in the end, more likely to be true. Bayesian statistics also allows us to update probability estimates in the face of new evidence. Silver notes that the probability of terrorists crashing a plane into the World Trade Center on September 11, 2001, was extremely low. So was the probability of such crashes based on previous accidents – it had happened only twice in the previous 25,000 days of aviation over Manhattan, or 0.005 percent. But once the first plane had crashed into the World Trade Center, the probability of the second crash shot to 38 percent – and the probability that the second crash was a terrorist attack was almost 100 percent.

How does this apply to life beyond Earth? The problem, of course, is in determining the prior probability. Unlike the cases of cancer, cheating spouses, and crashing planes, we have no data about the incidence of life beyond Earth. We have only preconceptions: for some of us, the Copernican and Darwinian preconceptions that life will arise wherever it can make it highly probable there is life beyond Earth. But if you do not accept those preconceptions and believe instead that life is a miracle created by God, life beyond Earth may (though not necessarily) seem much less probable, particularly if you are in a creationist frame of mind. I say "not necessarily" because if you are in a seventeenth-century natural theology frame of mind, it might seem very probable that God would create numerous worlds as a reflection of his omnipotence and glory, exactly the case for many natural theologians now and then. Before I am accused of theology-bashing, it must be said that on the scientific side, we cannot do much better in terms of prior probabilities. The entire Drake Equation for the number of communicating technological civilizations in the Galaxy may be seen as an attempt to produce a prior probability that such civilizations exist, and because the values of most of its factors are unknown, the prior probability itself is unknown. If, based on an array of assumptions for each factor, the answer is one in 1 million stars in the Galaxy have communicative civilizations, that would be a form of prior probability. The problem is that a more pessimistic set of assumptions might yield one in 400 billion, leaving ourselves as the only life in the Galaxy.

When it comes to the *impact* of finding such life under any given scenario, it might seem that we do have some data. But it comes not in the form of other cases when life has actually been discovered, but in historical cases where life was thought to have been discovered, or in the form of analogies such as culture contacts on Earth. As we just saw, Bayesian statistics can be applied to human behavior, but in our case, we cannot say that a certain percentage of humans behaved in a certain way in response to the discovery of life, since it has never happened. We might use our historical cases from Chapter 1: the Orson Welles broadcast in 1938 seemed to indicate that news of a close encounter on Earth would cause panic. But that was only one case, and scholars now believe there was very little panic. Or we might use analogy, culture contacts on Earth being a popular one. But far from the simple invocation of the largely disastrous European discovery of the Americas, one would have to conduct a much more extensive research program on culture contacts around the world and through time to come up with a valid prior probability – and most likely a very different one. And even then the circumstances of alien physical contact would not be entirely analogous. The bottom line is that in the absence of prior probabilities in connection with the impact of discovering life, we are reduced to methods such as history and analogy. As I have argued in this book, these are still valid methods that serve as guidelines to impact, but they do not rise to the level of statistical probability methods.[18]

Table 6.2 *Twenty-Five Years of Discussions on Societal Impact of Astrobiology, 1991–2016*

Meeting	Date and Place	Sponsor	Results
Cultural Aspects of SETI (CASETI)	1991–1992 Chaminade Conference Center, Santa Cruz, California	NASA	Billingham et al. (1999)
Many Worlds	November 22–24, 1998 Lyford Cay, Nassau, The Bahamas	John Templeton Foundation	Dick (2000c)
When SETI Succeeds	1999 Hapuna Prince Big Island of Hawaii	Foundation for the Future	Tough (2000)
Societal Implications of Astrobiology	November 16–17, 1999 NASA Ames	NASA	Harrison and Connell (2001)
Exploring the Origin, Extent and Future of Life	2003 American Association for Advancement of Science Washington, DC	NASA/ American Association for the Advancement of Science	Bertka (2009)
Astrobiology: Expanding our Views of Society and Self	May 2008 University of Arizona Biosphere 2 Institute	University of Arizona	Impey, Spitz, and Stoeger (2013)
Astrobiology and Society	February 2009 SETI Institute	NASA Astrobiology Institute	Race et al. (2012)
The Detection of Extra-Terrestrial Life and the Consequences for Science and Society	January 25–26, 2010 Royal Society in London	Royal Society of London	Dominik and Zarneki (2011)
The History and Philosophy of Astrobiology	September 27–28, 2011 Ven, Sweden	Pufendorf Institute for Advanced Studies, Lund University	Dunér et al., *Astrobiology* special issue, vol. 12 (2012); Dunér et al. (2013)
Preparing for Discovery	September 2014 Library of Congress	NASA/Library of Congress	Dick (2015b)
Social and Conceptual Issues in Astrobiology	September 2016 Clemson University	Clemson University	Kelly Smith, Forthcoming

Between imagination and prediction, then, lies the domain of reasoned scenarios constructed using history, discovery, and analogy, all within the general framework for societal impact discussed in this chapter. In contrast to imagining impact, we refer to this process as "deducing impact," not because it is based on rigorous deduction or even statistical probabilities, but in the sense that we use logic and reason to form a conclusion or an opinion based on known facts. Unlike Nate Silver's success with general election results in 2012, we have zero data about alien life, a point the critics of astrobiology never fail to make. But we do have abundant data about human behavior and reaction under a variety of circumstances, including historical events related to the discovery of life, the nature of discovery, and other areas of human endeavor where analogy can be applied. It is these data that give us some hope of progress in our problem. Indeed, a few serious attempts have already been made at deducing the impact of finding life beyond Earth, mostly relating to the discovery of intelligence (Table 6.2). We should not ignore these attempts, but rather examine their methods, conclusions, and recommendations, even as we evaluate their success.

A Gathering of Scholars

The question of the societal impact of extraterrestrial life first received high-level attention a quarter century ago when John Billingham, the head of NASA's SETI program at its Ames Research Center, convened a series of workshops on "The Cultural Aspects of SETI" (CASETI) on the eve of the inauguration of NASA SETI observations in 1992. When in October 1991 scholars gathered for the first of three workshops in Santa Cruz, California, they could not have known that exactly two years later the US Congress would cancel the entire NASA SETI program. We gathered at the Chaminade Conference Center confident, not that the search would be immediately successful, but that it would be an ongoing research program of long duration with some chance of eventual success – enough chance that we needed to gauge the cultural impact. Alas, petty politics intervened, and the search was halted in the same session of Congress that canceled the Superconducting Supercollider in Texas. But the intimate gathering of two dozen scholars (Figure 6.3) was a model of interdisciplinary brainstorming, with astronomers including Frank Drake and Jill Tarter, anthropologists represented by Ben Finney and Michael Ashkenazi, religious scholars and historians including Georgetown's Langdon Gilkey and Harvard's Karl Guthke, several representatives from media studies, and even two diplomats, represented by Michael Michaud from the State Department and Nandasiri Jasentuliyana, the director of the Office of Outer Space Affairs at the United Nations. The gathering was a de facto recognition that this was a broad-based problem not to be solved by scientists alone. While the publication of the results was delayed almost a decade by the cancellation of the scientific program, its recommendations are still valuable for contemplating the aftermath of any successful SETI program.

Figure 6.3 A rare photograph of the participants in the 1991–1992 interdisciplinary workshops on the Cultural Aspects of SETI (CASETI), published here for the first time. The meeting was held just before the inauguration of the NASA SETI program, and was the first of the meetings listed in Table 6.2. First row, left to right: David Milne, John Billingham (NASA Ames), education specialist Julia Koppich, Roger Heyns, Kent Cullers (SETI Institute), anthropologist Ben Finney. Second row: historian of science John Heilbron, JPL astronomer Michael Klein, SETI Institute CEO Tom Pierson. Third row: Vera Buescher (SETI Institute), historian of science Steven Dick, astronomer Julie Lutz, astronomer Jill Tarter, Alex Inkeles, Vivian Sobchak, lawyer Steve Doyle. Fourth row: Gary Coulter (NASA HQ), Amahl Drake, astronomer Andrew Fraknoi, anthropologist Michael Ashkenazi, John Rummel (NASA HQ), Harvard cultural historian Karl Guthke, student Alison Tucher, theologian Langdon Gilkey. Top row: astronomer Frank Drake, Bob Arnold, Michael Michaud (US State Department). Credit: Seth Shostak.

While the organizers failed to foresee the cancellation of NASA's SETI program almost before it became operational, the fact that the workshops were held at all was remarkably prescient. "From the time when NASA first began to plan a large-scale program of observations of the sky in search of signals from extraterrestrial societies," the workshop leaders wrote in the introduction to their final report, "the Agency has recognized that the social impact of unambiguous evidence of such civilizations would be substantial." Such recognition was one thing, but for a scientific agency to fund societal impact work was quite remarkable for the time, even though the original NASA Space Act called for studies of the societal implications of space exploration. In the end NASA enunciated three objectives in bringing together this eclectic group: to assess potential short- and long-term responses if its SETI project proved successful; to identify ways in which NASA could inform the public about possible outcomes of its search; and to recommend areas for further research regarding the implications of success.[19]

The scholars examined the problem from four perspectives: history, human behavior, policymaking, and media/educational issues. Not surprisingly, their methods reflected their diverse disciplinary backgrounds. The history group (of which I was a part) examined the history of the debate and found opinion about impact divided into two groups: millenarians and catastrophists. Millenarians (connected to the historical idea that a major transformation is coming, sometimes connected to the turn of a millennium) believe that the detection of life will have "profoundly beneficial" consequences for humanity. The early SETI pioneers were almost unanimously of this opinion. Catastrophists, on the other hand, point to the ill effects of culture contacts on Earth, or to the possible ill effects of receiving a vastly superior knowledge. Nobelist George Wald fell in this group, saying he could conceive of no worse nightmare, one that would jeopardize "the dignity and worth of man." He was joined by Nobelist Martin Ryle, who thought we should not reveal our position in the universe with purposeful radio transmissions – a viewpoint very much alive today with opponents of messaging extraterrestrial intelligence (METI). The historians did not come down on the side of either the optimists or the pessimists, but did suggest that future studies should be focused on analogies "based on the transmission of ideas within and between cultures, rather than on analogies based on physical encounters between cultures." As we have stressed in this volume, they cautioned that such analogs were "approximate guides" for impact scenarios, and not predictors of the future.[20]

The human behavior group reflected the complexity of the human condition and methodologies used to study them. The members of the group emphasized that human responses would depend on individual factors such as gender, race, and religion, as well as the factors of the culture in which they were embedded (language, belief systems, frames of reference, and so on). They foresaw a spectrum of reactions, ranging from indifference and mild curiosity through millennial enthusiasm or catastrophist anxiety and "full scale pronoia or paranoia." They identified the importance of education in cultivating positive impact outcomes, and they recommended research (such as polls on likely public responses to various discovery scenarios) on the factors underlying responses. Finally, they proposed that a "reaction team" of behavioral scientists be constituted to be used in the event of a discovery. Like the history group, the human behavior group also cautioned that "the reactions of individuals and groups depend upon interplays among factors, in ways that are not known or presently predictable."[21]

The policymaking and educational groups were different in that their goal was to provide guidance on "what to do" rather than to look at "what might happen." The policy group, headed by Michael Michaud, the US State Department Counselor for Science, Technology and Environment at the American Embassy in Paris, naturally focused on the likely important role of governments. Its recommendations included both preparing governmental bodies and international institutions, and establishing procedures for making humanity's responses more effective. More specifically, they suggested briefing appropriate governments and organizations, developing post-detection notification processes for governments and the public, and conducting analyses of the political consequences of detection.[22]

The media/education group emphasized the importance of media reaction, which might be a large determinant of public reaction. There is a long history of media determining public outcomes of the news it reports, not least in the Great Moon Hoax/Satire and the "War of the Worlds" broadcast. As I have already emphasized in previous chapters, the lesson should be well taken that the media will be a paramount factor in how the discovery is received. Since that warning from the CASETI group more than two decades ago, the role of the media has only heightened with the availability of social media. Keeping the media and the public informed prior to any detection will be essential. The group also emphasized that the public needs to be better informed and that "false alarms" and the inevitable controversy that will follow a reputed detection are a normal part of science. This is vividly displayed already in the detection of planets beyond our solar system.[23]

The Santa Cruz meetings were pioneering but not unique in the long run. In the wake of the first NASA Astrobiology Roadmap in 1998, calls were made for the study of cultural impacts of astrobiology, resulting in two meetings of scholars in 1999. The first was a seminar organized by Allen Tough, a Canadian futurist and social scientist, sponsored by the Foundation for the Future and held on the Big Island of Hawaii in conjunction with a much larger bioastronomy conference. The 16 participants illustrated the diverse backgrounds necessary to the discussion: John Billingham and Jill Tarter, both by this time representing the SETI Institute; social scientists including anthropologist Ben Finney and psychologists Albert

Harrison and Douglas Vakoch; physicists Paul Davies, Guillermo Lemarchand, and Claudio Maccone; astronomer Eric Chaisson; and me as the sole historian of science. The discussions concentrated solely on SETI, and particularly the long-term impact under the scenario where a message was deciphered. The report foresaw five possible long-term consequences: the receipt of practical information, answers to major questions, changes in our view of ourselves, cooperation in joint galactic projects, and long-term negative effects. The latter was a recognition of the potential "severe disruption" of human society.[24]

Three months after the Hawaii meeting, NASA's Ames Research Center also organized a workshop on the societal implications of astrobiology.[25] This time about 50 scholars ranging from futurists like Alvin Toffler to anthropologists, scientists, and journalists gathered to discuss the subject. Not surprisingly, the group emphasized the importance of its task: to encourage public understanding of this new science, to gauge public reaction to astrobiological discoveries, and to prepare for the future through policy decisions given "a possible sea of living worlds." More than a dozen recommendations were issued, including the importance of a multidisciplinary approach involving both scientists and humanists, studying the implications of a shift in our frame of reference from the Earth to a living cosmos, making "state-of-the-art preparations" for discovery of life, studying the ethical implications of discovering life, and implementing policy measures "to ensure the integrity of extraterrestrial life." They made a strong case for undertaking serious levels of research and outreach before the fact of discovery, arguing such research should be integrated into core science initiatives (as would soon be done with the Human Genome Project). "Science and society are deeply and irrevocably intertwined," they wrote, "and a mutual appreciation of the close relationship is vital to the integrity of both fields."[26]

Interest in the subject extended well beyond the United States, though still limited to the Western world. Particularly notable from the point of view of the sponsor and speakers was a two-day meeting of the Royal Society on "The Detection of Extra-Terrestrial Life and the Consequences for Science and Society," held in London in 2010.[27] It involved the president of the Royal Society, the president of the UK Institute of Physics, the French High Commissioner for Atomic Energy and recent president of the International Astronomical Union (IAU), the secretary general of the International Academy of Astronautics (IAA), and the director of the UN Office for Outer Space Affairs (UNOOSA). "The detection and further study of extra-terrestrial life will fundamentally challenge our view of nature, including ourselves," the organizers wrote, "and therefore the field of astrobiology can hardly be isolated from its societal context, including philosophical, ethical and theological perspectives." This declaration was followed up by a series of papers authored by stellar authors including Nobelists Baruch S.

Blumberg and Christian De Duve; astrobiologists Chris McKay, P. C. W. Davies, and Frank Drake; theologian Ted Peters; and social scientists Albert Harrison and Kathryn Denning, among others.

The most recent attempt to study the impact of discovering life beyond Earth again had a NASA connection. This was the symposium convened at the Library of Congress in 2014 as part of my duties as the Baruch S. Blumberg NASA/Library of Congress Chair in Astrobiology (Figure 6.4). My goal was to look at the humanistic aspects of astrobiology, in particular the impact of discovering life. This book is the ultimate product of that effort, but the issues were so far flung that it was necessary to engage scholars in a wide variety of fields. Thus philosophers, theologians, anthropologists, historians, and scientists gathered in Washington in the shadow of the Capitol, to discuss the big questions related to the societal implications of discovering life. Their goal was not to issue comprehensive recommendations, but to deepen the scholarly discussion of the issues. The product of their deliberations, *The Impact of Discovering Life beyond Earth*, stands as testimony to the richness of this embryonic field. The implicit conclusion was that the subject is amenable to serious discussion, and not only desirable but necessary if we are to prepare for such an important discovery. This conclusion was also borne out

Figure 6.4 The four initial Baruch S. Blumberg NASA/Library of Congress Chairs in Astrobiology, at the Library's John W. Kluge Center, September 15, 2016. The program was inaugurated in 2012 to explore the humanistic aspects of astrobiology. Left to right (and in order of succession): David Grinspoon, Steven Dick, Nathaniel Comfort, and Luis Campos. Credit: Travis Hensley, Library of Congress.

during congressional hearings on astrobiology in 2013 and 2014, where members of Congress wanted to know what should be done in the event of the discovery of extraterrestrial life.[28]

Other gatherings on the subject demonstrate how widespread the interest really is, spanning disciplines and cultures, including meetings sponsored by the Templeton Foundation in 1998, the American Association for the Advancement of Science in 2003, and the American Anthropological Association at its annual meeting over a period from 2004 to 2006.[29] The first two of these meetings focused on ethical and theological implications, and the latter on issues raised in an anthropological context. The problem is that there has been no synthesis of these discussions, and a field of "societal impact of astrobiology" is gaining momentum but has not yet gelled. As I discuss in Chapter 9 in connection with "astropolicy," a beginning has been made with an attempt to construct a societal impact roadmap.

Attempts at Synthesis

Individual efforts have also concentrated on aspects of the problem. Foremost among these are psychologist Albert Harrison's volume *After Contact: The Human Response to Extraterrestrial Life*, and American diplomat Michael Michaud's *Contact with Alien Civilizations: Our Hopes and Fears about Encountering Extraterrestrials*. Dividing the reaction into long and short term, Harrison viewed the initial reactions as depending on "the nature of ET, the information available to us, media coverage, personal predispositions, and social pressures." That reaction, in his view, will depend on three variables: the distance of the alien civilization, its perceived capabilities, and its perceived intentions. Since Harrison is a psychologist, some of his ideas about initial reaction are of particular interest. He believes that research on threat, stress, and coping suggests that we will not be overwhelmed psychologically. This applies, however, mainly to our Type 3 remote detection, since most people have been conditioned to expect contact by means of radio signals. Moreover, people will have more pressing things to worry about, like making a living and raising a family. Harrison also makes the important point that the likely incomplete and ambiguous nature of the discovery means that initial reactions will draw heavily on human expectations, hopes, and fears. This is all the more reason for the public to be educated about the history and science of the debate, and to encourage research on the social sciences in these areas.[30]

As for long-term consequences, Harrison believes contact with an extraterrestrial civilization will have a pervasive influence on philosophy and the arts, as well as science and technology. He sees three logical possibilities in the interaction among societies: intimidation and force, trade and exchange, and integration. These three possibilities could end in a golden age, in assimilation with the new cultures, or in the destruction of humanity. Lacking a knowledge of alien behaviors, we are unable to determine which outcome will occur in reality. But it is notable that in his discussion of alien cultures, Harrison employs James Grier Miller's Living Systems Theory, which finds commonalities and continuities between the biological and social sciences, ranging from organisms to societies and supranational systems. It is possible that knowledge acquired from such a theory could be useful in gauging impact.[31]

Michaud's *Contact with Alien Civilizations* is notable for its compilation of ideas about consequences of contact, examining both the pros and cons. But it is especially interesting because of Michaud's background as director of the US State Department's Office of Advanced Technology and Counselor for Science, Technology, and Environment at US embassies in Paris and Tokyo, roles in which he led US delegations in the negotiation of international science and technology agreements. More specifically, Michaud was deeply involved in developing post-detection protocols in the event of the discovery of extraterrestrial life (see Chapter 9). Of special interest here are his conclusions about the consequences of contact, described in the last and most original chapters of the book, the culmination of his work since 1972 on the political and social consequences of discovering intelligent life beyond Earth. Michaud believes the most probable scenario is a low-information contact, either in the form of an intercepted message or from an alien artifact found in our solar system. He argues that the contact will most likely be with an alien intelligence near our scientific and technological level, both because we cannot assume perpetual "progress" in these areas, and because our search and communication methods act as a filter of aliens that are more or less intelligent than we are. He broaches the possibility that such intelligence may be postbiological. In the end Michaud believes neither the optimists nor the pessimists have proven their cases about the consequences of contact, and that the most rational attitude may be ambivalence. Michaud concludes that "our speculations [about the consequences of contact] still rest on analogies with human experience, on our cultural and political contexts, and on our personal biases. They still reflect our hopes and our fears."[32] This is a lesson that we should take to heart, but it is no excuse for lack of thought and action on the subject.

The Significance of Detection

Finally, one of the more concrete results to emerge from the contemplation of the impact of discovering life is that methods have been devised to assess the significance of a detection, on the theory (as history has shown) that even a false alarm could have a considerable impact. This concept is a natural extension of ordinal scales developed for other phenomena, such as the Richter scale from 1 to 10 for earthquakes, the Fujita scale from 0 to 5 for tornados, and the familiar Saffir–Simpson scale from 1 to 5 for hurricanes. The SETI scales are direct descendants of the so-called Torino scale devised to assess the impact hazard of near-Earth objects. Astronomers Ivan Almar and Jill Tarter first presented what has become known as the Rio scale at the 51st International Astronautical Congress in Rio de Janeiro, and the SETI Committee of the International Academy of Astronautics adopted the scale in 2002. The scale runs from 0 to 15, ranging from no impact to extraordinary. Five years later the "San Marino scale" was devised to assess the hazard of deliberate transmissions from Earth, in other words, Active SETI, or what today has become known as METI.[33]

While the Rio and San Marino scales were devised for passive or active SETI, in 2010 Almar and Margaret Race first presented the broader London scale at the Royal Society meeting I described earlier. It is analogous to the Rio scale, but more encompassing in its discovery scenarios, intended to evaluate "the scientific importance, validity and potential consequences of an alleged discovery of ET life via various astrobiological methods and within the Solar System or Galaxy." In other words, the discovery scenarios could range from microbial life to advanced intelligence. The London scale combines factors of scientific importance and credibility and mathematically is simply expressed as LSI = Q × delta. Here Q is the sum of four more or less objective factors: lifeform, nature of evidence, method of discovery, and distance, while delta is the more subjective assessed credibility of a claimed discovery. The product of these factors can range from 0 (insignificant importance, validity, and impact) to 10 (extraordinary importance, validity, and impact). This scale thus explicitly incorporates what we have captured in Tables 2.1 and 2.2 showing discovery scenarios, namely that the method of discovery may be direct or indirect and the distance of the encounter may be near or far. In addition it takes into account the nature of the evidence (alive, fossilized, biomarker, and so on) and the type of life, defined as ranging from similar to Earth to completely alien and unknown to us – all possibilities within the boxes we have constructed in the tables in this chapter. The advantage of the London scale is that it quantifies impact, and thus may be very useful in policy discussions when it comes to reacting to the discovery of life beyond Earth (see Chapter 9). Moreover, the fact that the scientific factors are separated from the more subjective credibility factors means that the calculated value for any particular discovery may change with time as new evidence is assessed.[34]

Where We Stand

What are we to make of these far-reaching studies, ranging from the workshops to the individual studies of Harrison and Michaud? First of all, none of them devises or employs an impact framework of any kind, much less one similar to the Ladrièrean framework described earlier. This is certainly not surprising in the case of the group studies, which by their nature are gatherings where individuals voice their opinions that are then combined into a report, without any attempt at a framework. Nor is it surprising in the case of the few individual studies; though Harrison did employ a "living systems" framework in discussing what alien life and cultures might be like, the impact section of his book is guided by the social sciences rather than any coherent framework. Nonetheless, as with the imaginative approach of science fiction, many of these studies do support elements of the Ladrièrean impact framework, including belief in a profound impact resulting in a change in worldview after the destruction of old views and the building of new ones.

Second, many of the studies implicitly or explicitly employ two of our approaches laid out in the first three chapters: history and analogy. Only a few broach the third approach, discovery, which led to our insight about extended discovery and discovery scenarios. This suggests that, beyond the anatomy of discovery I laid out in Chapter 2, one way forward is the detailed study of the nature of discovery in the history of science and technology, and how any patterns there might be applied to the discovery of extraterrestrial life. At the same time, the finding that so many authors use history and analogy suggests that further study of their uses and abuses will also be important.

Third, almost all of the studies, ranging from the NASA-sponsored efforts to the individual efforts of Harrison and Michaud, largely address only contact with extraterrestrial intelligence, whether remotely through electromagnetic signals, or directly through interstellar travel. The 1999 NASA study did recognize that "very little work has been done on the consequences of detecting a single-celled life form elsewhere in the solar system," and urged more work in that area. The 2011 NASA-funded AAAS study also emphasized the need for more study of the impact of discovery of microbial life. Most recently Race has taken the lead in marshalling the astrobiology, social sciences, and humanities communities to address these issues in the context of the latest Astrobiology Roadmap (as of 2015 the NASA Astrobiology Strategy), with the support of the NASA Astrobiology Institute.[35]

Fourth, these studies are all a reflection of their times and cultures. This seems obvious, but it is an important point. Virtually all writings on the subject have been Western-centric, and one of the prime recommendations for further study must be that other cultures become involved in the conversation on a subject that will affect all of humanity. Other cultures, after all, have worldviews of their own that may illuminate astrobiology and its impact in ways the Western mind does not see. Work in this direction has only begun, and is one of the chief methods for furthering the conversation.

Fifth, these studies have most often fallen short of their goal of assessing long- and short-term responses to the discovery of life. They are more likely to issue recommendations for further study or action than to come to specific conclusions about encounters with alien life, microbial or intelligent. More definitive studies must end in definite recommendations that can inform policy in the event of a discovery of alien life, which could occur at any time.

Despite the sporadic nature of these studies, both natural and social scientists are increasingly serious about the "wild card" impact the discovery of extraterrestrial life might have on the future of humanity. The studies in this section point to some consensus on religious, political, and cultural impacts of discovering life, and a convergence with the imaginative conclusions played out in science fiction literature in the previous sections. This gives us some confidence that we are on the right track as we move on to the next section of this book about actual impacts. One thing is sure, the problem of encountering alien life is not going away. In envisioning impact what is needed is a more systematic effort, employing serious methods and with an eye toward an impact framework for which the Ladrièrean model is only a beginning. Michaud is right that both the optimistic and pessimistic scenarios of impact have largely reflected humanity's hopes and fears. To the extent we can, employing history, discovery, analogy, and an impact framework, using both imagination and deduction, and doing our best to transcend anthropocentrism, we need to move beyond hopes and fears to a comprehensive assessment of impact. We attempt that daunting effort in Part III.

Part III

Impact!

7

Astroculture

Transforming Our Worldviews

I can think of nothing so positively transforming of human consciousness as the discovery, study, and conservation of life somewhere off the earth.

J. Baird Callicott[1]

O be prepared, my soul!
To read the inconceivable, to scan
The million forms of God those stars unroll
When, in our turn, we show to them a Man.

Alice Meynell[2]

Come, my friends,
'T is not too late to seek a newer world.
Push off, and sitting well in order smite
The sounding furrows; for my purpose holds
To sail beyond the sunset, and the baths
Of all the western stars, until I die.

Tennyson, Ulysses

It has been one of the great honors of my life to be associated with NASA, the premier agency for exploration in the world. During my time as chief historian, wherever I went around the country and the world, it was clear that people were excited about seeking new worlds, revealing things never before seen, and perhaps in the future, discovering life. At the same time I liked to remind them that through space exploration we are creating a new world right here on Earth. During the course of the twentieth century, and especially in its second half with the beginning of the Space Age, our eyes were increasingly turned outward, toward the heavens. Humans have, of course, contemplated the stars for millennia, but only in the twentieth century did they begin to understand how those roiling cauldrons work, and to realize that we might one day travel across that sea of suns, at first vicariously, but eventually in person. The inspirational images from the Hubble Space Telescope and other spacecraft brought the universe to our doorstep. Knowledge of our place in space, and our place in time in the course of 13.8 billion years of cosmic evolution, has greatly increased our cosmic consciousness, completing the Copernican revolution. That consciousness is only increasing in the twenty-first century. My colleague, astronomer David Grinspoon, likes to emphasize that for the first time Earth is now in human hands. This is true not only for climate change and other aspects of what some call the Anthropocene Era but also for ourselves and the cultures we build together, which are changing on a scale never before seen. When and if life is discovered beyond Earth, that cultural change will only accelerate. But exactly how?[3]

Given all I have said in previous chapters about approaches to the impact of discovering life beyond Earth, and about the need to transcend anthropocentrism, to examine the universality of our knowledge claims, and to attempt systematically to envision impact, I finally arrive at the ultimate question of this book: what is the *real* impact likely to be if we find life? Will it be philosophical, practical, or both? Will it be gradual or Earth-shattering? Should we fear or embrace the consequences? I have already cautioned that the impact will very much depend on the scenario, whether we discover microbial or complex life, near or remote, friendly or hostile. And I have warned that we cannot predict the impact. But the discussion of history, discovery, and analogy, aided by attempts to think out of the box, can provide guidelines and likely scenarios of impact that allow us to anticipate and assess what might happen. The consequences are likely to be a mix of hope and fear, and while we cannot succumb to wishful thinking, the very act of studying impact gives us a better chance of controlling our own future, and perhaps even tilting the balance in favor of hope rather than fear. In the best-case scenario, such studies may help us to optimize the outcome for all involved, just as other societal impact studies aim to do.

In this chapter I explore the overall impact on humanity of finding life beyond Earth, using as my framework the concept of worldviews. I have already broached that idea in Chapter 3 from the point of view of analogy, arguing that past changes in worldviews such as the Copernican and Darwinian might provide guidelines to the impact of discovering life. And in the previous chapter I used the rise and fall of scientific worldviews as a possible framework for analyzing societal impact. In this chapter I adopt that framework to explore how changes in worldview at various levels might play out in more concrete terms. If indeed the deceptively simple Ladrièrean scheme of preimpact, impact, and postimpact, of worldview destructuring and restructuring, holds with regard to the discovery of life beyond Earth, then this chapter explores those phases, in both the short and long terms.

A Hierarchy of Worldviews

The changing nature of worldviews is in a sense a notorious subject, to the extent they are taken as coextensive with the "paradigm changes" advocated more than

50 years ago by scientist, historian, and philosopher Thomas Kuhn. That concept, which energized the fields of history, philosophy, and sociology of science, and much else, is now hopelessly theory-laden with baggage such as the debate over whether science progresses, the relation of normal science to revolutionary science, the science wars, postmodernism, and so on. For our purposes here I return to a broader, more holistic, meaning of worldview in the sense of the German *Weltanschauung*, defined as how we see the world from a variety of perspectives, and carrying with it the implication that worldviews have an impact on our daily lives. Each of us sees the world in a different way, our individual worldviews a mix of political, cultural, religious, philosophical, and scientific components, among others. There are also national and cultural worldviews; we can speak of an American worldview and more broadly a Western civilization worldview grounded in democratic thinking and Christianity, although there is now greater diversity of religious thought – or lack of it – than in the past. We can speak of an Islamic worldview grounded in theocracy and the Koran, or an Indian worldview grounded in Hinduism. The number of worldviews humanity has developed is quite impressive, and how individuals, nations, and cultures develop, maintain, and act on their worldviews is an area of active and important study.[4]

How to make sense of all these interacting worldviews for the purposes of our study? In this chapter I argue that cosmological and scientific worldviews preside at the apex of the hierarchy of worldviews each of us holds, consciously or unconsciously, followed in order of decreasing generality by philosophical and religious worldviews, with cultural worldviews at the base of a broad pyramid (Figure 7.1). Numerous cross-currents among these worldviews make our study at once richer and more difficult. But our cosmological and scientific worldviews are grounded in nature, and, I believe, should lay claim to grounding our overall worldviews in reality. This is a controversial statement these days; while some might claim the primary direction of influence in Figure 7.1 is upward because worldviews are socially constructed, I claim the primary influence is downward. Trickle-down economics demonstrably does not work, but I am arguing trickle-down worldviews do. Thus, cosmological and scientific worldviews should inform our worldviews at all levels, including the theological and cultural.

Given this framework, our task of exploring the impact of discovering life beyond Earth then becomes an exploration of its impact on current worldviews and the generation of new worldviews. The three worldview levels in Figure 7.1 thus provide the organizing principle for this chapter: the cosmological, the theological, and the cultural. Furthermore, I view this discussion as an aspect of what has been called "astroculture," the extent to which our culture is increasingly being transformed by all things related to outer space. Space exploration and astronomy have already deeply affected not only science but also music, literature, and the

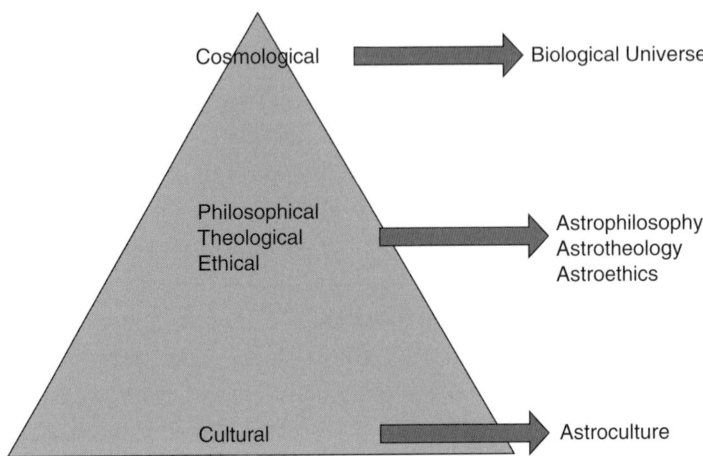

Figure 7.1 A hierarchy of worldviews, in which cosmological worldviews trickle down and profoundly affect various levels of human culture. The discovery of life beyond Earth may result in a transition from our current physical cosmology to one based on a biological universe. Philosophical, religious, and ethical worldviews may be generalized and transformed into astrophilosophy, astrotheology, and astroethics. And the transformation of human culture to astroculture may be accelerated beyond the initial thrust of the Space Age. Credit: author.

arts. Who can forget Arthur C. Clarke's *2001: A Space Odyssey*, David Bowie's "Space Oddity," or the thousands of other artistic expressions inspired by space? In short, our increasing consciousness of the heavens is affecting our general cultural worldviews. In the same way, surely the discovery of alien life will magnify and potentially transform our worldviews even more.

Cosmological Worldviews: The Rise of the Biological Universe

Throughout most of history, our cosmological worldviews have been largely physical. Theology aside, planets, stars, and eventually galaxies were both the physical and the metaphysical basis of our understanding of the universe. They were, and are, the fundamental units of the cosmos, which, combined with changing ideas of space and time, constitute the fabric, the weave and warp, of all scientific cosmologies. Ancient civilizations in Egypt, Babylonia, and Greece separated the wandering planets from the stars based on their motions. It was only in the twentieth century that a third "kingdom" of astronomy was added – galaxies demonstrated to be far beyond our own Milky Way system. Within astronomy's three kingdoms, by my estimation some 82 classes of objects have been discovered within the past four centuries since the invention of the telescope. Taken together, these kingdoms and classes form the rich tableau of modern astronomy.[5]

What matters most in the present context, however, is not the colorful and sometimes spectacular details of these objects and their discovery, but how they were ordered into the world systems we know as cosmologies. Throughout history there have been a number of these systems, ranging from the ancient atomists with their infinite number of worlds, to the Aristotelian/Ptolemaic cosmology with the Earth at its center, and the Copernican heliocentric worldview that made the Earth a planet and the planets potential Earths. The short-lived Cartesian system with its vortices first gave credence to solar systems beyond our own, and the Newtonian universe ruled by gravitational law often imagined other solar systems like ours. These cosmologies have risen and fallen with new discoveries, or in the case of the Newtonian universe, been subsumed in the worldview of Einstein's space-time and general relativity. Whether we realize it or not, these cosmological worldviews have historically played a major role in our everyday thinking. Dante's *Divine Comedy*, with its moral lessons played out in a cosmological worldview of nested spheres with the Earth at the center and the sphere of fixed stars at the periphery, provided the backdrop for medieval life. Within this ordered framework, both a mental and moral model as well as a physical system, all human actions took place. "In every age the human mind is deeply influenced by the accepted model of the universe," C. S. Lewis reminded us in his book on medieval and Renaissance literature, *The Discarded Image*. The model, he insisted, "is also influenced by the prevailing temper of mind." As historian of science Nasser Zakariya has demonstrated in detail in his book *A Final Story*, the same is true in the modern world, where the sweeping scientific epic of cosmic evolution (Figure 7.2) embraces the cosmological, the biological, and the social, and provides the background for modern life – including the biological universe.[6]

From the Physical World to the Biological Universe

When the geocentric framework was gradually destroyed as the heliocentric theory of Copernicus was accepted in the course of the sixteenth and seventeenth centuries, the impact on society was profound. The "discarded image" also discarded ideas of humanity's central place in the universe. We are now at a turning point similar to that of the sixteenth and seventeenth centuries. Over the four centuries since Copernicus, the idea of a purely physical universe has gradually given way to the idea of a universe filled with life, what I call the biological universe (Figure 7.3). Those championing this idea hold that planetary systems are common, that wherever conditions are favorable life will originate and evolve, and in its strongest form, that this evolution will likely culminate in intelligence. All three of these ideas over the past four centuries were preconceptions not yet proven, assumptions utilized as a working hypothesis about the universe based on analogy and

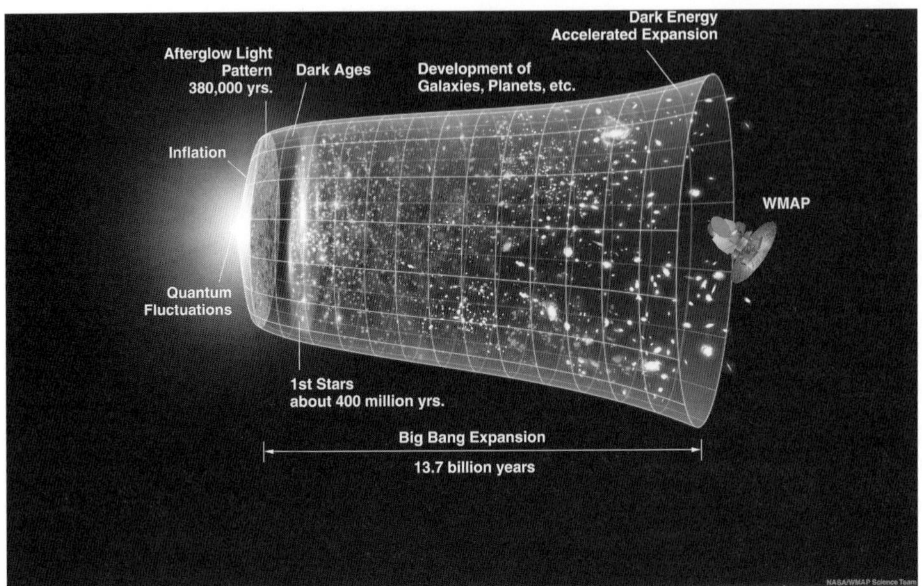

Figure 7.2 The scientific epic of cosmic evolution provides our modern worldview. This master narrative of the universe is now known to have occurred over the past 13.8 billion years, beginning with the Big Bang (at left), continuing with the formation of stars within the first few hundred million years, followed by the development of galaxies, planets, and life. Credit: NASA/WMAP Science Team.

other largely non-empirical factors. In the past two decades, however, astronomers have finally demonstrated that planetary systems are extremely common, a normal by-product of stellar evolution, even if they come in an immense variety of forms. The other two preconceptions about life and intelligence remain just that. But in my view, and in that of most astrobiologists, the general uniformity of nature and nature's laws, even in the midst of its diversity of planets, stars, and galaxies at a lower level of generality, indicates these preconceptions will also eventually be proven true.[7]

Glimpsing this vision of a biological universe, some now fear with John Donne that "Tis all in peeces, all cohaerence gone." For the biological universe concept makes claims about the large-scale nature of the universe, especially that life is one of its basic properties. It is therefore more than an idea, more than another theory or hypothesis. It is sufficiently fundamental and comprehensive to qualify as a worldview of its own, one that has implications for all humanity. Moreover, because it is testable, it is a scientific worldview perhaps best described as a cosmology. And because it combines the biological with the physical, it may accurately be termed "the biophysical cosmology." Nor is this idea one we need impose on history. In the mid-twentieth century Harvard astronomer Harlow Shapley called a universe full of life the "Fourth Adjustment" in humanity's view of itself in the universe

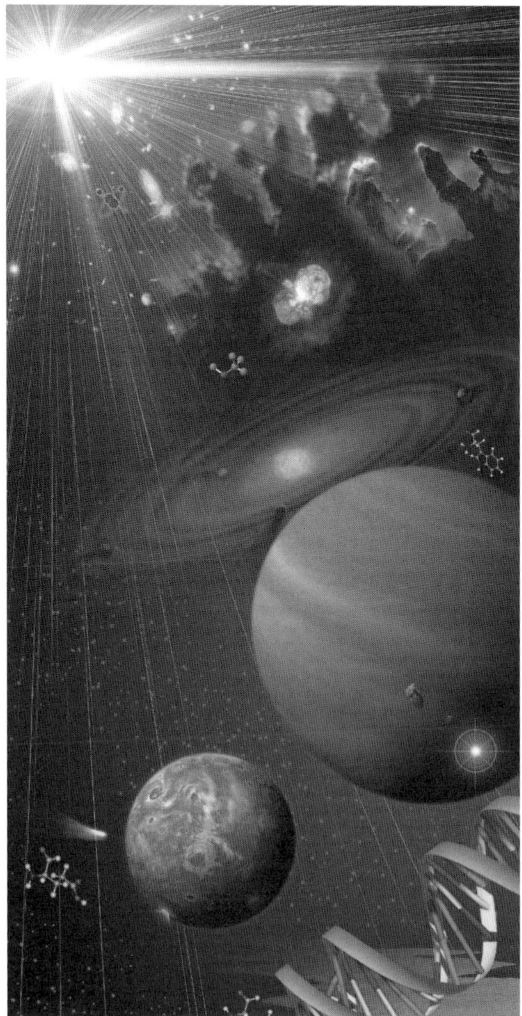

Figure 7.3 Whether life is widespread in the universe is the main concern of the discipline of astrobiology. This view of cosmic evolution, as it appeared in the roadmap for NASA's Office of Space Science Origins theme in 1997, foresees cosmic evolution ending in a microbial biological universe. The image reflects the cancellation of the NASA SETI program in 1993, but it is possible we live in an intelligent or anthropic biological universe. Credit: NASA.

since the time of the ancient Greeks. At about the same time astronomer Otto Struve saw it as a revolution comparable to the Copernican cosmology of the sixteenth century and the galactocentric cosmology in the early twentieth century, whereby the solar system was seen to be at the periphery of the galaxy, a situation now known to be true. And NASA SETI pioneer Barney Oliver even went so far as to use cosmological terminology, speaking of it as a "biocosmology."[8]

The rise of the biological universe as a new worldview did not come easily or quickly. Grounded in Copernicanism, by the middle of the eighteenth century, it was widely accepted as invoking a grand analogy, but without proof. By the middle of the nineteenth century, it was given scientific underpinnings with the spectroscopic discovery that the same elements were found throughout the universe as on Earth, and Darwin's insight that life develops in a natural way through evolution by natural selection. And in the middle of the twentieth century the proto-science of exobiology was founded, as the Space Age provided the means for in situ investigations of life on planets in our solar system. These and related investigations have now grown into the robust field of astrobiology.[9]

It is notable that the worldview of a biological universe is widely accepted even without proof, with those preconceptions I mentioned earlier continuing to play a prominent role. And it is clear that even in the absence of proof, the destructuring of a worldview represented by the purely physical universe and its restructuring into the biological universe is well under way. It is also remarkable that this destructuring and restructuring have gone on simultaneously and over an extended period. Some argue that since this worldview is already so firmly entrenched, the discovery of alien life will have little impact. But in my view, this is dead wrong. Possibility is not the same as actuality; the idea is not the same as reality. Replacing preconceptions with real scientific data and conclusions will accelerate the building of the new worldview – and accelerate its implications as well.

Forms of the Biological Universe

The potential impact of this biological universe is best approached by distinguishing several possibilities. As many scientists have argued, it is entirely possible the universe is full of microbes, but lacks complex life or intelligence. Such a *microbial biological universe* would be a fundamental discovery in the history of science, though arguably not as exciting as the discovery of intelligence, where the objects of our discovery may talk back to us. If found to be of independent origin, the discovery of such microbes would open the possibility of a universal biology in the same way we now believe we have a universal physics. It is tempting to say that the impact of a microbial biological universe would be primarily scientific. But the reaction in 1996, when nanofossils were claimed in the Mars rock ALH84001, clearly indicates the impact will be much broader than that, at least over the short term. After all, where there is smoke, there may be fire; if microbial life is common, it may indicate that, given enough time as we have had on Earth, intelligence may develop. But the discovery of microbial life in itself would be the greatest discovery in the history of science.

An *intelligent biological universe* is one where intelligent life is a common feature. We can argue about the meaning of "common," or in its reverse "rare Earth" formulation, about the meaning of "rare." But in a universe so large that even if one in a million planets have intelligence, millions of planets with intelligence would still occur even in our own Galaxy, the semantic argument over the definition of common or rare does seem rather pointless. More urgent is a discussion of what it means to be sentient, intelligent, or superintelligent, what interaction with such entities implies in terms of ethical standards, and beyond that, diplomacy. As seen in the next chapter, our interactions with microbial life also have their ethical problems, but they are surely less complex than when Klaatu or the Overlords arrive.

The *anthropic biological universe* view holds that life and mind are intimately connected with the physical universe. The physical and biological universe are obviously already connected in the sense biology cannot exist without a prior physical substrate. But scientists and philosophers alike are increasingly aware of possible deep connections between the physical universe of planets, stars, and galaxies, and the biological universe of life, mind, and intelligence. The more we learn about the universe, the more we find that it is even more mysterious than we thought – that not only is the universe subject to mathematical laws, but also finely tuned for life. In its deepest structure and most fundamental properties such as the physical constants, the universe seems tailor-made for life. Or, to put it another way, the universe is "biofriendly," as witnessed by our existence, and the possible existence of other life beyond Earth.

The logical force of this idea, known as the "anthropic principle," has been growing over the past century. In 1913 Harvard biochemist Lawrence J. Henderson, marveling at the fitness of the terrestrial environment for life, made a bold statement in the final paragraph of his book *The Fitness of the Environment*, where he extended his argument to the universe at large. "The properties of matter and the course of cosmic evolution," he wrote, "are now seen to be intimately related to the structure of the living being and to its activities; they become, therefore, far more important in biology than previously suspected. For the whole evolutionary process, both cosmic and organic, is one, and the biologist may now rightly regard the universe in its very essence as biocentric." What we might call Henderson's "biocentric principle" expressed here is a much better term for the fitness of the universe for life than "anthropic," since it does not imply that terrestrial life, including humans, could be the only form of life. To put it bluntly, and non-anthropocentrically, the universe was fine-tuned for life, not humans. But it of course includes beings of higher consciousness such as we humans fancy ourselves to be. By whatever name, the idea is powerful. As cosmologists John Barrow and Frank Tipler wrote in their landmark volume *The Anthropic Cosmological Principle* (1986), "The realization that the possibility of biological evolution is strongly dependent upon the global structure of the Universe is truly surprising and perhaps provokes us to consider that the existence of life may be no more, but no less, remarkable than the existence of the Universe itself."[10]

Finally, we have already broached the high possibility of a postbiological universe, the product of cultural evolution in which one of several forms of postbiology may be created. Of these, we have argued that artificial intelligence (AI) is the most likely. While there is a growing body of literature on the promise and dangers of AI on Earth, we need also to consider what the societal impact might be if AI is discovered beyond the Earth.

Impact of Astrobiological Worldviews on Culture

The microbial, intelligent, and anthropic versions of the biological universe, and the postbiological universe as well, each represent different worldviews with different implications for humanity. They can also be viewed in terms of human destiny, usually seen in theological terms, but also in more secular terms. In a microbial biological universe, aside from its scientific value we may be dealing primarily with the instrumental rather than intrinsic value of life, ethical distinctions I elaborate on in the next chapter. Instrumental value in this case implies that humans might be mainly interested in the utility of extraterrestrial microbes for medicines and other purposes, just as has been the case with newly discovered extremophile organisms on Earth. That would be a subject of major ethical and practical discussion. In such a microbial biological universe, human destiny would be to explore the universe as the lords of creation – at least until sentience or intelligence was discovered near or far.

In an intelligent biological universe, by contrast, human destiny is to interact with extraterrestrials, whether in struggle or to join the galactic club. This is the stuff of science fiction, but in case of an actual discovery it would very quickly become a practical problem. As I have emphasized in earlier chapters, the outcome would very much depend on the scenario. In an anthropic biological universe, the questions turn more to the philosophical. The obvious question is, why is the universe so biofriendly? Could it be just a coincidence? For most people who have thought deeply about the problem, this seems too good to be true. Could it be that a supernatural intelligence fine-tuned the universe, the God hypothesis? Many would say so, but it depends on whether the supernatural is part of your metaphysics. Could a highly evolved *natural* intelligence have fine-tuned the universe? This possibility has received some attention, but is difficult to prove. Or perhaps a "multiverse" exists, an ensemble of universes, and we happen to live in a universe that is suitable for life. Whatever version of the anthropic universe we accept, and whatever explanation for it, the anthropic biological universe is a cosmological worldview as new and expansive as any in the past. Finally, the postbiological universe has barely been discussed in the terrestrial context, much less the extraterrestrial.[11]

If after many centuries, it turns out we find none of these versions of the biological or postbiological universe, then we will have determined that we live in a physical universe in which humans are unique or nearly unique. In that case, human destiny will be to explore this universe and perhaps to spread humanity throughout its vast spaces. We need not invoke or repeat the sins of manifest destiny in doing so. If we are indeed alone, we need not worry about harming the "natives." But there will be many questions about exploiting and harming the space environment for human purposes.

In summary, the impact of actually discovering life – as opposed to conjecturing and theorizing about it based on preconceptions – will be to strengthen and consolidate the new worldview called the biological universe, or its cultural extension, the postbiological universe, all as a part of the epic of cosmic evolution. Whether it is microbial, biological, anthropic, or postbiological, or whether some mix of these or something yet to be discovered, remains to be seen. But in any case, with the discovery of alien life our worldview will have been transformed in proportion to the complexity of life discovered, and in proportion to the degree to which the universe is anthropic or biocentric.

Along with the mysteries of dark energy and dark matter, the abundance of or lack of life is one of the most fundamental *cosmological* discoveries yet to be made. As physicist Freeman Dyson put it in slightly different fashion:

> The prospects are bright for a future-oriented science, joining together in a disciplined fashion the resources of biology and cosmology. When this new science has grown mature enough to differentiate itself clearly from the surrounding farrago of myth and fiction, it might call itself "cosmic ecology," the science of life in interaction with the cosmos as a whole. Cosmic ecology would look to the future rather than to the past for its subject matter, and would admit life and intelligence on an equal footing with general relativity as factors influencing the evolution of the universe.[12]

Fine-tuning and the multiverse are two concepts that may prove central to this task. In such a "cosmic ecology," life and intelligence do indeed play a central role in the evolution of the universe, no less than its physical laws. If the universe indeed "saw us coming," life may be its ultimate purpose for existing, returning to physics a kind of teleology that had been banned from science. This lends all the more importance to the current disciplines of astrobiology and SETI, which will determine just how abundant life is in the universe – or (perhaps in the very long run!) the multiverse. The biological universe may be not the result but the cause of the physical universe, part of its very fabric, and more surprisingly fundamental than we have ever realized.

What does such a seemingly esoteric cosmological theory have to do with the man or woman in the street, you might ask? The common person in the sixteenth century might have asked the same question when hearing about some obscure theory about the Sun being at the center of the cosmos. We now know the answer is "everything." Not only did that idea spark the scientific revolution in the seventeenth century associated with the names of Galileo, Kepler, and Newton it also sparked the popular sense of dislocation embodied in Donne, Milton, and Pope, with all that meant for literature and the arts. The discarded image of the Earth-centered cosmos also undermined the ordered meaning of life as it had existed for millennia since the geocentric cosmos was first broached among the ancient Greeks. The discovery of an intimate connection between cosmology and life implied by the anthropic biological universe – or even the discovery of an intelligent and possibly microbial biological universe – would do the same in ways whose details we cannot yet fathom. More broadly, such a cosmic connection with life, the ultimate form of the biological universe, should counter any feelings of alienation from the universe brought about by the great decentralizations of the past four centuries. It should make us feel at home in the universe.

Theological Worldviews: The Rise of Astrotheology

Below the cosmological level in our framework of Figure 7.1 looms a perplexing variety of theological worldviews, a dominating feature of human cultures – whether for better or worse is a subject of perennial debate. Whatever the other effects on cultural elements at a level below the theological, the impact of the discovery of life beyond Earth on these theologies is likely to be pervasive, foundational, and personal (Figure 7.4). Such impacts will vary depending on the nature of the discovery and the particular religion involved. Because theological worldviews are so deeply held by such a large part of the population, changes in worldviews at this level – whether those worldviews are objectively true or not – are sure to have profound effects. Conversely, however, they also have the potential to help Earthlings cope with the discovery in whatever form it takes.

The actual impacts of changes in theological worldview must remain speculative, but in examining the possibilities we can draw on at least three sources: historical discussions of the religious impact of possible inhabited worlds; theological reactions to putative discoveries of life beyond Earth; and empirical surveys of theological and public opinion on the subject. All of these streams of thought have resulted in what is now a considerable scholarly discussion of the subject, known variously as exotheology, astrotheology, and cosmotheology. Such terminological diversity is not unusual in the early stages of a new intellectual endeavor.

Figure 7.4 "The Creation of Adam," a fresco painted by Michelangelo for the Sistine Chapel ceiling around 1510. It depicts a one-to-one relationship between God and man, but astrotheology raises the fundamental question of how unique this relationship really is. This painting was conceived within the Christian tradition; other religious traditions will be affected differently. Credit: John Parrot/Stocktrek Images/Getty Images.

History and Astrotheology: A Planet-Hopping Jesus?

The effect of what we have called the biological universe on theologies, particularly Christian theology, has been the subject of sporadic debate since Copernicus made the Earth a planet and the planets potential Earths. In the sixteenth century Philip Melanchthon, Martin's Luther's faithful supporter, argued against the Copernican theory since "it must not be imagined that there are many worlds, because it must not be imagined that Christ died and was resurrected more often, nor must it be thought that in any other world without the knowledge of the Son of God, that men would be restored to eternal life." That statement opened a Pandora's box of implications that has reverberated through history and has not been closed even today.

In subsequent centuries three choices were logically open to Christians who pondered the question of other worlds: they could reject other worlds, reject Christianity, or attempt to reconcile the two. Historically, all three of these possibilities came to pass in the eighteenth and nineteenth centuries. The scriptural and doctrinal problems of the issue had been widely discussed throughout the seventeenth century in the context of various cosmologies, only to be overwhelmed by natural theology, which held that a universe full of life better reflected the magnificence and glory of God. But the doctrinal problems would not go away. No one more forcefully expressed the continuing difficulties for Christianity of what was then known as the "plurality of worlds" debate than Thomas Paine writing during the Enlightenment. In his influential *Age of Reason* (1793), Paine bluntly stated that "to believe that God created a plurality of worlds at least as numerous as what we call stars, renders the Christian system of faith at once little and ridiculous and scatters it in the mind like feathers in the air. The two beliefs cannot be held together in the same mind; and he who thinks that he believes in both has thought but little of either." Pointing to the doctrines of redemption and incarnation and to the absurdity of a planet-hopping Savior, Paine rejected Christianity. Few would go that far in the wake of Paine's diatribe, but to this day the doctrines of redemption and incarnation remain two of the central astrotheological issues in the context of Christianity.[13]

Theological Reactions to Astrobiology

The relation of alien life to theology – most often Christian theology – has also been discussed as events have unfolded in the Search for Extraterrestrial Intelligence (SETI) and as new discoveries or reputed discoveries have been made in NASA's astrobiology program. In the wake of Frank Drake's Project Ozma in 1960, for example, Catholic priest Daniel C. Raible wrote, "Yes, it would be possible for the Second Person of the Blessed Trinity to become a member of more than one human race. There is nothing at all repugnant in the idea of the same Divine Person taking on the nature of many human races. Conceivably, we may learn in heaven that there have been not one incarnation of God's son but many." A few years later an editorial in the Catholic journal *America* suggested that "today's theologians would welcome the implications that such a discovery might open – a vision of cosmic piety and the Noosphere even beyond that of a Teilhard de Chardin." The same flexibility was expressed in Jewish thought, which need not worry about redemption and incarnation. Rabbi Norman Lamm, citing research in exobiology as documented in Walter Sullivan's *We Are Not Alone* and Harlow Shapley's cosmological decentering of humanity popularized in *Of Stars and Men*, argued anyone would be "profoundly mistaken" in assuming the number of intelligent species in the universe had any relation to the significance of humanity in theological terms. Judaism, he concluded, "could very well accept a scientific finding that man is not the only intelligent and bio-spiritual resident in God's world," and that God could just as easily exercise providence over 10 billion worlds as one.[14]

Thirty years later, in the wake of the claims of fossil life in the Mars rock ALH84001 in 1996, the same kinds of issues were raised. As we saw in Chapter 1, along with the scientific aspects they were reporting, newspapers were full of speculation about the impact of the discovery on theology. There was some consensus that the discovery of alien life in any form would confirm the expansive nature of God – the seventeenth-century natural theology argument – even as many cautioned that Christianity would have to rethink some of its central dogmas. The discussion reached the highest level of government, as represented in the meeting hosted by Vice President Gore four months after the announcement, described in Chapter 1. That meeting included not only scientists (Stephen Jay Gould and Lynn Margulis, among others), a historian (me), and a journalist (Bill Moyers) but also Catholic theologian John Minogue (president of De Paul University) and Carol Brown representing the National Council of Churches. A few months later the subject was still fresh, as Jesuit Vatican astronomer Christopher Corbally presented a paper at the American Association for the Advancement of Science (AAAS) on the religious implications of possible fossils in the Mars rock. In the end it was a false alarm, but for those who think the discovery of microbial life would not have a theological impact, this served as a lesson about the reaction when even possible fossilized life was found.[15]

The subject has only grown more intense in the two decades since planets have been discovered beyond the solar system, giving way to more substantial discussions of what is now increasingly termed "astrotheology," or, in one of its many possible flavors, "cosmotheology."[16] Under whatever name, these terms embody the radical idea that our theologies may need to be expanded or modified in rather wholesale fashion to encompass the new worldview that cosmic evolution has resulted in abundant planets and perhaps abundant life and mind beyond Earth. An early indication of increasingly organized interest was a meeting on theology and alien life in 1998. Sponsored by the Templeton Foundation, which fosters dialogue between religion and science, I still vividly remember it as an intimate meeting of about a dozen scholars held in the Bahamas at the home of Sir John Templeton. Several papers at this meeting independently elaborated on issues that had been raised before. Ernan McMullin, a Catholic priest and eminent philosopher at the University of Notre Dame, argued that "the Creator of a galactic universe may well choose to relate to creatures made in the Creator's own image in ways and on grounds as diverse as those creatures themselves." For him, Christian doctrine in the context of aliens boiled down to three issues: original sin, soul and body, and incarnation. He speculated that an omnipotent creator might want "to try more than once the fateful experiment of allowing freedom to a creature," a freedom that Adam and Eve had failed in the Earthly Garden of Eden. He pointed to the possibility that aliens might or might not have souls; if they did, God might or might not elect to become incarnate. And regarding incarnation, McMullin suggested that conflicting theological interpretations of that central doctrine of Christianity left wide open the question of Christ's incarnation on other worlds. George Coyne, at the time the director of the Vatican Observatory, also offered no definitive answers, but suggested that theologians would have to "rethink some fundamental realities within the context of religious belief." In other words, neither McMullin nor Coyne were at all sanguine about the rapid adjustment of Christianity to extraterrestrials, but they believed it would be accommodated in the long term. By contrast, astronomer Jill Tarter, a pioneer in the field of SETI, argued that an extraterrestrial message might be "a missionary campaign without precedent in terrestrial history, replacing our diverse terrestrial religions with a universal religion." Alternatively, a message that indicated long-lived extraterrestrials with no need for God or religion might undermine our religious worldview completely.[17]

Aside from me as a participant and editor of the proceedings, the roster included luminaries such as Nobelist Christian de Duve; physicists and astronomers such as Paul Davies, Freeman Dyson, and Martin Rees; and evolutionary biologist Richard Dawkins, a skeptic if ever there was one when it came to dialogue between science and religion. The meeting was lively. As the theological deliberations proceeded, I recall Dawkins more than once interrupting a speaker to ask in his British accent, "what on Earth are you talking about?," with emphasis on a long and drawn-out "Earrrrth." There was a lively discussion among believers and skeptics – not so much when it came to the possibility of life, but when it came to the possible theological implications. And I vividly remember George Coyne commenting on my paper on cosmotheology by saying, "there is a special place in hell for those who think God isn't supernatural!" – whether tongue in cheek or not I was never sure.

Precisely because of new results in astrobiology that kept rolling in – thousands of planets in the next 20 years, evidence of oceans on the Jovian and Saturnian moons such as Europa and Enceladus, the tenacity of life seen in extremophiles – the problem of adapting theologies to current knowledge has only grown more urgent. In what can only be seen as a laudable forward-looking stance to determine the impact of one of its own programs, NASA itself has sponsored programs that include the theological implications of astrobiology. In conjunction with NASA and the Templeton Foundation, in 2003 and 2004 Constance Bertka, a planetary scientist who also headed the Dialogue on Science, Ethics, and Religion program at the AAAS, convened a series of workshops in Washington that included ethical and theological perspectives. In 2011 NASA established the Baruch S. Blumberg NASA/Library of Congress Chair in Astrobiology specifically to examine humanistic aspects of astrobiology, including theological aspects. NASA's astrobiology program also funded, somewhat controversially, research at Princeton's Center of Theological Inquiry on the societal implications of astrobiology.

With all this as background, astrotheological books began to appear with increasing frequency, all by authors with dual astronomical and theological interests. Not surprisingly, three of those authored by theologians focused almost exclusively on Christianity. Dominican theologian Thomas F. O'Meara, situated, like extraterrestrial life historian Michael J. Crowe, at Notre Dame University, argued that Christianity offers "no veto against a history of free intelligence on another planet," and that "bringing together intelligence, matter, and divine presence may find realization in multiple ways in the galaxies." Methodist minister David Wilkinson, who holds PhDs in both astrophysics and systematic theology, expanded the scope of discussion by examining the meaning of SETI and astrobiology for creation, incarnation, revelation, and salvation. He argued that theologians need to take these issues seriously, that SETI can be of value to Christian theology by expanding its perspectives, and that Christian theology can be of value to SETI by its long tradition of discussion about what makes us human, about the nature of good and evil, and a variety of other issues. By contrast Jesuit astronomer and Vatican Observatory director Guy Consolmagno focused on a single issue guaranteed to spark public interest: "Would you baptize an extraterrestrial?" His answer was, "only if it asked," a considerable improvement from sixteenth-century conversion attempts during the European first contact with the Americas. Discussions are also under way in theological journals, such as a special issue of *Theology and Science* in 2017, and *Zygon: Journal of Religion and Science*. Taken together, these books and articles constitute the beginnings of a new intellectual endeavor, one sure to accelerate when an actual discovery is made.[18]

With this expanding literature, as well as articles from a variety of authors both secular and religious, we can begin to parse the theological questions more finely, both in terms of the impact of microbial versus intelligent life aspects and in terms of the impact on different religions. Not surprisingly, almost all scholarship has dealt with the implications of alien intelligence rather than microbes. But as the Mars rock episode indicates, even the discovery of microbes will certainly raise theological questions. This is largely because the microbial discussion quickly transforms into a discussion about intelligence because the discovery of simple microbes overcomes one large hurdle toward intelligence – the origins of life. Absent that connection, it seems unlikely microbes in and of themselves will raise difficult theological issues except in the area of origins of life. With intelligence, however, the issues are clear and have focused on three areas: creation, incarnation, and the effects on human dignity. After examining theological opinion on the creation issue, Berkeley theologian Ted Peters thought it "safe to forecast" that contact with extraterrestrials would expand the existing biblical vision that all of creation "is the gift of a loving and gracious God." As for incarnation, opinions are mixed, especially when it comes to the question of "one incarnation or many." And in terms of human dignity, there is some consensus that expanding the concept of dignity to extraterrestrials is not only possible, but desirable, and doing so would help us understand our own concepts of human dignity.[19]

Typically and unsurprisingly most of these discussions continue to focus on Christianity. Although some outside the Christian tradition scoff at the need for discussing theological issues such as incarnation and baptism, for 2 billion Christians, these will be important questions in the event of discovery. But what about other religions? Unlike Christianity, no substantial historical tradition exists in the other two "Adamist" religions analogous to the Christian tradition on this subject. In his Jewish exotheology, Rabbi Lamm did point to some historical precedents, and he expressed his viewpoint that extraterrestrials would not be problematic, much less fatal, for the Jewish religion. In the Islamic tradition, second only to Christianity with its 1.5 billion adherents, astronomer David Weintraub points out that some particulars of Islam would prove problematic in its geocentric and anthropocentric aspects: the importance of Mecca for prayer and travel; praying five times per day on worlds with potentially vastly different day lengths; and in general following the teachings of Muhammad, who was born and lived on Earth. Thus, while Muslims do not have to deal with incarnation, they still have some of the same problems brought about by the anthropocentric nature of their religion. On the other hand, Allah is the creator of the entire universe, and learning about extraterrestrials would seem to be part of Muslims' goal of reading the book of Nature to strengthen their faith.

Beyond the Adamist traditions, there has been even less discussion, but Weintraub stands out for his ecumenical interests, touching on more than a dozen religions in his book *Religions and Extraterrestrial Life.* In his view, the least impact on these would be to Hinduism (900 million members) and Buddhism (400 million members). Each of these, Weintraub finds, would easily absorb, and even welcome, the discovery of extraterrestrial life. Smaller religions would have to each deal with its own problems. Some, such as Mormonism, even have extraterrestrials as part of their doctrine, an artifact of the time of their founding. Belying the seemingly simple question of the effect of astrobiology on theology, these findings point to one main conclusion, in some ways obvious from the outset: each religion will have to deal with the discovery of extraterrestrial life in its own way. There will not be one effect, but many.[20]

The View from Popular Culture

One final source helps us to gauge impacts: empirical surveys of theologians and the public. In my book *The Biological Universe*, published just as the first exoplanets were being announced, I posited based on historical evidence that within various religious traditions the consensus was that terrestrial religions would adjust to the discovery of extraterrestrial life, while those external to religious traditions thought they could not survive. Moreover, I hazarded the opinion that religions such as Hinduism and Buddhism would be less affected than the Abrahamic religions of Judaism, Christianity, and Islam, since the latter had to deal with dogma related to the godhead. These conclusions can now be refined by a variety of surveys, which must, however, be used with caution. Among the earliest of these, conducted by sociologist William Sims Bainbridge in 1983, surveyed almost 1,500

American college students and focused on the question of whether humanity should attempt communication with extraterrestrial intelligence. Thirty-eight percent of Protestants answered yes, as did 42 percent of Jews, 44 percent of Catholics, and 50 percent claiming no religious affiliation. Moreover, the more conservative the Protestant in this survey, the more likely the answer was to be no.

In 1991 Michael Ashkenazi of Ben Gurion University surveyed 21 theologians and found that on the spectrum of religions, Eastern traditions such as Hinduism and Buddhism were more receptive than Christian fundamentalists. In 1997 Douglas Vakoch and Yuh-Shiow Lee compared American and Chinese college student reactions to the possibility of extraterrestrial contact and found that the "more anthropocentric students from both countries were … less open to the existence of extraterrestrial life than were their less human-centered counterparts." And in 2009 Ted Peters conducted the "ETI Religious Crisis Survey," which received 1,300 responses worldwide from individual religious traditions, including Catholic, Protestant, Jewish, Mormon, and Buddhist. The conclusion was that these religious traditions believed overwhelmingly that the discovery of extraterrestrial intelligence would not provoke a crisis in their particular religion. Of those that did, only 22 percent of Catholics, 17 percent of Jews, and 14 percent of evangelical Protestants thought that it might cause a crisis, while only 3 percent or less of Mormons, Buddhists, and nonreligious thought it might.[21]

As always, we need to be careful about surveys, since so much depends on how the question is asked. Had Peters, for example, asked Christians, "do you believe in a planet-hopping Jesus?," which, as we have seen, is really the central issue, the responses might have been quite different. Bertka, for one, argues cogently that responses to the discovery of life beyond Earth would be more varied than scholarly and theological opinion would indicate, in the same way that the American public's responses to evolution are varied. The public response might have more of an emotional than a rational component. She points out that generalizing about Christianity's response to aliens is therefore very difficult, is individually and culturally dependent, and may "at minimum be a long and convoluted process with more than one likely outcome." In other words, and as history shows, the Ladrièrean destructuring and restructuring of theological worldviews will likely be a messy process.[22]

In addition to scholarly discussion, extraterrestrial theological scenarios continue to be explored in some detail in science fiction, ranging from such classics as C. S. Lewis's cosmic trilogy (1938–1944), including *Out of the Silent Planet*, and James Blish's *A Case of Conscience*, to Mary Doria Russell's more recent novels *The Sparrow* and *Children of God*. While Lewis uses extraterrestrials to paint from a celestial viewpoint a vivid picture of an Earth in which humanity is in a spiritual struggle from which it must extract itself, both Blish and Russell use

Jesuit missions to inhabited planets to make their points. Blish portrayed Jesuit Father Ramon Ruiz-Sanchez and a biologist who must deal with the religious implications of the discovery of intelligent bipedal reptilian inhabitants on the planet Lithia. The Lithians have no religion, but an innate sense of morality, raising questions of the proper sources of morality. Russell, who uses her anthropology background to good effect in her two "Sparrow" novels, takes four Jesuits to the planet Rakhat following a SETI detection of music emanating from the planetary system surrounding Alpha Centauri. There Father Emilio Sandoz is raped, mutilated, and enslaved, all of which are considered honorable activities on Rakhat. Since the gospel of Matthew tells us not even a sparrow falls to Earth without God's knowledge, the question is, how does God allow such things in the universe – again a question of morality, a cosmic variation of the theme "why do bad things happen to good people?" These fictional scenarios represent deep thought about a problem that has now been with us for centuries or even millennia in hypothetical form, and that will be given greater urgency as soon as a discovery is made.[23]

If there is any consensus from these sources of thought about astrotheology, it is that in the short to medium term current terrestrial religions would adjust to extraterrestrials. Even physicist Paul Davies, the popular author who has written widely on science and religion and who was pessimistic about the ability to adjust, changed his mind in 2003 when he argued that "the discovery of extraterrestrial life would not have to be theologically devastating."[24] Professor of Christian ethics and theology Cynthia Crysdale emphasized that Christianity will need to adapt in ways that are already foreseeable. If it does not, "to the degree that the religious public has a certain narrative about how the world has unfolded, a narrative that includes divine determinism and/or intervention, and especially one that sees human life and salvation as the apex of meaning in this unfolding, the discovery of life elsewhere will cause huge problems."

In fact there is some evidence that those outside traditional religions believe many religions, particularly the Adamist religions, would be decimated by such a discovery. Philosopher Roland Puccetti, for one, argued that monotheistic religions involving a personal relationship between a creator and its creatures would be in trouble. On the other hand he believed that for the Oriental faiths, such a discovery would be a far less serious matter, "since in their 'higher' forms at least they teach salvation through individual enlightenment and conceive the supreme Reality in strictly impersonal terms." Or as E. M. McAdamis puts it in his excellent study, the more any religion places humanity at the core purpose of the universe, in other words, the more anthropocentric and teleological it is, "the more potential there will be for religious resistance to astrobiological evidence and the possibility of life beyond Earth."[25]

Astrotheology or Astrophilosophy? The Case of Cosmotheology

In the longer term, theologies could radically change, even to the extent that some might call them astrophilosophies rather than astrotheologies. A case in point is my own personal formulation of what I call cosmotheology, which may be seen as a particular brand of astrotheology bordering on astrophilosophy. Cosmotheology, which I first conceived in 1997 as an offshoot of the same AAAS session where Jesuit Chris Corbally presented his reaction to the Mars rock, is in its most general sense a theology that takes into account what we know about the universe based on science. It is therefore a naturalistic cosmotheology, but it is not coextensive with scientism because it does not imply that science is the only way to understand the world. In my formulation, cosmotheology has six bedrock principles: 1) humanity is in no way physically central in the universe; 2) humanity is not central biologically, mentally, or morally in the universe; 3) we must take into account the probability that humanity is near the bottom in the great chain of beings in the universe; 4) cosmotheology must be open to radically new conceptions of God, not necessarily the God of the ancient Near East, nor the God of the human imagination, but a natural God grounded in cosmic evolution; 5) cosmotheology must have a moral dimension, extended to embrace all species in the universe – a reverence and respect for life in any form; 6) cosmotheology can and should transform our ideas about human destiny. As emphasized at the end of the previous section, although human destiny has most often been couched in divine terms, as in Reinhold Niebuhr's *The Nature and Destiny of Man* (1941) or Pierre Lecomte du Noüy's best-selling *Human Destiny* (1947), or, indeed as in the entire Christian theology, it need not be linked to the supernatural. Rather, it can be linked to the process and endpoint of cosmic evolution.[26] Taken as a whole, this version of astrotheology can also be seen in the tradition of religious naturalism, but specifically formulated to take into account the realities of cosmic evolution and the many possibilities inherent in life in the universe.

Such a religious naturalism as expressed in cosmotheology raises an important question: is it not really a form of astrophilosophy rather than astrotheology? Perhaps, but history suggests humanity is a long way from giving up supernatural theologies in favor of secular natural philosophies, no matter how "true" the former are in an objective sense, nor how much the latter tend to spiritual needs. Moreover, many religious naturalists would disagree that their formulation is not theological. But it may be that over the course of centuries astrotheology will transform into astrophilosophy, as the natural gods of cosmic evolution and the biological universe replace the supernatural God of the ancient Near East. At the same time other secular forms of philosophy may also be transformed. As we see in the next chapter, such developments may parallel the formulation of a "cosmocentric" ethic where the cosmos and its inhabitants are increasingly valued as our cosmic consciousness increases.

Cosmotheological precepts are unlikely to gain wide acceptance among Earthlings in the near term. Science and religion have a long history of conflict, and the religious reaction to past ideas that threaten centrality of humans is not encouraging – we need only think of the fate of Darwin, even among the most advanced societies on Earth, most especially the United States. I like to tell my friends who are skeptical of evolution that if you don't believe in evolution, you don't believe in biology; if you don't believe in biology, you don't believe in science; and if you don't believe in science, welcome to the Middle Ages. But the past need not be prologue, and as more theologies become astrotheologies in the future, they might help us cope with the discovery of alien life.

There is another reason to think more carefully about cosmotheology. For non-Christians and those outside the Adamist/Abrahamic tradition in general, the chances that aliens will have heard of redemption, incarnation, baptism, Jesus, or Mohammed seem virtually zero. They may have come up with their own versions of these ideas, but it seems much more likely that they will have some kind of naturalistic cosmotheology, if they deal in theology at all. Perhaps they will have an astrophilosophy that deals with issues such as morality that we normally think of in theological terms. If that is the case, a philosophy grounded in cosmic evolution may serve as a common subject for communication, rather than a source of tension as with so many theologies on Earth. This would indeed be a magnificent dis-analogy compared to culture contacts on Earth.

Cultural Worldviews: The Rise of Astroculture

The cosmological and the theological are major worldview components of any given culture. The cosmological worldview will usually pervade any given culture: the Western geocentric worldview belonged not only to Ptolemy and the Aristotelians but also to Dante and his readers, just as the cosmologies of other civilizations ancient and modern mirrored the heavens and the Earth. These worldviews were thus woven into the framework of the daily lives of the literate populace. By contrast, theological worldviews may be more diverse, especially in the modern multicultural world. Any given culture may be dominated by a particular theology, as Christianity in the West, Islam in the Middle East, or Hinduism in India, but that same culture may also harbor other theological worldviews. Indeed, different worldviews in any culture are often a source of serious conflict. But there are other parts of culture than the cosmological and theological, and we would be remiss if we did not recognize how, at the broad bottom of the pyramid of worldviews of Figure 7.1, culture is at an ever-accelerating pace being transformed into astroculture, in its broadest sense an increasing awareness of our place in the universe.

More specifically, astroculture is a relatively new umbrella concept used to describe the array of images, events, and media reactions that "ascribe meaning to outer space while stirring both the individual and the collective imagination." In a landmark international meeting on the cultural history of outer space held in Germany in 2008, German historian Alexander Geppert introduced the concept, which was then given concrete form by more than a dozen authors. Geppert's goal was to examine whether a unique Western European perspective on space existed in the three decades following World War II, in the same way that American historians had examined the cultural history of space in the United States. We can expand the concept to argue that, while different perspectives on space may exist in different cultures, humanity as a whole is increasingly creating and immersed in an overarching astroculture that transcends national boundaries. Call it *global astroculture*. This transformation is much broader than the possibility of life beyond Earth, but the idea of alien life has for some time been a major component of culture (at least Western culture), and is sure to accelerate if a discovery is made of life beyond Earth.[27]

As I emphasized when I gave the keynote address at the meeting in Germany, and as shown in Chapter 4, culture itself is a slippery concept, but one that anthropologists have had to grapple with over the past century. Although there are different conceptions of culture in different times and places, it is my contention that human knowledge and attitudes toward life are increasingly creating, and created by, astroculture. To be sure, different individuals are affected in diverse ways. For me, my lifelong fascination with what we now call astroculture began with a movie I saw in Germany at age 11. *Raumschiff Venus antwortet nicht* [*Spaceship Venus Won't Answer*] (Figure 7.5), based on the first novel of the great science fiction writer Stanislaw Lem, aroused my youthful imagination with its story of a spaceship sent to Venus, where scientists find the remains of a warlike civilization that perished in a nuclear war. All that survives are advanced machines, programmed to carry out the goals of the original Venusians. In conjunction with the real space programs launching in the early 1960s, this movie helped to create my intellectual background even as I have played a small part in creating astroculture through a career in astronomy and space exploration. It is the combination of such individual experiences, along with joint human experiences such as the first Moon landing, that give rise to the overarching astroculture.[28]

The Impact of the Space Age: The Cosmic Connection

The rise of astroculture has been long coming, and these changes in humanity's worldview have increased our cosmic consciousness especially in the past half-century. In particular the space program for the first time has given us stunning

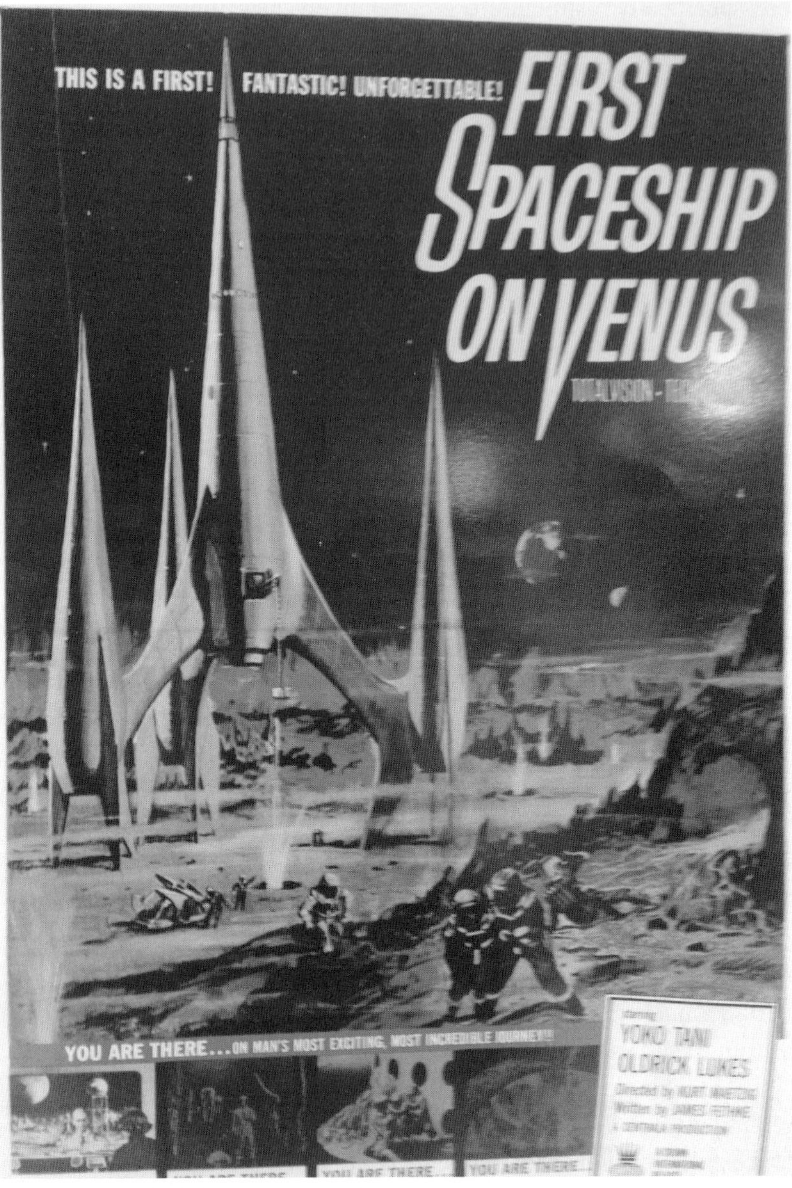

Figure 7.5 Poster for *First Spaceship on Venus*, the American version of *Der schweigende Stern*, directed by East German director Kurt Maetzig in 1960. The movie, released in the United States in 1962, was based on Stanislaw Lem's first novel, *The Astronauts* (1951). Science fiction has had a major effect in generating astroculture. ©DEFA-Stiftung.

images of our home planet, beginning with Earthrise first seen from *Apollo 8* as the American astronauts circled the Moon on Christmas Eve 1968 (Figure 7.6). The "Blue Marble" image of the full Earth, taken by the *Apollo 17* astronauts in 1972, became one of the most widely reproduced images in history, followed by many other "full Earth" images that show a world without national borders. The "Pale Blue Dot" image, taken at the behest of Carl Sagan as the *Voyager 1* spacecraft turned its eye toward the Earth in 1990, revealed the Earth not as a large and colorful world, but as a small and insignificant dot against the blackness of space. "Our posturings, our imagined self-importance, the delusion that we have some privileged position," Sagan wrote, "are challenged by the point of pale light. Our planet is a lonely speck in the great enveloping cosmic dark."[29]

For some people these images give rise to what American author Frank White calls "the overview effect." White documents how some spacefarers have felt a series of positive mental experiences as a result of seeing the Earth from space, giving them a sense of unity and even renewed purpose. He uses the analogy of a fish jumping out of water and perceiving a new objective reality for the first time, if only briefly. More generally, he argues we all become "terranauts" as we vicariously view the images the astronauts saw firsthand. The idea of such an overview effect has been criticized as fundamentally religious (even though it does not invoke supernaturalism), methodologically flawed, and dependent on a given

Figure 7.6 Earthrise, image taken on December 24, 1968, during *Apollo 8*, the first circumlunar flight. The iconic image, taken by astronaut William Anders, was an early step toward humanity's cosmic consciousness. Credit: NASA.

person's state of mind. As such critics caution that its widespread use for ideological purposes to support a future in space is overrated and perhaps dangerous. Be that as it may, surely the consciousness of humans in general has been affected by the new knowledge of our place in space in ways writ large and small, including the well-documented effect of the *Apollo* program on the environmental movement. It is no accident that the first Earth Day was held in 1970 in the midst of the *Apollo* flights to the Moon. "Earth Day and the environmentalism of the 1970s owed much to ideas about Earth as a living organism in which humanity and the physical environment had to coexist," wrote space policy analyst W. Henry Lambright, citing the *Apollo* images of Earth.[30]

More general and perhaps more esoteric than those direct images, but no less powerful, is a series of related ideas that gradually entered human consciousness and culture during the course of the twentieth century: that we are part of a cosmos billions of years old and billions of light-years in extent; that all parts of this cosmos are interconnected and evolving; and that the stories of our galaxy, our solar system, our planet, and ourselves are part and parcel of the ultimate master narrative of the universe, a story we now collectively term "cosmic evolution." Even as in some quarters of popular culture, heated debate continues over Darwinian evolution 150 years after the idea was published, over the past 50 years the much more encompassing idea that Carl Sagan embodied in the phrase "the cosmic connection" has become more and more a part of our daily lives, and will even more in the future as our cosmic consciousness increases. The cosmic connection of which Sagan spoke in 1973 embraced not only cosmic evolution, life beyond Earth, and the exploration of space in 100 different forms, it also emphasized the most intimate connection of all: "All of the rocky and metallic material we stand on, the iron in our blood, the calcium in our teeth, the carbon in our genes were produced billions of years ago in the interior of a red giant star. We are made of star-stuff." The gospel of humans as star-stuff was spread far and wide by his *Cosmos* television series in 1980, absorbed in 1,000 ways into popular culture, and reinforced in the twenty-first century by Neil DeGrasse Tyson's *Cosmos II*. This gospel not only has the indisputable value of being true, but also, in Ann Druyan's words, imparting "a new sense of the sacred." It is part and parcel of the epic of cosmic evolution, an epic that has been developing over the past century and more, as described in Nasser Zakariya's *A Final Story* mentioned at the beginning of this chapter.[31]

The cosmic connection has only increased over the past several decades as we have viewed spectacular imagery from the Hubble Space Telescope and other observatories, as well as absorbed the results of other spacecraft that have revealed our place in space and time (Figure 7.7). Not without reason has Hubble has become known as "the people's telescope," to the extent that the American people reacted strongly when the last Hubble-servicing mission was canceled, only to be restored after both popular and congressional protest. Nor was the uproar over the demotion

Figure 7.7 The Hubble Space Telescope has provided thousands of dramatic and now iconic images that have increased our sense of cosmic connection. Seen here are the "pillars of creation," located in the Eagle Nebula some 7,000 light-years from Earth. The gas and dust here are in the process of creating new stars, which we now know will eventually result in planets as a normal by-product of star formation. Courtesy of NASA, ESA, and the Hubble Heritage Team (STScI/AURA).

of Pluto as a planet devoid of cultural meaning. Signs were posted, protests were staged, songs were written, and schoolkids were depressed at the thought of losing Pluto's planethood. In a larger sense, the scholarly and popular movement known as "Big History" now views human history not as a series of episodes over the past few thousand years, but as the latest adventure in 13.8 billion years of cosmic evolution. Such a view allows us to appreciate the emergent properties of culture in the same way as the emergent properties along the earlier path of cosmic evolution. In short, an appreciation of cosmic evolution reintegrates humans with the long history of the cosmos whence they sprang, giving people increasing reason to care about the universe of which they are a part.[32]

Captured by Aliens

It is within this grand context of cosmic evolution that the search for life in the universe finds its place, becoming in the process a dominant part of astroculture. While some UFO enthusiasts (and SETI/METI opponents) worry about hostile aliens invading the Earth, others emphasize that our minds have already been invaded by aliens. Historians of the extraterrestrial life debate have documented how the idea of extraterrestrials has become increasingly prevalent in Western culture over the past four centuries, until today it plays a major part in both science and popular culture. In the felicitous phrase of Joel Achenbach's book, we have become "captured by aliens," in the sense that aliens have become ubiquitous in cultural manifestations ranging from UFOs, alien abductions, and science fiction, to NASA's astrobiology program and policy concerns in the US Congress. As anthropologist Debbora Battaglia argues in her study of communities centered on a belief in aliens, some of them even represent "extraterrestrial (E.T.) culture as lived experience." The Heaven's Gate cult, the Raelian religion, and the most ardent of UFO enthusiasts and abductees are examples. These experiences, she argues, also reflect the culture in which they are embedded.[33]

From the viewpoint of this chapter, astrocultural phenomena such as science fiction and UFOs may be seen as attempts to work out the worldview of the biological universe in popular culture. It is no coincidence that H. G. Wells's invaders in *War of the Worlds* (1897) came from the Mars of Percival Lowell, whose Martian canal controversy had just reached England. Similarly the science fiction novel of German historian and philosopher Kurd Lasswitz features a Mars straight out of Lowell, crisscrossed with canals. His *Auf Zwei Planeten* [*On Two Planets*] (1897) was a bestseller throughout Europe, and a young Wernher von Braun "devoured this novel with curiosity and excitement" on his way to a brilliant career in rocketry, both engineering war and conquering space. Once out of the bag, the alien was off to a successful career, which has only become more pervasive with every discovery of astrobiology. Strikingly, the first wave of sightings in what we now call the UFO phenomenon also came in 1896–1897, along with the idea that they might have come from space. It is more difficult, however, to trace this directly to Lowell, who himself never offered an extraterrestrial hypothesis for the phenomenon. But again, ideas about life beyond Earth and the opening of the Space Age saw a peak in UFO reports, and the idea that aliens might be piloting spaceships in the skies of Earth has not abated, an apparently enduring part of astroculture.

The discovery of life in any form will also impact broader aspects of culture. Given the impact that even the possibility of alien life has had on science fiction, the effect on literature in general will likely be enormous if extraterrestrial life is discovered. In the same way as literature was affected in the wake of the

Copernican theory with John Donne and other writers, surely the biological universe in any of its forms will do the same in ever more expressive form. The discipline of history will be affected, as our attempts at "universal history" in the tradition of Spengler, Toynbee, and Fukuyama will become only subsets of a much larger and truly universal history. Beyond theological-philosophical worldviews, philosophy will also change, as venerable questions such as the nature of objective knowledge are finally answered, or at least advanced or expanded. In short, the transformation of culture to global astroculture will witness the slow dismantling of one worldview and the reconstruction of a new one. In the process, new values will also emerge, giving rise to a whole field of study known as astroethics.

The Gutenberg Effect and the Internet Revolution as Analogy

It is difficult to envision exactly how the implications of a biological universe will unfold once it is verified. But in closing this chapter I once again deploy several analogies, among many that could be studied in a more comprehensive research program. I have already indicated how changes in cosmological worldview might model the impact of fully embracing the biological universe. More down-to-Earth analogies are also illuminating. For example, Gutenberg's invention of the printing press has widely been seen as heavily influencing, if not triggering, the Renaissance, the Reformation, and the rise of modern science (Figure 7.8). Scribes went the way of the dodo, as the slide rule and typewriters would with the coming of the computer 500 years later. In their place came fifteenth-century typesetters, proofreaders, paper suppliers, and salesmen. Ideas spread like wildfire, most notably religious ideas, but also astrology, alchemy, and a variety of other esoterica. The pace of cultural evolution in the West was vastly accelerated. As one scholar put it, "Gutenberg's invention completely and radically restructured the trade in intellectual products and thereby, even more radically, enlisted a vastly increased number of minds recruited to the challenge of formulating more and better solutions to the ever-constant problems arising in human society. This new army of minds forever altered the tempo and directional evolution of the culture of the West."[34] Note the "restructuring" language, used again and again in connection with what we might call "the Gutenberg effect:" not just assorted and disconnected elements, but "the entire infrastructure of cultural debate … was massively reconstructed."

In Ladrièrean terms, in the decades and centuries after Gutenberg the old world order was gradually dismantled, and the new one rose with every passing year. Canadian philosopher of communication theory Marshall McLuhan has described the resulting sea change in culture and human consciousness in his book *The Gutenberg Galaxy: The Making of Typographic Man*, wherein he analyzes

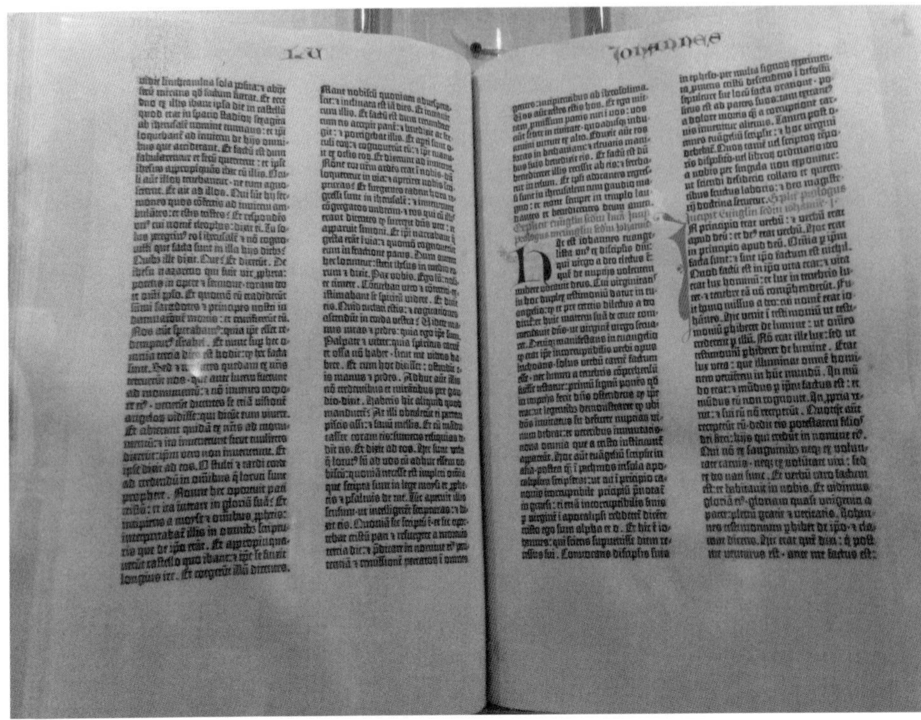

Figure 7.8 The Gutenberg Bible, the first great book in the Western world printed from movable metal type, was completed in Mainz, Germany, in 1455. This Library of Congress copy is displayed near the "Giant Bible of Mainz," created by a scribe and completed in 1453. The two Bibles thus span two cultural eras. Gutenberg's printing press set off a series of events that represent a turning point in the transition from the Middle Ages to the modern world. The discovery of extraterrestrial life may precipitate a similar transition. Photo by author.

the effect of mass media on human culture and consciousness. The "Gutenberg Galaxy" referred to the accumulated knowledge of humanity. In astrobiology we are potentially referring to the accumulated knowledge of multiple extraterrestrial civilizations, and in the long run the accumulated knowledge of the Galaxy and billions of galaxies. Perhaps in some way, at least in terms of knowledge, the unexpected impact of the printing press will foreshadow the kind of change we may see with the discovery of an intelligent biological universe.[35]

Nor need we limit ourselves to the aftermath of printing in the fifteenth century. If printing inaugurated the biggest cultural change since the invention of the alphabet and writing, the computer age, also dubbed the Information Age, represents the biggest and broadest change since Gutenberg's revolution.[36] We are living through it now, but when we ask how this revolution is changing society, we begin to see the difficulties of gauging the societal impact of any large-scale change. Right before our eyes, the internet and the World Wide Web are changing culture in ways large and small never anticipated. More and more people are able to access more and more information, affecting social, political, and cultural life. In attempting to assess the effects, scholars and commentators have also turned to the Gutenberg analogy as one way of illuminating impact. A detailed Rand study reveals its methodology in its title "The Information Age and the Printing Press: Looking Backward to See Ahead," based in part on Elizabeth Eisenstein's pioneering book *The Printing Press as an Agent of Change*.[37] The Rand study points out that both the Gutenberg and internet eras represent communication breakthroughs enabling unprecedented new ways of generating and disseminating knowledge, ranging from social interaction to classroom instruction. And, just as with the printing press, the Information Age is having unintended consequences unfolding every day, including fraud and security concerns on the negative side, and the joys and frustrations of social media. By some accounts, it is changing our very brains.[38]

It might seem presumptuous to claim that all of humanity will be affected, or affected in the same way, by the discovery of a biological universe. Yet just as we speak of the effect of science and technology – or parts of it such as medicine, computers, and the internet – on culture, so we may speak about the biological universe, astrotheology, and astroculture as worldviews that will pervade everything humans do. This is not an elitist statement. It is unfortunately true the world is divided into the haves and have-nots, and that they may be affected in different ways by contact with alien life. The alien component of astroculture may be weak or dominating depending on the discovery scenario, and the astroculture component of any culture may be weak or dominating depending on its circumstances. But it seems to me that astroculture and its elements will be not only persistent in human thought, but most likely growing at an accelerating rate in future years. In the wake of the discovery of life beyond Earth, it will surely accelerate even more.

8

Astroethics

Interacting with Alien Life

The arc of the moral universe is long, but it bends towards justice.

Martin Luther King[1]

Ethics is nothing other than Reverence for Life. Reverence for Life affords me my fundamental principle of morality, namely, that good consists in maintaining, assisting and enhancing life, and to destroy, to harm or to hinder life is evil.

Albert Schweitzer[2]

Nothing within these Articles of Federation shall authorize the United Federation of Planets to intervene in matters which are essentially the domestic jurisdiction of any planetary social system.

Star Trek, The Prime Directive

The Golden Rule may be as universally true as is the theory of relativity.

Holmes Rolston III[3]

A biology-seeking spacecraft lands on Mars, and after a series of experiments and an extended period of analysis finds strong indications of microbial life. A long-sought signal of intelligent origin is received on Earth from a planet 50 light-years away, and the signal appears to bear an information-rich message. An unmistakable transmitting active artifact is found in the outer solar system, left by a civilization long gone. An alien spacecraft lands on the White House lawn and friendly aliens emerge promising world peace. Each of these scenarios is, thus far, the stuff of science fiction, as we have seen in earlier chapters. But each of them could happen, some with more likelihood than others. And each raises all sorts of ethical questions. Does Mars belong to the Martians, even if the Martians are only microbes? What do we say in response to an alien message, and who speaks for Earth? How do we treat an alien in a "close encounter of the third kind"? In short, whether we discover alien microbes or advanced alien life, we will immediately be faced with the problem of how to interact. Welcome to the world of astroethics – the contemplation and development of ethical standards for a variety of outer space issues, including terraforming the planets, resource utilization, near-Earth asteroid threats, space exploration, planetary protection – and the discovery of extraterrestrial life.

Before we can act in any situation that involves life, it is first important to assess the moral status of the organisms involved. This is no easy task, since we are ambiguous about relations with animals on Earth, on the one hand sheltering them as beloved pets, on the other rather arbitrarily hunting, eating, and exterminating them. But over the past few centuries, a good deal of thought has been given to the subject of moral status of Earth organisms and the idea of intrinsic value on which it is often based. Contemplating encounters with alien life tremendously expands our ethical horizons. In the case of intelligent aliens, it also encompasses not just the problem of how we might treat them, but also how aliens might act or react. In other words, it is not just a question of our ethics. What about their ethics? Is there any basis for inferring whether alien intelligence might be good or bad? Might there be such a thing as a universal ethics, perhaps in the form of the Golden Rule as Holmes Rolston suggests, or simply a reverence for life as Albert Schweitzer famously taught? Or is *Star Trek*'s "Prime Directive" of nonintervention a naïve one-way street, a recipe for our own extinction? In short, to expand one of the best-known statements of Martin Luther King to an extraterrestrial context, does the arc of the moral universe indeed bend toward justice? Or perhaps more precisely, does the moral arc of the universe bend toward justice? This is the domain of altruism, which surprisingly has also been the subject of considerable thought in an extraterrestrial context, though with no decisive conclusions.

The answers to these questions will inform our actions in real-world contacts with alien life under different scenarios. I argue that by contemplating these issues, and certainly by putting them into practice in the event of the discovery of life beyond Earth, we will not only address an important "wild card" problem in our near or far future but also transform our thinking by moving from an anthropocentric ethic toward a "cosmocentric" ethic, one that establishes the universe and all or part of its life as a priority in a value system rather than just humans or even terrestrial life in general. This is compatible with certain brands of astrotheology, including the cosmotheology I proposed in the previous chapter, which includes as its fifth principle a moral dimension, extended to include all species in the universe – a reverence and respect for life that we find difficult enough to foster on Earth. For both our Earthly and celestial concerns, a cosmocentric ethic will surely be a shift in worldview as radical as any of those I discussed in the previous chapter.

The Moral Status of Alien Organisms

What exactly is ethics and what are its sources? Jesuit theologian and astronomer William Stoeger argues that ethics is "about recognizing and respecting relationships. It is about how intelligent, freely choosing agents and societies can live and act in harmony with the lower- and higher-level networks, ecosystems, organisms, and communities upon which they depend and in which they exist and function."[4] Over most of the course of human history, this formulation has applied to relations among lifeforms on Earth. But ethical horizons have expanded over the past century even beyond lifeforms. In the 1940s, ecologist Aldo Leopold in his *Sand County Almanac* formulated a "land ethic," so influential in the environmental movement that it is now usually called environmental ethics: "All ethics so far evolved rest upon a single premise: that the individual is a member of a community of interdependent parts … The land ethic simply enlarges the boundaries of the community to include soils, waters, plants and animals, or collectively: the land." In other words, Schweitzer's reverence for life would not be inclusive enough for those concerned about environmental ethics. It is more in tune with the "deep ecology" philosophy formulated in the 1970s and 1980s by Arne Naess and others, which states that all nonhuman life has intrinsic value independent of its value to humans. With environmental ethics and deep ecology as implicit or explicit background, astroethics "enlarges this perspective beyond our planet to the extraterrestrial systems and communities on which we depend and of which we are a part."[5]

Where Does Ethics Come From?

Although both environmental ethics and astroethics encompass more than just life, here we focus on alien life. At the basis of our actions toward alien life, we must first wrestle with the problem of its moral status, often seen as the basic unit for ethical analysis. And in order to do this we must focus on the question of the foundation for ethics, or "ethical value." The most common answer to this question among the general public would undoubtedly be that the foundation for ethics is religion. In Judaism the Torah, in Christianity the Bible, and in Islam the Koran promulgate ethical principles, often converging on the idea of compassion. But that is only one point of view. Sam Harris, in his book *The Moral Landscape*, espouses a secular ethics expressed in his subtitle *How Science Can Determine Human Values*. He argues that values can be defined as the well-being of conscious creatures, which in turn is based on the facts of world events and brain states; in other words, values are the subject of science in its broadest sense as verifiable knowledge. There is no Christian or Muslim morality, he contends; rather morality is an undeveloped science, one that he begins to develop in his book. In doing so, Harris explicitly denies the so-called naturalistic fallacy, the idea (dating back to our philosopher friend David Hume mentioned in Chapter 5) that we cannot infer from what *is* to what *ought to be*.[6]

As he is wont to do, Harris is bucking popular opinion here. But this is not only the opinion of an author who has notoriously written against religion. Despite his Jesuit background, Stoeger emphasizes that "there are strong philosophical foundations for ethical principles and norms that do not rely on religious belief … much of ethical reasoning and development relies on philosophical, not theological, considerations," even if they are bolstered by theological considerations. Although many schools of thought inform ethics, he adopts two principles in a seminal article he wrote on astrobiology and ethics shortly before his death in 2014: first, that ethics "relies in large part on the way reality is organized," and therefore must be informed by our scientific knowledge of the world around us; and second, that while ethics is not reduced to science, it is heavily informed by it, a principle he views as a modified naturalistic position. Although he does not go as far as Harris, Stoeger explicitly rejects the idea that there is an insurmountable barrier between science and values.[7]

In fact, there is some consensus among philosophers and other scholars who have addressed astrobiological concerns that the naturalistic fallacy is not an impermeable barrier, is in itself fallacious, and that secular ethics derived from the universe around us is the optimal route to determining moral status. Thus young Belgian philosopher Clément Vidal argues, "we can and must use insights from science to build ethical theories," since social or supernatural alternatives are less objective and cannot naturally be extended to cosmology, which he sees as an important goal. He then develops a whole series of values such as maximizing fitness by evolutionary trade-offs, including egoism-altruism, stability-adaptability, specialist-generalist, exploration-exploitation, and competition-cooperation. In the end Vidal proposes an ethic that values "the infinite continuation of the evolutionary process" as an "ultimate good." Of course, one can agree or disagree with this goal, but as we see throughout this chapter, the idea that science informs values is increasingly accepted among both secular and religious ethicists. It is implicit, for example in the idea of cosmological theories of value as pioneered by NASA engineer and philosopher Mark Lupisella. And it is explicit even in many Christian authors, including philosopher/theologian Nancey Murphy and cosmologist George F. R. Ellis. In their book *On the Moral Nature of the Universe: Theology, Cosmology, and Ethics*, Murphy and Ellis argue that "morality has an 'objective' basis in the nature of reality and that the discipline of ethics can be construed as the scientific study of this moral order." Their theory of ultimate reality, however, is deeply teleological and theological, a theory "about the purpose for which God created the universe." This leads them to a controversial pacifist thesis that involves a renunciation of all force in international politics.[8]

From a historical point of view, scholars have pointed out that the idea of the naturalistic fallacy has become too broad since Hume first broached it in the eighteenth century, that in reality we simply cannot separate humans from nature. With Stoeger, we are not necessarily saying that ethics can be fully reduced to nature (that would be Harris's version of the naturalistic fallacy, and if misused might lead to programs like social Darwinism), but that it can be strongly informed by nature. This has important implications for astrobiology, since other lifeforms might adopt ethics based on the same scientific facts, if indeed they see the world in the same way.[9]

Finally, in the same pioneering and thought-provoking collection of essays in which Stoeger's views appeared, several authors from a variety of fields implicitly or explicitly adopt a secular ethics approach or a mix of the theological and secular. In systematically examining an ethics for astrobiology, philosopher Carol Cleland and political scientist Elspeth Wilson argue that because of the pluralist nature of religions, secular approaches have advantages over theological ones with regard to theorizing about the moral status of such organisms. In other words, because it is highly unlikely that aliens will have any of the religions of Earth (if they have any at all), a secular or nature-based ethics is more likely to be shared with us. Theologian Ted Peters, in an article on "Astroethics," argues that at least for aliens who are our peers, we should invoke the Golden Rule and the related concept of dignity, both of which have secular and theological roots. In other words, Kant's eighteenth-century Enlightenment categorical imperative, from which, in his view, all morals flow, resonates with the gospel of Matthew: "In everything do to others as you would have them do to you." You don't have to be a Christian to accept that. It seems to me, then, that we can draw on both theological and secular perspectives when it comes to astroethics, realizing that the latter may well be more universal. Ethicist Robin Lovin, for example, has emphasized that the concept of dignity has a long history of discussion in the Christian tradition, and that this tradition should be useful in ethical issues arising with extraterrestrial life. We should recognize, however, that since dignity is often associated with rational capacity, defining that concept may well be problematic to say the least in the context of alien life. In short, it seems that while theology can provide potentially universal principles such as compassion and dignity that will be useful in the context of astroethics, the problematic naturalistic fallacy should not stand in the way of secular ethics playing an important and perhaps predominant role.[10]

What Do We Value?

Whether secular or theological, the most important question about the foundations for ethical value turns on the distinction between intrinsic value – value independent of a valuing agent – and instrumental value – value in relation to something else like the needs of humans. A human will always have intrinsic value by our standards, but a tool like a wrench can have only instrumental value, and even then only under those circumstances when it is usable. Humans may also have instrumental value, but that is secondary to their intrinsic value, which is to say inherent worth. The real heart of the question for us is why do humans have intrinsic value as opposed to other life? What else might have intrinsic value, and why? Philosopher J. Baird Callicott – a leading proponent of Aldo Leopold's "Sand County Almanac" ethic, has emphasized that the idea of intrinsic value in environmental ethics turns on a non-anthropocentric attitude, which is in turn related to intrinsic value. This, it seems to me, is the general attitude required for astroethics dealing with alien life. Callicott points out that environmental ethics as espoused by Leopold does not completely apply since it is based on things that have "evolutionary kinship and ecological community." All living things on Earth have this kinship, but not all extraterrestrial life does, as far as we know. Nor, he says, is the idea of "sentiency" (the capacity to experience pleasure and pain) broad enough. Instead he argues that Albert Schweitzer's reverence for life can provide the common basis for the moral consideration of life. To put it another way, Callicott proposes that a "life principle/reverence for life ethics" be the standard for intrinsic value or what he calls "moral considerability," that is, those things that have "conscious wishes, desires and hopes; or urges and impulses; or unconscious drives, aims and goals, or latent tendencies, directions of growth and natural fulfillments." This is what we might call a "biocentric" ethic.[11]

Another point of view comes from philosopher Kelly Smith, who points out that environmental ethics as an attempt to extend intrinsic value from humans to the environment is problematic. He argues that the ability to reason (ratiocentrism) is a better approach than trying to extend intrinsic value to the environment, as it has been in some astrobiological issues such as terraforming Mars. He later broadened his criteria for intrinsic value to include not only rationality but also sociality and culture, which may coevolve with rationality. He is at pains to emphasize that ratiocentrism, or its sociality-rationality-culture triad, is a broader category than both intelligence and anthropocentrism, including some animals that have awareness of self and their environment, symbolic thinking, and an idea of rules of conduct.[12] Ratiocentricity or its triad are considerably more restrictive than environmental ethics in terms of moral status. But Smith's more important point is that intrinsic value may be too blunt an instrument to be applied when it comes to moral status and practical ethical situations. After all, in a more hierarchical moral ranking we may accord some ethical status to our pets, but if we are in a zero-sum game or a life-or-death situation, the pet will be the first to go. Mark Lupisella goes in the opposite direction, arguing for cosmocentricity, which in one of its forms argues that cosmological intrinsic value could be based on the universe realizing its nature of relatedness and connectedness, compatible with Vidal's emphasis on the cosmic evolutionary process. In that case, the nature of the universe grounds intrinsic value.[13]

Thus we have an entire spectrum of frameworks or theories for moral status, an ethical landscape summarized in Table 8.1, an extension and modification of a scheme first presented with broad brushstrokes by Swedish philosopher Erik Persson.[14] He distinguishes four ethical theories of moral status, in order of increasing inclusivity: *anthropocentrism*, in which only humans have moral status; *sentientism*, in which all and only sentient beings have moral status; *biocentrism* in which all and only living beings have moral status; and *ecocentrism*, in which all living beings, species, and ecosystems, and perhaps even nonliving matter, have moral status. In the expanded scheme of Table 8.1, many religious people would fall in the anthropocentrist camp if moral status is expressed in terms of a soul, and so would Kant if expressed in secular terms; Smith would be a ratiocentrist; Schweitzer and his many followers would be biocentrists; Callicott would teeter somewhere between biocentrism and ecocentrism; and Leopold, Naess, and Rolston would be ecocentrists. Persson himself proposes that sentientism may be the most plausible theory to apply to extraterrestrial life.

The difficulty with these categories is in the definition of their distinguishing terms, which imply bright lines, neat boundaries that evolution arguably does not create. In determining the nature of sentience, for example, Persson describes it as "a subjective perspective from which things can be good or bad," one measure of which is feeling pleasure or pain. Although the definition of sentience is open to question even on Earth, the practical implication is that "extraterrestrial life will have, or not have, moral status on exactly the same terms as Earth life," so that "we will be able to use the same rules and theories for moral conduct toward extraterrestrial life as we do toward Earth life." This, of course, does not bode well for octopuses (unless we grant them sentience) or bees, much less for Martian bacteria, although we should emphasize that just because things are not granted moral status does not mean they cannot be protected under special circumstances, including our ecological entanglement.[15] Moreover, moral status should not be confused with scientific importance. The discovery of a blade of grass on Mars would be the greatest discovery in the history of science, one well worth protecting. But it does not necessarily follow that the blade of grass should be found to have intrinsic value and granted moral status.

Table 8.1 *Theories of Moral Status**

Theory	Explanation	Representative Proponents	Implications for Life Beyond Earth
Anthropocentrism	Only humans have moral status	Most religions Zubrin and Wagner (1996)	Alien life has only instrumental value; protection possible
Ratiocentric Social-Reason-Cultural Triad	All organisms that have reason, or the social-reason-culture triad have moral status	Smith (2009 and 2014)	Alien microbes and some complex alien life have only instrumental status
Sentientism	All and only sentient beings have moral status	Persson (2012) Peters (2013b)	All and only sentient aliens have moral status
Biocentrism	All and only living things have moral status	Schweitzer (1960) Callicott (1986b) McKay (1990) Sagan (1980)	All living alien life has moral status
Ecocentrism (Environmental Ethics; Deep Ecology)	All living beings, ecosystems, and perhaps nonliving Nature have moral status	Leopold (1949) Naess (1973) Devall and Sessions (1985) Rolston III (2014)	Even nonliving Nature has moral status
Planetocentrism	All planets have intrinsic value, especially with life	Sullivan (2013a)	All planets have intrinsic value
Cosmocentrism	The entire cosmos and its constituent parts have moral value		
Basic	Cosmic consciousness	Vidal (2014) Dick (2009) Hart (2013)	Some physical or metaphysical aspect of the universe has priority in a value system, and provides a justification for intrinsic value
Enhanced	Derived from alien contact		
Strong	Value from physical or metaphysical aspect of the cosmos	Haynes (1990) Lupisella and Logsdon (1997) Lupisella (2016)	
Anthropic	Intimate connection between cosmos and life		

*Note: Modified from Erik Persson (2012), by permission.

What Is the Moral Status of Aliens?

Although clearly these ideas about intrinsic value greatly affect how we treat alien organisms, the more general question is how to choose among them to determine the moral status of alien organisms. Agreeing that our ethics are too anthropocentric, Cleland and Wilson argue that a clarifying exercise in the absence of alien life is to look more closely at the moral status of nonhuman animals. They point out what animal rights activists have emphasized for a long time, that many animals "display characteristics associated in humans with moral status." In particular the great apes, including chimpanzees, gorillas, and orangutans, are highly social, can solve problems, display some of the characteristics of culture, display a range of emotions, and may even have self-awareness. A ratiocentric approach might grant them moral status, though not as much as humans. But they argue further that elephants are highly social and even rodents show some behavior suggestive of moral status; that might be logical if one takes a sentientist approach. Dolphins and whales have sophisticated cognitive abilities, and so might even qualify in the ratiocentric framework. The fact that many humans are reluctant to grant moral status, much less ethical significance, to these animals, Cleland and Wilson point out, does not bode well for humans recognizing moral status in alien life.[16]

Cleland and Wilson also point to the utility of science fiction in illuminating these issues. Hoyle's *Black Cloud* is granted moral status because of its cognitive ability, and also the Horta on *Star Trek* because of the behavior of a mother defending its children. One might argue this still displays anthropocentric bias about intrinsic value because it is based on what *we* value, but it is hard to imagine a mother not defending her children under any normal circumstances, barring complete neglect. Perhaps better examples are real-life animals found on Earth, but even more alien than the ones discussed earlier. Cleland and Wilson point to cephalopods (octopuses and cuttlefish) on the one hand, and Hymenoptera (bees, ants, wasps) on the other, as cases that expand our moral categories and ethical reasoning. While octopuses are asocial compared to humans and do not exhibit much emotive behavior, they are "intelligent," or at least sentient in terms of our moral frameworks, in the sense of exhibiting problem-solving behavior, curiosity, and sometimes even deception. The heptapods in *Arrival* were clearly granted moral status the moment communication with them began, their somewhat repulsive morphology notwithstanding. Bees display collective behavior resulting in complex hive societies whose individuals show division of labor and that depend on mass cooperation. The question of whether we are willing to grant these lifeforms moral status directly relates to our ideas of the moral status of the diverse and probably weird lifeforms we may find in outer space.[17]

Since we have not decided these questions for Earth, much remains to be done. As Cleland and Wilson conclude, although the moral status even of terrestrial animals at various levels remains an open question, they challenge our concept of moral status in ways that will be relevant when we discover lifeforms beyond Earth. In a broader sense, our view of the moral status of alien life depends on what we choose as the source of our ethics, whether secular or religious, on our criteria for intrinsic value, and on our theories of moral standing. In the last section of this chapter I return to how these choices affect the ethical significance of moral status for specific cases, a status that guides us on what we owe alien life and how we treat it. In doing so I will explain the planetocentric category of Table 8.1, and flesh out the several possible forms of cosmocentric ethics. But first we need to look at the problem from the point of view of alien life.

On Good and Evil: Altruism in the Universe

Astroethics is a two-way street when it comes to alien intelligence, where our options at once both narrow and expand. If we are dealing solely with intelligence, we no longer have to worry about the rights of microbes and lower forms of life (unless that intelligence has its own microbiome, which is entirely possible!). Instead, we now have to pose questions such as this: How will aliens treat us? Is there any basis for inferring whether alien intelligence might be good or bad? What do we know about good and evil on Earth, and specifically the evolution of altruism on Earth? Is it possible there is a universal theory of ethical value when it comes to intelligence? If so, what should be the basis for such a theory? As with many other issues I have dealt with in this volume, these are all questions of the philosophy of ethics, or metaethics – not surprising when we are dealing with the foundational issues that astrobiology inevitably raises. And like those other issues, this one must remain speculative, since we have no consensus on theories of ethics. Our discussion is, however, grounded in reality, admittedly *our own* reality, plus any universal principles we might infer.

A Spectrum of Possibilities

Extraterrestrials in popular culture tend to be advanced (*2001: A Space Odyssey*), combative (*Star Wars*), brutish (*Aliens*), and occasionally friendly (*E.T.*). In short, they have been envisioned along two main prototypes: good and evil. This is undoubtedly a reflection of what we find on Earth, yet a moment's thought reminds us that humanity itself is a disturbing mix of good and evil, often found in the same individual. So in outer space, is it to be friendly E.T. who seeks only to "go home," the drooling and deadly creature in Ridley Scott's *Alien* series, or something in between? The powerful yet compassionate aliens in *The Day the Earth Stood Still* or the destructive monsters of *War of the Worlds*? Both views have a long pedigree in science fiction literature and film, which has ingrained these extremes in the popular mind. But it is important to remind ourselves that these Hollywood prototypes may bear no relation to reality.

At the same time, these prototypes do lay out the starkest possibilities at either end of the spectrum of behaviors that is likely to stretch between them. Astronomer and SETI scientist Seth Shostak has memorably captured these polar alternatives in a two-page short story, "The Second Signal," where a long awaited signal from extraterrestrial intelligence arrives on Earth to much fanfare, and an enthusiastic reply is sent. Later, a second signal is received, from a second civilization, warning Earthlings not to reply to the first message. It turns out the second civilization is dead, having been destroyed by the first civilization. They have set up a beacon to warn others, a purely altruistic motive from which they will gain nothing. So which is it in the universe, altruism or destruction? If terrestrial history is any guide – and wishful thinking about the "progress" of old and "advanced" civilizations aside – probably both in equal parts. If so we have an interesting time ahead of us.[18]

How can we productively approach such a seemingly intractable problem as the balance of good and evil in the universe? Science fiction and theology aside for the moment, there has been no lack of opinion on the subject among philosophers, as well as natural and social scientists. One trigger for discussion was a 2010 statement in which renowned astrophysicist Stephen Hawking famously pronounced, "If aliens ever visit us, I think the outcome would be much as when Christopher Columbus first landed in America, which didn't turn out very well for the Native Americans." We have seen that this is an analogy to be viewed with suspicion, since historians find that particular situation much more nuanced than many think and it represents only one example of culture contact among thousands in Earth history. In any case, many scientists consider physical contact unlikely due to the difficulties of interstellar travel.

Nevertheless, Hawking's statement, taken seriously more because of his celebrity than for any expertise on history or the nature of aliens, went viral in the media and elicited a number of considered responses. Robert Wright, most famous for his book *Non-Zero: The Logic of Human Destiny*, invoked philosopher Peter Singer's 1981 book *The Expanding Circle*, where Singer noted the striking moral progress humans have made since the days of ancient civilization, acting on evolutionary impulses and nurtured by reflection, reason, and intellectual momentum. This view was later supported in Steven Pinker's controversial book *The Better Angels of Our Nature: Why Violence Has Declined*, which analyzes the historical forces that have caused human violence to decline over both the long and short terms. This seems counterintuitive given today's incessant headlines about terrorism, war, and refugees, but its statistical basis seems solid and is supported by many others, even if the conclusion has its critics. In addition, Wright argued, moral progress has been driven by pragmatic self-interest. If we were in an optimistic mood, we might project this claimed trend of human moral progress to alien intelligence. But in and of itself the claim of moral progress is no less suspicious than other claims about linear progress in evolution, and it seems a thin reed on which to hang such an overarching conclusion that the arc of the moral universe bends toward justice.[19]

Evolutionary Ethics

Nevertheless, despite suspicions about linear progress, evolution is the closest thing we have to a universal principle in biology. Might it not shed light on behavior and ethics also? And to the extent that evolution operates universally (in my view, a very good bet or at least a good working hypothesis), might that in turn illuminate the nature of extraterrestrials? Here we need to tread carefully, since social Darwinism – often associated with ruthlessness, force, and violence – immediately raises its head. But we need not throw out the baby with the bathwater. As philosopher Michael Ruse has pointed out, "Social Darwinians may have been wrong in execution but were certainly right in conviction, that evolution – specifically Darwinian evolution – has to be important in the understanding of morality."[20]

The application of evolution to behavior and ethics, and more particularly the field of evolutionary ethics, has its basis in the concept of universal Darwinism. "Darwin's dangerous idea," as philosopher Daniel Dennett puts it, holds that evolution by natural selection applies not only to biology, but also in modified form at many other levels – mind, language, knowledge, and ethics. If one accepts this idea, the challenge is in the details of "Darwinizing culture," and elucidating how genes and culture, including human behavior and ethics, may coevolve. This is the realm of the controversial fields of sociobiology, evolutionary psychology, behavioral ecology, memetics, and gene–culture coevolution, all fields on which evolutionary ethics can draw for insights into the sources and development of ethics. But the bottom line in all cases is that in the view of universal Darwinism, "only an evolutionary analysis of the origins – and transformations – of ethical norms could ever properly make sense of them."[21]

Such evolutionary analyses have been undertaken in many fields of terrestrial cultural evolution, resulting in much controversy, ranging from scholarly treatises to more popular treatments such as Michael Shermer's *The Science of Good and Evil*. Among the most relevant to our discussion is E. O. Wilson's work on sociobiology, which in his words, is "simply the extension of population biology and evolutionary theory to social organization," including human behavior. In his Pulitzer Prize-winning book *On Human Nature*, Wilson argues that evolution has molded characteristics such as aggression, religion, and altruism that make us human. In the end Wilson holds that the cultural evolution of higher ethical values is controlled by genes. In his image, genes hold a very long leash on culture, but in any case, "inevitably values will be constrained in accordance with their effects on the human gene pool." Human behavior is a technique for keeping human genetic material intact. To Wilson, "morality has no other demonstrable ultimate function." One might infer that alien intelligence would be similarly constrained based on its genetic material. To the extent that alien genetic material is different, does this imply a different basis for morality? Or is altruism beneficial to maintain any genetic material?[22]

Those questions cannot be answered at this time, nor is this the place to delve into the details of the many theories about Darwinizing ethics. But it seems obvious that, despite all our differences, and beneath all our diversity, humans (and in some cases animals) have a capacity for working together socially. Some philosophers such as Ruse argue this in an innate capacity that "manifests itself at the physical level as a moral sense," and that morality "is hard-wired into humans. It has been put there by natural selection in order to get us to work together socially or to cooperate." It is this innate capacity generated by natural selection that is at the basis of evolutionary ethics, in the same way it was the basis for evolutionary epistemology, as I discussed in Chapter 5. Both are examples of evolutionary naturalism, cases of "taking Darwin seriously." Although some would admittedly counter that it is taking Darwin too seriously, because of its likely universal character evolutionary naturalism is one of the few ways forward when it comes to analyzing the possibilities of alien behavior.[23]

One of the manifestations of this moral sense is altruism, a seemingly selfless concern for the welfare of others and the opposite of selfishness. It has often been viewed as a puzzle inconsistent with evolution by natural selection, which is seen as acting on individuals for the preservation of individuals. In general, two factors are seen as motivating altruism – kin selection, where altruistic behavior is triggered for close kin with the idea of spreading one's genes, and reciprocal selection, where altruism is triggered with the expectation that helping someone else will result in them helping you later. Both kin altruism and reciprocal altruism are well documented within species on Earth; beyond the kin group (obviously the case with alien intelligence, unless there's something we really don't know) the situation becomes more complicated. Some scientists also argue that altruism occurs at the non-kin group level in order to benefit the group, a mechanism known as group selection. For example, evolutionary biologist David Sloan Wilson has proposed a much-discussed evolutionary theory of morality and religion, in which both are seen as culturally evolved adaptions to make society more cohesive. The mechanism of multilevel group selection is to select for cooperation, cohesive groups, and altruism.[24]

Discounting kin altruism (unlikely in a literal sense but not impossible in a long-term sense if we are somehow related), could reciprocal altruism be applied to contact with extraterrestrials? Could group selection work at the level of the galactic club, making those who belong to such a club more cohesive and likely to survive? Undertaking such evolutionary analyses at the terrestrial level has proven controversial enough, but if universal Darwinism is truly universal, it seems it should apply to alien life no less than terrestrial life. Remarkably, embryonic efforts have already been made in studying these possibilities, most notably in a volume edited by Douglas Vakoch, *Extraterrestrial Altruism: Evolution and Ethics in the Cosmos*. The volume was undertaken with a specific practical question in mind – should the Search for Extraterrestrial Intelligence (SETI) be extended to Messaging Extraterrestrial Intelligence (METI), and would transmitting such directed messages – giving away the position of our home planet and potentially much more – be dangerous? Also in the background was Stephen Hawking's warning about the dangers of contact, and two of the authors in the volume argue that interstellar travel is very difficult and physical contact is therefore unlikely, rendering Hawking's Columbus analogy impotent. Still, even remote contact, whether by SETI or METI, raises issues of the nature of any intelligence and their intentions, whether altruistic or otherwise.[25]

Many of the articles in Vakoch's volume deal with the problem of how to encode altruism if we send a message, or how to determine if an incoming message indicates altruism. But several of the articles address the prospect of universal altruism. One of the leaders of evolutionary psychology, Jerome Barkow, argues that just as gravity constrains body size and shape, so "selection pressures that result in intelligence and cultural capacity constrain the psychologies a species can develop." Adopting convergent evolution as a working assumption, he argues further that the psychology of creatures that can produce high technology implies a social species capable of cooperation, and that for this reason we can hope to cultivate reciprocal altruism. But he urges caution: for example, predation is likely necessary for intelligence, so we should avoid sending information that might provoke a xenophobic response, instead communicating our art, music, and scientific knowledge. Anthropologist Douglas Raybeck agrees that predation is one path to intelligence, and also urges caution. We can begin to see why it is crucial to involve social scientists in any decision about whether and what to send in response to an alien signal – or a signal we initiate ourselves.[26]

In the same volume, animal ecologist Abhik Gupta points out that because aliens are likely totally unrelated to us, "notions of kin selection would not guide their relations with humans." Reciprocal altruism might also be inoperative given the distances involved in remote contact; at the very least any reciprocal altruism would take place over a long period of time. However, Gupta sees hope in our altruistic behavior toward anything that we consider to have value, especially but not exclusively life. The value (hopefully intrinsic and not instrumental) we place on life, which Gupta terms "biophilia" borrowing a term from E. O. Wilson, could provide the basis for cooperation between humans and extraterrestrials. Taking another approach, Mark Lupisella argues that biological altruism on Earth is now supplemented by biocultural altruism, and that a more advanced "cosmocultural altruism" is possible, one that would nurture a broad-based respect for the universe and all its life. Such cosmocultural altruism transcends biological self-interest, and could be cultivated by two mechanisms: first, specific efforts at increasing "caring capacity" by encouraging increased incentives for caring via social organization, cultural constructs and technological advances, and second by highlighting the concept of cosmocultural evolution, which could result in respect for the universe and all its beings. He sees this model as generally applicable to all beings that first evolved via natural selection, with an emphasis on group selection followed by robust cultural and philosophical "evolution," including possibly machine "collectives." In this view, the ultimate in group selection and philosophical evolution, extraterrestrial civilizations that belonged to a kind of galactic or cosmic club (not to be confused with a military alliance) would indeed have a better chance of surviving and thriving and being good cosmic neighbors.[27]

Finally, several of the articles in the Vakoch volume address the possibility of a universal theory of ethics. If we have found reason in past chapters to doubt that science is universal, wouldn't universality be even more problematic in ethics? Philosopher Holmes Rolston III – whom we saw in the previous section in connection with environmental ethics – points out that moral agents preceded science by a long time. Rolston argues that humans have discovered values "that are objective enough to urge on moral agents from whatever extraterrestrial origin … One can plausibly venture the claim that if there are moral agents anywhere, anytime, they have not matured until they have reached the capacity for altruism." Among these values are not lying, cheating, stealing, or hating, or put more positively, having the capacity for love. "The Golden rule may be as universally true as is the theory of relativity," he suggests. In this he follows Kant, who argued that such fundamental moral precepts are "categorical imperatives" that must apply to all moral agents. With their theological and teleological theory of ultimate reality, Murphy and Ellis suggest that everywhere ethics evolves, it would proceed along similar lines.[28] So whether through cosmocultural altruism, similar motives, categorical imperatives, or certain theories of ultimate reality, a universal ethics seems at least possible. It must be said, however, that we are a long way from reaching this goal on Earth.

The bottom line is that it is not unrealistic that altruism could be operative among intelligent species in the universe. But even the optimists urge caution, and there are those who are vociferously pessimistic in this regard, especially when it comes to deliberately sending messages as part of METI programs. Among the leaders of this group are astronomer and science fiction writer David Brin, who presents a contrarian perspective in both his fiction and nonfiction writings. His novel *Existence* – with its vast graveyard of destroyed robot starships tens of millions of years old, which were followed by the more efficient but no less deadly crystal artifacts that effectively function as viruses – presents a cosmos more dangerous than we might think. Warning against extrapolating any terrestrial altruistic trends into a natural law that might apply universally, Brin suggests in his nonfiction writing that we not send messages "until we know a lot more" about what's out there. Like Hawking he raises the culture contact analogy on Earth with its often disastrous outcomes. He also points out the possibility of disease, and of "berserker probes" that destroy life. He argues there may be serious problems even with remote contact in which information is exchanged, in the form of "infectious" information (as in *Existence*), resonating with the infectious memes of Richard Dawkins. If a message were relevant to various religions, wars might be fought over its interpretation, just as they have been over various scriptures on Earth. Thus, all kinds of policy questions are raised about the dissemination of information, and ultimately who speaks for Earth. Rather than evolution, analogy is the tool Brin uses, and he concludes that "quid pro quo" is more likely than pure altruism. Of course, other analogies might lead to the opposite conclusion.[29]

Approaches other than evolution and analogy might also illuminate the nature of the alien. We might ask in an archaeological context "what is it like to be human?" as expressed in the Smithsonian Institution's Hall of Human Origins exhibit at the National Museum of Natural History. We might take a philosophical approach and ask "what is it like to be an extraterrestrial?" in the same way philosopher Thomas Nagel famously asked "what is it like to be a bat?" – a question he found to be problematic and an answer even more elusive. Or "what does it mean to be a person?" as in Roland Puccetti's *Persons: A Study of Possible Moral Agents in the Universe* (1968), which extends the idea of "persons" beyond the human context. Still others might incline toward a theological approach, favoring Thomas Aquinas's argument that the soul, the "first principle of being," embodies what it is to be human. But it is hard to see how to proceed in these directions in a way that makes progress, although the archaeological approach also emphasizes the evolutionary fact that human nature developed gradually over time.[30]

I close this section by returning to the two prototypes in science fiction. Chinese science fiction writer Cixin Liu has argued in his popular trilogy *The Three Body Problem* that the universe operates on two axioms: 1) survival is the primary need of civilization, and 2) civilization continuously expands, but the total matter in the universe remains constant. Since any cosmic civilization cannot know at the outset whether another one is benevolent, this sets up a chain of suspicion in which one must destroy the other no matter what its own morality and social structure. "The universe is a dark forest," he concludes after a METI signal results in invasion of Earth. "Every civilization is an armed hunter stalking through the trees like a ghost, gently pushing aside branches that block the path and trying to tread without sound. Even breathing is done with care. The hunter has to be careful, because everywhere in the forest are stealthy hunters like him," which must be eliminated. Cixin sees this as the picture of cosmic civilizations, the explanation for the Fermi Paradox of why we do not see other civilizations. With such a picture, it is easy to see why some think that METI projects, and possibly even SETI projects, should be halted. The universe is an evil place in this Darwinian survival-of-the-fittest scenario.[31]

On the other hand, many novels view the universe in a more positive light, one in which life is essentially good, even if sometimes at the expense of humanity. Arthur C. Clarke's *2001: A Space Odyssey* series features not battles but renewal, and the title of his *Childhood's End* speaks for itself, even if humanity eventually loses its identity and culture as a new Ladrièrean-type worldview takes shape. (One need not go as far as Clarke in evaporating the Earth!) Carl Sagan's *Contact* features a message, a journey, and the discovery that intelligence is built into the universe itself. Examples could be multiplied of the many science fiction scenarios where contact with aliens is beneficial, uplifting, or transforming.

All of these scenarios are obviously projections of the hopes and fears of particular individuals, times, and cultures. More objectively, in the final analysis evolution seems to be the key, or at least an important key, in discussing the nature of alien behavior. At least it is the best we can do at this point in our own evolution. Although it may seem a cop-out, the evidence such as it is points to the conclusion that extraterrestrial intelligence will exhibit a mix of good and evil, depending on whether aliens have gone through a predatory phase, whether they have developed cosmocultural altruism, and whether they have a biophilic ethic. Longevity is no guarantee of advanced morality, unless aliens have developed a cosmocentric/biophilic ethic that emphasizes the intrinsic value of life. The surprising evolution of altruism on Earth, and its evolving extension to members outside the group, gives reason for hope about the nature of extraterrestrials, and may tip things slightly to the "good" side in our estimation of alien morality, but that conclusion is far from definitive. Should we, then, apply a "Precautionary Principle," not undertaking METI projects, essentially doing nothing because we know nothing about aliens with certainty – a version of "Buridan's ass" in which the donkey starves because it can't decide which bale of hay to eat? Or should we attempt communication and encode altruism in any messages we send to aliens? A prudent answer at this point would seem to be that caution is in order, but not paranoia. There may be a way forward through consultation and consensus. That is a policy question I address further in Chapter 9.[32]

How Should We Treat Alien Life?

Having explored in our first section some of the foundations for ethics and its concepts of moral status and intrinsic value, and having seen some of the difficulties of applying various frameworks of morality even to terrestrial life, how can we apply this knowledge to extraterrestrial life? In short, how should we treat alien life? We have just seen in the second section that in the case of intelligence, it may well depend on alien concepts of ethics and how aliens treat us. That will certainly be a dominant and commonsense factor in our view of the alien. But in all cases, these questions raise practical and profound issues, most notably which of the ethical frameworks in Table 8.1 we should apply to extraterrestrial life. In tackling this problem, let's look at how it has been treated thus far in the four domains in Table 4.1: microbial, complex, intelligent, and superintelligent life, the latter possibly postbiological.

Do Martian Microbes Have Rights?

We should start with microbes, which many consider most likely to be the first discovery of life beyond the Earth. Microbes have been a focus of attention in the context of Mars exploration – an important driver in NASA's space program from its inception. The *Viking* project, which landed two spacecraft on Mars in 1976 with the goal of searching for microbes, raised these issues in a direct and urgent way. NASA's long-standing planetary protection program has one guiding ethical principle: first, do no harm, both in terms of the contamination of Mars and back contamination of Earth. In contemplating exobiology already in 1960, Nobel geneticist Joshua Lederberg wrote that "the human species has a vital stake in the orderly, careful, and well-reasoned extension of the cosmic frontier … The introduction of microbial life to a previously barren planet, or to one occupied by a less well-adapted form of life, could result in the explosive growth of the implant … The overgrowth of terrestrial bacteria on Mars would destroy an inestimably valuable opportunity of understanding our own living nature."[33] The consensus from the beginning of the American space program has been that it simply would not do to contaminate Mars, nor, of course, would it do to "back contaminate" the Earth, either from returning *Apollo* astronauts, lunar sample return, or sample returns from cometary bodies or planetary bodies like Mars.

This is not the place to give a history of NASA's planetary protection program – that has been well done in Michael Meltzer's book *When Biospheres Collide*. Instead I focus on the practical question of what should we do if we actually find microbes. Simply and operationally put, how do we treat microbes? The answer on Earth is that we often "treat" them, pathological ones at least, with antibiotics or extermination. On the other hand we now know the human microbiome is essential to our viability, so we ought to treat some microbes with respect. When it comes to microbes on Mars, the opinions vary widely. In *Cosmos* (1980), Carl Sagan famously wrote "If there is life on Mars, I believe we should do nothing with Mars. Mars then belongs to the Martians, even if they are only microbes."[34] In terms of Table 8.1 this implies a biocentric approach, in which all living things have moral status – but we are left with the inconsistency that here on Earth we destroy microbes pathological to humans. Do Martian microbes have more value than terrestrial microbes? How can we grant them moral status on Mars, but not on Earth? At the opposite extreme, in his book *The Case for Mars: The Plan to Settle the Red Planet and Why we Must* (1996), engineer and Mars exploration enthusiast Robert Zubrin framed the argument as a choice between human exploration versus indigenous life forms, resulting in an anthropocentric approach. He saw back contamination of Earth as one of five "dragons" impeding human exploration of Mars, including radiation hazards, zero gravity, human factors, and dust storms. One by one he attempted to slay these dragons. In the case of back contamination, he argued, not entirely persuasively, that any life on Mars would already have been brought to Earth by Mars rocks, that any such organisms would not be harmful to Earthlings, and that in any case, life "almost certainly" does not exist on the surface of Mars. In his view, contamination issues are unimportant as far as astronauts or Earthlings are concerned. On the other hand, Zubrin did recognize that the discovery of Martian organisms would be a "find beyond price," because it would help us understand the nature of life, and because of its utility to humans for genetic engineering, agriculture, and medicine. In other words, Zubrin seems to come down on the side of Martian organisms as having instrumental value because of their utility to us rather than intrinsic value, and he would presumably argue they should be treated accordingly. And human exploration of Mars, his life-long aspirational goal, should proceed apace.[35]

Is the answer, then, that Martian organisms have scientific but not moral importance? Such a view did not go unanswered for long. In fact, NASA planetary scientist Chris McKay had argued the opposite in the context of terraforming Mars, where he claimed that Martian life has rights, and that any attempt at Martian terraforming needs to consider those rights. In an article a few years later provocatively titled "The Rights of Martians," NASA engineer Mark Lupisella argued that an international body should consider these issues, and suggested it may be necessary to move away from the Western model of exploitation and colonization. Lupisella wrote just as the famous ALH84001 Mars rock controversy was in full force with regard to the possibility of Martian nanofossils. Although that result has proven spurious in the eyes of most astrobiologists, the question of microbes will undoubtedly arise again in our exploration of the solar system, perhaps even in the context of Mars exploration.[36]

McKay has since considerably extended his argument by suggesting that the long-term goal of astrobiology should be to enhance the richness and diversity of life in the universe, a sweeping and bold ethical choice if ever there was one. He summarizes his approach as "Do no harm and provide help as needed." It is the second part of this approach that goes a significant step beyond current planetary protection protocols. In McKay's view, if Mars never had life and doesn't appear suitable for life under its present circumstances, we could ecologically terraform Mars with the goal of making it habitable with Earth lifeforms. If it has life with a common genesis with Earth, he sees no ethical barriers to mingling Mars and Earth life. But if it truly represents a second genesis, McKay recommends focusing on "the resurrection of the Martian life as necessary and the restructuring of the Martian environment so as to allow that life (and only that life) to expand to produce a global biosphere." McKay also advocates only biologically reversible exploration. He points out that because of the revision of NASA protocols in the wake of *Viking* showing harsh surface conditions, spacecraft are no longer sterilized. Hundreds of thousands of terrestrial microorganisms have thus likely been carried to Mars aboard landing spacecraft. Any exterior microbes exposed to sunlight would be killed, but any interior microbes would need to be killed by dismantling and removing all landing craft – thus biologically reversible.[37] This is one set of ethical choices with values to guide future exploration. It seems to be a limited form of biocentrism, since it kills microbial organisms from one planet in order to save those on another.

In contrast to McKay, astronomer Woodruff T. Sullivan III argues we need to preserve extraterrestrial life and keep it untrammeled by extending and adapting a rigorous environmental ethic, a deep ecology view based on Holmes Rolston. He further argues we need a "planetocentric ethic," where planets themselves are held to have intrinsic value, "especially if they might potentially harbor life." This means a hands-off attitude similar to designated wilderness areas on Earth, in other words, the antithesis of any terraforming. His guiding principle is "cause neither physical nor biological harm to any planetary body and its ecosystems." Sullivan suggests a Planetocentric Solar System policy modeled on the Antarctic Treaty of 1961, in particular its Protocol on Environmental Protection added in 1998. In this way, we might avoid some of the exploitation problems of European contact with the Americas during the Age of Discovery, even where life is not involved.[38]

These differing views offer stark ethical choices for practical problems involving microbes. They are choices that will not go away. As of 2017 Zubrin was still pushing his view of exploration over planetary protection at the International Development and Space Conference, sponsored by the National Space Society and dedicated to the settlement of space. He argued that the planetary protection community was overzealous in many of its demands, which were stifling exploration. Concerns will also grow with exploration of Mars led by private companies: Elon Musk's company SpaceX intends to send spacecraft to Mars as early as 2020, and has indicated planetary protection policies are too stringent.[39] In the end it comes down to very large questions for humanity's relationship to space: do we assert hands off, play a more active role to extend and diversify life, or exploit outer space? It depends on which of the theories of moral status we choose, a policy question I return to in the next chapter.

Of Animals, Aliens, and Postbiologicals

If such conundrums arise with microbes, we should brace ourselves for the astroethical problems that arise with complex and intelligent life, categories that overlap with the ratiocentrism and sentientism that may be found in some animals. In tackling the question of how we should treat extraterrestrial intelligence (which is not equivalent to rationality), theologian Ted Peters suggests we begin by dividing such intelligence into three "astroethical categories" or "moral communities" based on how advanced they are: inferior, peer, and superior. Exactly how to parse these categories may well be problematic, especially when it comes to defining intelligence. Even doing this in a general way based on evolutionary development is problematic, since greater longevity does not necessarily imply greater intelligence. The dinosaurs were around a lot longer than *homo sapiens*, but give every evidence of not having been as smart as we are; at least, we are still here, though admittedly perhaps by good luck. Peters suggests three complementary tests for assessing these levels of intelligence: the technology test, the Turing test, and the naming test. The first is based on the sophistication of alien technology compared to ours; the second is the Turing test similar to that used to test the level of artificial intelligence; and the third is the "naming test," the extent to which sentient aliens name things. The latter is based on the idea that human beings name animals, not the other way around, a fact related to the emergence of language and symbolic representation.[40]

Given these moral communities, Peters suggests that our engagement with the inferior category might be analogous to our responsibilities toward animals on Earth. Of course, this engagement is ambiguous and problematic, but a great deal of thought has been given to it that might be useful in assessing our responsibilities toward what we consider inferior aliens. This is the same point Erik Persson made in arguing for a sentientist framework of morality. Moreover there are lessons to be learned here with regard to human culture contacts; the European contact with the Americas saw the "Indians" designated as inferior, much to their detriment and the disgrace of the Europeans. In ethical terms, we will have to decide whether extraterrestrials designated as "inferior" have intrinsic or instrumental value. Peters suggests that we model our responsibilities toward such life based on our concern for animals: to protect such life from suffering and enhance its experience of well-being.[41]

When contemplating ethical issues with equal or superior intelligence, we are at once faced with the possibility that our peers may be hostile or peaceful, and that our superiors may offer salvation from our problems, situations that will affect our attitudes and moral responsibilities. For peer aliens, Peters suggests we invoke the Golden Rule and our concepts of human dignity. But the options always come back to the two prototypes: the Darwinian survival of the fittest, and the altruism perhaps fostered by group selection, as discussed in the previous section. We simply cannot choose at this point without projecting our individual hopes and fears.

If we adopt philosopher Kelly Smith's sociality-rationality-culture triad as the basis for intrinsic value moral status and apply this framework to the problem of how to treat aliens, we might utilize his three moral categories for the triad: advanced, limited (similar to Peters's inferior category), and none. The advanced category includes humans and "any aliens with scientific knowledge sufficiently advanced to contact us." The limited category includes those beings that can "participate" in societies, including most animals on Earth. Those creatures totally lacking the sociality-reason-culture triad include "microbes, plants and lower animals with only rudimentary neural systems" – likely to be most of the creatures in the universe. He therefore lumps humans in with much more advanced aliens, a possible weakness in the scheme in terms of interactions, but perhaps sufficient for the purposes of only determining intrinsic value.

Because his advanced category has intrinsic moral worth, Smith argues they should never be treated instrumentally for our use and they should be all treated equally, in accordance with the ethical considerations that humans use among each other. The limited category possesses some intrinsic value and in his view should not unnecessarily be treated instrumentally, with exceptions such as scientific research. In the case of humans and canines starving in a lifeboat, he suggests it is not difficult to decide which should survive. Creatures completely lacking rationality have no intrinsic value, Smith suggests, and can be used as we wish, subject to negotiation in specific cases like strip-mining Mars with indigenous microbes. Microbes' potential to develop into more complex life, as happened on Earth, should not confer intrinsic value on them. This is quite a different view from the McKay–Sullivan scenario, nor presumably would Carl Sagan be happy. But it does seem to be a practical solution more likely to actually be implemented, followed, and enforced by any future United Federation of Planets.

Other approaches to the problem of how to treat alien intelligence are possible outside of the scheme of Table 8.1. In a pioneering essay in 1985, for example, Michael Ruse simply asked, "is rape wrong on Andromeda?" Ruse's controversial answer was no, because "although the immorality of rape is a human constant, we cannot thereby assume that it will be a constant for other organisms, including extraterrestrial intelligent organisms. This is certainly a non-anthropocentric solution, but beyond that it is hard to categorize in terms of ethical frameworks. It brings to mind the stark conundrums of James Blish's *A Case of Conscience* and Mary Doria Russell's *The Sparrow* and *Children of God.*"[42]

Finally we arrive at the idea of superintelligence, most often envisioned in the form of postbiological artificial intelligence, an idea I broached in Chapter 4. Given the timescales involved in the universe, our relatively recent emergence, and what I call the "Intelligence Principle" as a driver of cultural evolution, there is good reason for suspecting that most intelligence will not be biological, as I have argued in Chapter 4 and at length elsewhere. This kind of scenario is also in the spirit of transcending anthropocentrism urged in Chapter 4.[43] Again, the idea of human interactions with artificial intelligence has been explored in science fiction, and Isaac Asimov's Three Laws of Robotics, introduced in 1942, immediately come to mind as an early example: 1) A robot may not injure a human being or, through inaction, allow a human being to come to harm; 2) A robot must obey the orders given it by human beings except where such orders would conflict with the First Law; 3) A robot must protect its own existence as long as such protection does not conflict with the First or Second Laws.

These rather hopeful laws were anthropocentric since they were applied to human interaction with human-built robots. To apply them to postbiologicals beyond the Earth, which are out of human control, would be Pollyannaish in the extreme, even if we believe such intelligence could in some sense be altruistic. A postbiological might just as easily act on a law stating "a robot must protect its existence at all costs," especially if it was on a mission. Whether advanced versions of superintelligence can ever have consciousness, free will, emotions, and morality is the subject of a vast literature, with no consensus. In his book *Is Data Human: The Metaphysics of Star Trek*, philosopher Richard Hanley argues that artificial personhood with these qualities is "a distinct possibility," and that an anthropocentric test of humanness is not a fair test for a postbiological. Accordingly, "we ought to consider these individuals to be persons, and treat them accordingly – which is to say, we should consider their interests equally with the interests of human beings." In short, we should grant them intrinsic value and moral status. This means that whatever moral frameworks of Table 8.1 we choose, we should extend them to artificial intelligence, or more broadly to superintelligence in whatever form. Given scenarios like those expressed in Ray Kurzweil's *Singularity* and James Gardner's

Intelligent Universe, as well as in the movies *Her* and *Ex Machina*, we may well confront this scenario on Earth before we do in the universe. It behooves us to be prepared.[44]

Toward a Cosmocentric Ethic

The questions we have been asking go to the very core of the concepts of intrinsic value, moral status, and their meaning for practical ethics. They raise the issue of whether an anthropocentric ethic is enough for an astroethics dealing with alien life, even when extended to environmental ethics and deep ecology, or to one of the more general ethical categories in Table 8.1. Do we need something even broader, a "cosmocentric ethic," as Mark Lupisella and space policy analyst John Logsdon suggested two decades ago? By cosmocentric ethic in its strongest sense they mean one that suggests some physical or metaphysical reason why the universe itself might be the priority in a value system, which then provides a justification for intrinsic value and perhaps even an objective measurement of value. This ethic could be purely secular if it avoids the supernatural, or tend toward the theological if the divine is considered part of a given metaphysics, as in pantheism where the divine is a basic aspect of the universe. Such an ethic would have practical application; in fact, Lupisella and Logsdon came to their idea in light of the Mars rock controversy when they asked this question: "Should human space exploration and/or settlement of Mars take priority over preserving possible indigenous extraterrestrial life of even a primitive nature?"[45] Their answer – contrary to Zubrin – was that it should not, but that what is needed is to take a cautious approach that does not compromise possible Martian life (e.g., perhaps achieving some kind of coexistence), an approach, they suggested, that may require a more cosmocentric ethic.

The question of whether we really need a cosmocentric ethic depends to a large extent on how it is defined. So let me establish some categories of cosmocentrism. A *basic cosmocentric ethic* might stipulate that we should invoke the increasing cosmic consciousness I spoke of in the previous chapter, our new-found cosmic perspective, requiring us to consider our place in the biological universe when we make ethical judgments. We are, after all, part of the cosmos and perhaps not the most important part when it comes to life – the central question of astrobiology. In this view, when we ask about the rights of Martian life, or how to treat alien intelligence, we should certainly avoid an anthropocentric stance that only humans have moral status. Whether we should invoke any of the other moral theories of Table 8.1, ranging from ratiocentrism to planetocentrism is open to question, though to the extent environmental ethics is based on evolutionary relationships it may not be relevant to extraterrestrial life, unless we go all the way back to common cosmic evolutionary relationships. Whatever moral theory we chose, a basic cosmocentric ethic would urge us to consider that humans, and all life on Earth, are only one local outcome of cosmic evolution in a biological universe. And in order to have a cosmocentric ethic rather than just a cosmocentric perspective, such a framework would have to confer intrinsic value and moral status on specific objects or organisms.

An *enhanced cosmocentric ethic* might arise when we actually discover other moral agents in the universe, and can learn about their ethics. Whatever that ethic is, it would most certainly not be anthropocentric and perhaps not any of the other categories of Table 8.1, although some version of a cosmocentric ethic would be most likely to align with ours since aliens inhabit the same universe we do. To a degree, this also depends on the prevalence of altruism as discussed in the previous section, and altruism, as we suggested, may be linked to genetic factors. In any case, the discovery of alien moral agents holds the prospect of greatly enhancing our concepts of ethics, and perhaps our own ethics insofar as we treat each other and life on Earth. Surely this mingling of ethical systems, whether for good or ill, would be a landmark worthy of another level of cosmocentrism.

What I call a *strong cosmocentric ethic* would hold that the cosmos itself confers meaning in some way, and is the basis for a universal ethics that we should apply when considering practical questions about how to treat life in the universe, whatever its complexity. This is the mode of cosmocentrism that Lupisella and Logsdon favor. The question here is, can the universe itself provide a justification for inherent worth and thus a basis for ethics?[46] If so, exactly what aspects of the universe could provide a basis for a universal ethics? Kelly Smith has argued that the unfolding of complexity produced by the social-rational-cultural triad "may be the best means to realize the manifest destiny (manifest complexity) of all life, which could provide an ultimate, metaphysical foundation for ethical value." Clément Vidal has offered the infinite continuation of the cosmic evolutionary process itself as the ultimate good. Related to that view, Mark Lupisella has gone even further, arguing that the metaphysical basis for ethics might be "connectedness," which gives rise to relationship and interaction, diversity and complexity, and cosmic evolution itself, all of which are necessary for life. The knowledge gained from an enhanced cosmocentric ethic through alien contact could shed light on this, the ultimate foundation for universal ethics.[47]

Finally, an *anthropic cosmocentrism* would hold that life and the universe are intimately connected in the now familiar sense of the elegantly misnamed anthropic principle. In other words, if the fundamental physical constants such as the gravitational constant were different, life as we know it would not exist.[48] The word "anthropic" is unfortunate, since the anthropic principle is really biocentric, and not anthropic, and both words are unfortunate in terms of the present discussion, since "biocentric" in Table 8.1 has a more specific meaning. Nevertheless, "anthropic principle" is the term now in common usage, and I use it to raise the possibility of a deep connection between the universe and life, a cosmocentrism that could provide the ultimate basis for ethical choices. In a sense, anthropic cosmocentrism encompasses all other frameworks in Table 8.1, since life would not exist without the universe and (more controversially) the universe would not exist without life.

Cosmocentric perspectives give rise to a whole spectrum of philosophical worldviews, ranging from what Lupisella calls pragmatism (including anthropocentrism and ratiocentrism) to cosmocentrism (Figure 8.1). The latter includes what he calls "bootstrapped cosmocultural evolution," in which value, and hence some meaning and purpose, in the universe is bootstrapped as life emerges and increases as cultural evolution takes place, "making rational cultural beings and the universe 'co-priorities' in the overall evolution of the cosmos."[49] Although a strong or anthropic cosmocentric ethic may seem dangerously close to the naturalistic fallacy, we have seen that inferring from "is" to "ought" may not be a fallacy after all if one takes into account what we know about the cosmos, without necessarily implying that ethics is totally reduced to physicality. If the naturalistic fallacy is itself fallacious, strong or anthropic cosmocentrism – which could include cosmic

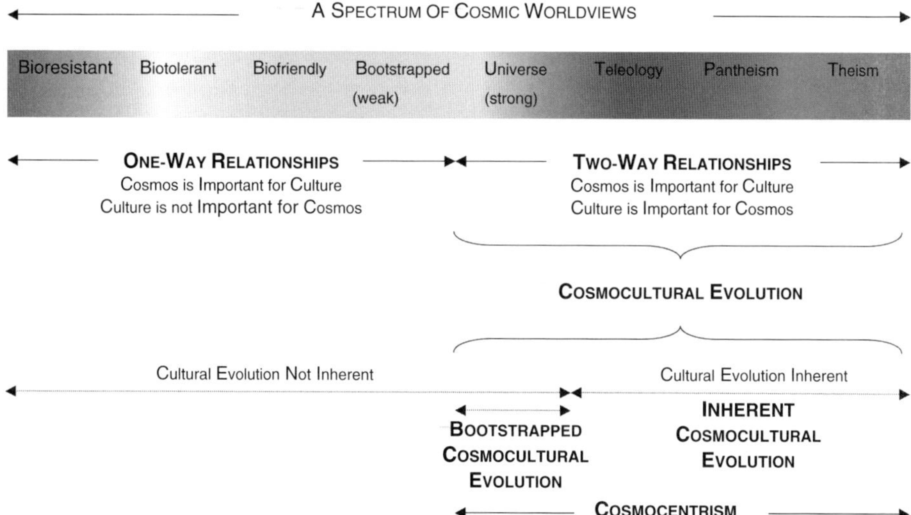

Figure 8.1 A spectrum of possible relationships between cosmos and culture. The spectrum ranges from bioresistant on the top left, whereby life emerges against substantial odds in a hostile environment, to (at upper right) a theistic universe with a transcendent personal God who is active in that universe. Intermediate between these extremes are a biofriendly universe where the nature of the universe tends to produce life, and a strong bootstrapped universe in which the universe raises itself into the realm of value via cultural beings such as ourselves, without theism. From Lupisella (2009a), by permission of Mark Lupisella.

teleology suggesting purpose in the universe – may be the ultimate goal in our search for the source of value in the universe.

How some of the ethical frameworks might apply in practical situations has been discussed by Jacob Haqq-Misra in the context of planetary engineering, in particular with regard to terraforming, geoengineering, planetary protection, and space exploration in general. Adopting an environmental ethic that may in principle extend to the cosmocentric, he develops an "ecological compass" (Figure 8.2) in which one axis extends from empty space to intelligence, and the other axis from instrumental to intrinsic value. The four quadrants thus produced can be applied to different practical cases. This is not the place to go into detail, but the fact that such schemes are being developed is just what we would expect as our exploration of space continues and our cosmic consciousness expands.[50]

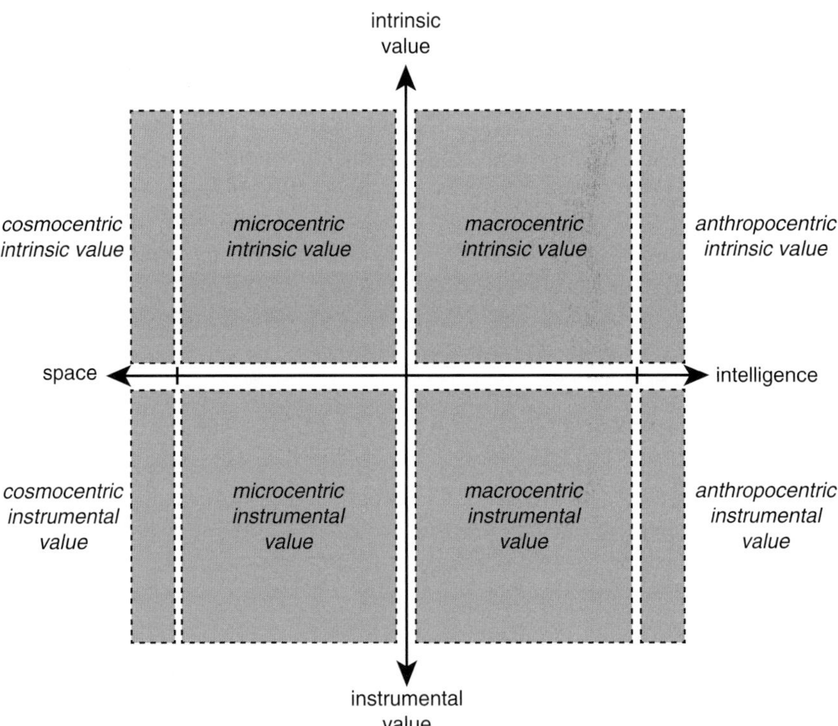

Figure 8.2 An ecological compass, according to Jacob Haqq-Misra. By plotting a spectrum of values ranging from instrumental at the bottom to intrinsic at the top, and a spectrum of objects that have moral consideration ranging from the universe itself on the left to intelligence on the right, the compass creates four quadrants in which the different theories of moral status in Table 8.1 can be plotted and discussed. Between anthropocentrism on the right and cosmocentrism on the left, lay the moral theories including the ratiocentric, biocentric, and planetocentric. By permission of Jacob Haqq-Misra.

Some forms of the cosmocentric ethic suggest the cosmos is valued in proportion to the complexity it produces. The practical result would seem to be that higher complexity is valued more than lower complexity, which means trouble for microbes. But in a broader sense, it values the universe that produced life, especially if there is a deep connection between the universe and life, as some versions of the anthropic principle suggest. I would argue that at a minimum we need a basic cosmocentric ethic, even as the discovery of alien intelligence would give rise to an enhanced cosmocentric ethic, and perhaps together we continue our search for an objective criteria for strong and anthropic cosmocentric ethics.

What would a cosmocentric ethic look like? Elements of it are found in theologian John Hart's book *Cosmic Commons: Spirit, Science, & Space*. Hart is a professor of Christian ethics at Boston University School of Theology, so it is no surprise that the book is Christian-centric. It also, in my opinion, takes too seriously UFOs and alien abductions, which are mixed in with more serious endeavors of astrobiology and SETI. But whatever his motivations Hart nevertheless has a cosmic perspective that encourages a kind of cosmocentric ethic. "Planetary, interplanetary and universal well-being require that humans develop a new sense of their place in, and integrality within, Earth and cosmic contexts," he writes. Only when people have a cosmic consciousness and foundation for a terrestrial-extraterrestrial ethics (what I term "astroethics"), he says, can we explore the universe within a well-developed ethical framework. The result, he writes rather optimistically, "will be responsible scientific research and outreach, respectful terrestrial–extraterrestrial engagement, and religious, philosophical, and ideological reconciliation with the implications and impacts of Contact. If this were to be the case, the likelihood for cosmic interrelationship, integration, and interdependence would be enhanced."[51] Such a cosmocentric ethic is in line, for example, with ideas expressed in Ursula Goodenough's *The Sacred Depths of Nature*, Stuart Kauffman's *At Home in the Universe*, and even older books like Pierre Teilhard de Chardin's *The Phenomenon of Man* – in short, books that find religious and ethical implications in the cosmos and the process of cosmic evolution.

In my view, the need for a basic cosmocentric ethic is already growing in proportion to the growth of our cosmic consciousness and the development of astroculture as discussed in the previous chapter. At one level, the novel ethical challenges as we explore the extraterrestrial environment argue for a cosmos-wide ethical view. At another level, we need such an ethic because the cosmos potentially grounds our ethics in some objective way, or at least serves as a shared frame of reference with other intelligent beings, or even because the cosmos and life are intimately entwined. The answer turns on how constrained we consider our terrestrial ethics, but a cosmic perspective is surely in order as we expand our views of the space environment, including (and especially) life. Such a view will not happen overnight, but perhaps humanity will increase its awareness in stages as we encounter the universe in increasingly intimate ways that will become a basic part of what it means to be human, or post-human.

We have come a long way in our discussion from anthropocentric ideas of intrinsic value to more universal moral standards and ethical practice, in the process even entertaining extraterrestrial altruism. How will astroethics transform our thinking? Clearly it already has to some degree, as evidenced by the considerable amount of work that has gone on in the subject over the past three decades, much of it in scattered workshops and conferences sponsored by NASA and other forward-looking groups. As in other areas, astroethical thinking makes us expand our ethical horizons, just as environmental ethics has expanded our terrestrial horizons. Ultimately, astroethics may transform our thinking by leading us toward a cosmocentric ethic. Even if no life is found, as Ruse has suggested, by thinking about such matters from a novel perspective, "we see more clearly the nature and extent of our own knowledge. This is true both in the area of epistemology and in the area of ethics."[52] In the same way that Kant decentered ideas about the validity of human knowledge, so will astroethics decenter our ideas about ethics. What remains is to transform a set of rather loose ideas into a systematic policy that can be implemented in the event of discovery.

9

Astropolicy

Preparing for Discovery

Confirmed discoveries of single-celled life fossils on Mars or simple life forms on Europa could have profound effects on worldviews and religious beliefs, and will raise many ethical and practical issues. We need formal post-detection protocols for single-celled organisms as well as for advanced technological civilizations.

NASA Report, 2001[1]

The interesting question is what do we do when we find life on another planet … What's the plan? Do we announce it to the world? Do we research more to determine if these are friendly or collaborative? What do we do when we make the discovery?

Congresswoman Suzanne Bonamici (D-Oregon)
Congressional hearings on Astrobiology, May 21, 2014[2]

Fortune favors the prepared mind.

Louis Pasteur

With all of the foregoing discussion of rational approaches, critical issues, and potential impact of finding life beyond Earth, I have yet to address two practical questions: how do we prepare for the discovery of life, and what should we do if we find it? As indicated earlier, these questions have been sporadically raised at the highest levels of government, usually in the context of astrobiology programs at NASA, but also at other space agencies such as the European Space Agency. During congressional testimony before the House Science Committee several years ago I raised issues of astrobiology and society and asked, "how do we balance planetary protection and stewardship with the human exploration imperative, of which astrobiology is a significant part?" Members of Congress had their own societal impact questions, only a few of the many that will require policy decisions at multiple levels. For astrobiology, such issues have been more often addressed outside of government, in institutional settings and individual studies. In either case, these issues place our subject squarely in the arena of space policy, which itself has been an active area of science policy for more than a half-century of space exploration. In short, an important part of the impact of discovering life is what I will call astropolicy – preparing for, managing, and embracing the discovery of life on other worlds. Space policy in general is a scholarly and sometimes lively endeavor in journals such as *Space Policy* and *Astropolitics*. But astropolicy as defined here and as a subset of space policy is sure to become an even more lively issue in the event of actual discovery. Arguably, preparing for astrobiological discovery and its aftermath is no less urgent than other science policy issues with the potential to impact all of society.

The questions here are legion, and potentially Earth-shaking. Who should take the lead in preparing for discovery? What do we do if life is actually discovered, microbial or intelligent, near or far? Should national governments be in charge, international political and scientific institutions, scientists and social scientists, ethicists and theologians, or some mix thereof? How do we prevent contamination of potential microbes on Mars, Europa, Enceladus, or other habitable sites in the solar system, and (perhaps more urgently from most Earthlings' point of view) how do we protect our planet from back contamination in the event of the discovery of microbial life? If a message is received as a result of a successful Search for Extraterrestrial Intelligence (SETI) program, should we answer? If so who speaks for Earth? Should we initiate messages as part of a Messaging Extraterrestrial Intelligence (METI) program? If so what should we say, and who, if anyone, should control what is said? These questions are only the leading edge of the many decisions that will have to be made once alien life is actually discovered. And, as I have repeatedly emphasized throughout this volume, each scenario will have its own unique problems and solutions.

While we can debate how likely any extraterrestrial life discovery scenario might be, in the past few decades developments in astrobiology have made the discovery of life beyond Earth more and more feasible, to the extent that in 2013 the World Economic Forum named the discovery of alien life as one of five "X factors" – emerging concerns with scientific grounding but unknown consequences. Clearly, the time is ripe for serious discussion of strategies and policies in the event of discovery. Here I look at policy issues in three time domains: preparing for discovery, managing discovery, and embracing discovery. I do so keeping in mind the two primary scenarios I have distinguished throughout this book: the discovery of microbial life and intelligent life, each with its multiple modes of near and remote discovery and direct or indirect impacts. I also do so mindful of the sentiment of both Eisenhower and Churchill that "plans are worthless, but planning is everything." While policies do not always survive contact with the actual events, it is indisputably prudent to contemplate them ahead of time, especially when the implications for Earth are so all encompassing as in the discovery of alien life. Flexibility will be the watchword in the event of any discovery, but as the World

Economic Forum concluded, "looking forward and identifying emerging issues will help us to anticipate future challenges and adopt a more proactive approach, rather than being caught by surprise and forced into a fully reactive mode." This chapter argues that we need not only to "look forward" but also to develop policies to achieve the goal of being fully proactive – and to have the best chance of having a positive impact on society when and if alien life is discovered under a variety of scenarios.[3]

Preparing for Discovery: Policies, Protocols, and Strategies

It is one thing to contemplate the consequences of discovering life beyond Earth, but quite another to develop policy for what to do when the event occurs. "Policy" can have a variety of meanings at a variety of levels, but its importance is evident in the fact that virtually all institutions have policies. They codify expected behavior, ways of acting under certain circumstances, and courses of action adopted by organizations and governments with the goal of implementation. Policy is closely allied to strategy, protocols, and roadmaps, all terms used in the domains of astrobiology and SETI. It is probably counterproductive to distinguish precise differences among them, although strategies may be general, protocols more specific, and policies very specific. The several studies discussed in Chapter 6, such as NASA's workshops on the cultural aspects of SETI, do not constitute policy, though ideally the ideas generated there should feed into policy. The same is true of NASA's early Astrobiology Roadmaps and its current Astrobiology Strategy. The important point is that all are agreed-upon ways of proceeding in the event of discovery. Short of legislation, none of them can compel or prohibit behavior with legal sanction; rather, they are meant as guidelines toward a desired outcome. As former US State Department officer Michael Michaud has pointed out in the context of extraterrestrial intelligence, because the impact of life beyond Earth has not yet captured the sustained attention of governments at the executive or legislative levels, the contemplation and development of strategies, policies, and protocols has thus far come largely under the purview of science organizations.[4]

Developing Policy for Astrobiology Impact

How do we develop strategies, protocols, and policies to prepare for discovery? To start with, those contemplating astrobiology impact policy can learn lessons from the recent development of astrobiology science policy. For astrobiology science in the past two decades, roadmaps, strategy, and policy have been trickle-up rather than trickle-down processes. After astrobiology grew out of exobiology in the mid-1990s at NASA, and after three years of preliminary workshops, and at the invitation of NASA, in the summer of 1998, 150 scientists from many disciplines met at NASA Ames Research Center to develop an astrobiology roadmap for the next 20 years, with emphasis on the first five. In its original form released in 1999, the roadmap included three fundamental questions, four operating principles, 10 goals, and 17 objectives for astrobiology. This is not the place to detail the roadmap's science content. The point is that when NASA adopted the roadmap strategy as a significant part of its astrobiology program, it became a course of action carried out by the astrobiology program office at NASA Headquarters in Washington, implemented by the NASA Astrobiology Institute it supports and by the many scientists who seek funding under the NASA astrobiology research announcements that are published each year. The National Research Council of the National Academy of Sciences also provided input through meetings and reports, but the astrobiology roadmap gained force from the broad base of scientists who formulated it and from the funding NASA provides to carry it out. The astrobiology roadmap, known in its latest version as the "astrobiology strategy" (Figure 9.1), is occasionally reviewed and revised, and is subject to oversight by the relevant committees of Congress, especially the House Science Committee, which holds hearings on the subject. When the NASA authorization and appropriations bills are written in Congress, funding for astrobiology may be increased, decreased, or zeroed out completely; the latter rarely occurs in science projects, but it did happen in 1993 for the NASA SETI program. This bottom-up process, which is very different for bigger projects like the space shuttle or the International Space Station where politics is involved up to and including the presidential level, has nevertheless worked well for astrobiology and other relatively small NASA programs. The latest Astrobiology Strategy process involved some 800 members of the astrobiology community, who used meetings, white papers, and webinars to create its final document. A similar bottom-up process was followed to create the first scientific roadmap for European astrobiology, published in 2016.[5]

What does this tell us about astrobiology impact policy? Importantly, two of the four roadmap operating principles that emerged from the original roadmap process were related to astrobiology and society issues, one in encouraging planetary stewardship by emphasizing planetary protection and avoiding contamination, and another by recognizing "a broad societal interest in our subject," including the discovery of extraterrestrial life and engineering new lifeforms adapted to live on other worlds. Inspired by this formal recognition of societal interest, a small group of participants at the 1998 meetings proposed that a multidisciplinary approach be used to understand the consequences of the search for life beyond Earth. This splinter group, which included scientists, social sciences and humanities scholars as well as futurist Alvin Toffler, met with considerable opposition of the "two cultures" type, claiming the social sciences and the humanities had no place in NASA,

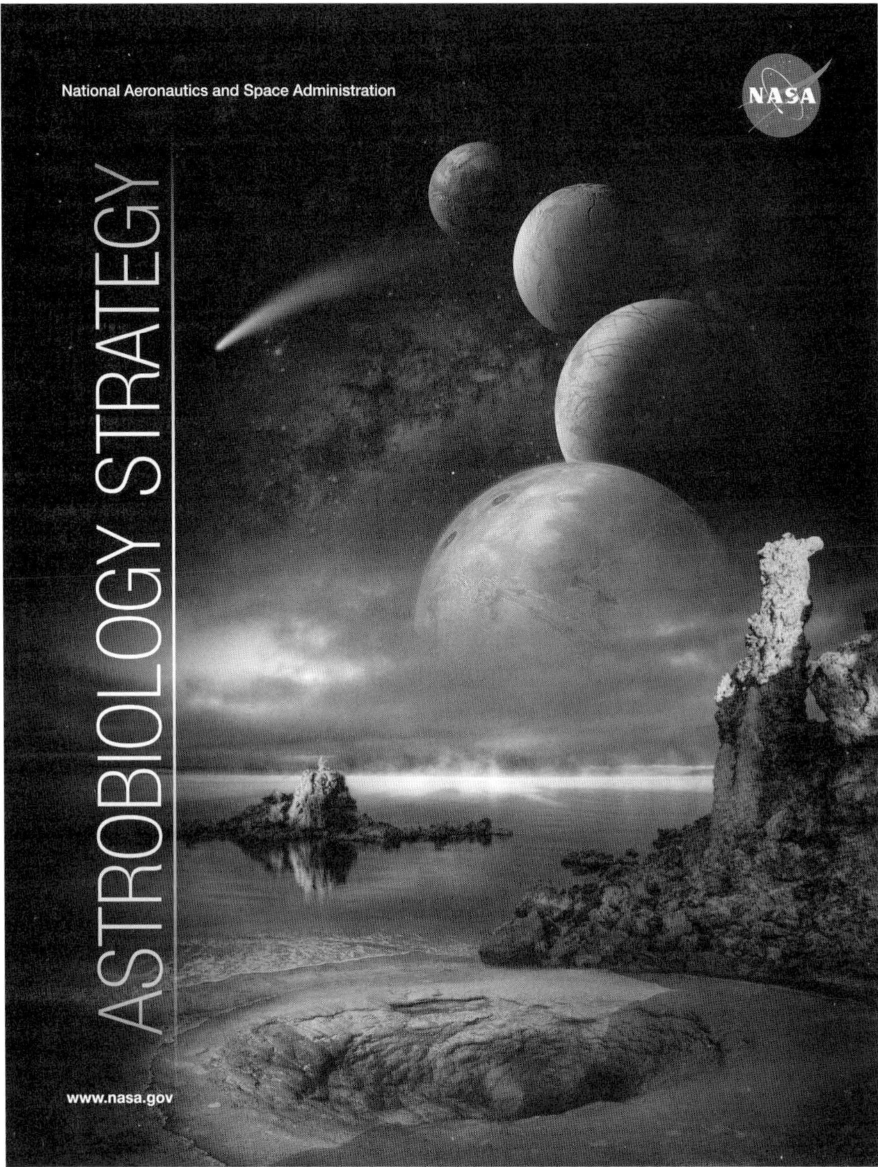

Figure 9.1 The process used to develop NASA's astrobiology science strategy may provide a model for developing strategies and policies to prepare for the societal impact of discovering life. Credit: Jenny Mottar, NASA.

especially if it were going to divert funding from *real* science. Others argued that to a large extent philosophical questions were the intellectual drivers behind astrobiology, and that it was incumbent on the scientific community to work through the issues of astrobiology and society. In the end the three societal goals the group proposed were not included in the final roadmap, which meant that they were not priorities to be acted on with specific objectives and funding.[6]

Nevertheless the broad societal interest stated in the roadmap's operating principles encouraged further discussions, and as discussed in Chapter 6, in 1999 an intrepid and much larger and diverse group of scholars met at NASA Ames Research Center for a workshop on the societal aspects of astrobiology. That meeting resulted in a report with four recommendations, four action items, and eight areas for further research. Ten years later, under the auspices of the NASA Astrobiology Institute, 43 invited scholars met at the SETI Institute and developed an "Astrobiology and Society" roadmap, fully aware of the astrobiology science process. Unlike the science roadmap, however, the societal impact roadmap has not yet become policy backed up by sustained funding. But the work continues at a basic level, and the process seems to be following its companion science roadmap in percolating from the bottom up with minimal funding and the hope of eventually becoming a more recognized and funded policy. That will require the two cultures to work together, and a good start has been made at the biennial Astrobiology Science Conferences, which sometimes include discussion of issues related to society. It is encouraging that the Introduction to the latest 2015 Astrobiology Strategy document still lists a goal to enhance societal interest and relevance. "Astrobiology recognizes a broad societal interest in its endeavors," it states, "especially in areas such as achieving a deeper understanding of life, searching for extraterrestrial biospheres, assessing the societal implications of discovering other examples of life, and envisioning the future of life on Earth and in space." The document also includes as an Appendix a humanities and social sciences section. Finally, NASA's establishment of the Baruch S. Blumberg NASA/Library of Congress Chair in Astrobiology in 2011, specifically to address the humanistic and societal aspects of astrobiology, is a de facto recognition of the importance of these issues.[7]

Far from initial skepticism about a role for the social sciences and humanities, there is now considerable consensus that the problem of the impact of discovering life in any form is important, and should not be left to scientists alone. When the Royal Society of London sponsored a meeting on the detection of extraterrestrial life and the consequences for science and society, and a satellite meeting in 2010 seeking a scientific and societal agenda on extraterrestrial life, the organizers wrote that "while scientists are obliged to assess benefits and risks that relate to their research, the political responsibility for decisions arising following the detection of extraterrestrial life cannot and should not rest with them. Any such decision will require a broad societal dialogue and a proper political mandate. If extraterrestrial life happens to be detected, a coordinated response that takes into account all the related sensitivities should already be in place."[8] As we have seen, scientists and scholars in the humanities and social sciences can and do play a major role in formulating policy, but because there are many other stakeholders in society, they may not be the final arbiters of policy, much less the major players in implementing it. Thus it is important not only to continue to develop strategies and policies but also to have the political mandate to implement them when the time comes.

Issues and Policy for Microbial Life Impact

In astrobiology as elsewhere, policy begins by identifying issues and desired outcomes, exploring alternative courses of action, and deciding on the best course of action. So we begin by asking what are the policy issues and desired outcomes bearing on the impact of finding life beyond Earth, what has been done to prepare for them, and what remains to be done before discovery? Another important practical question is how do we generate a proper political mandate to ensure that policies and protocols have force when the event occurs? The issues are numerous, but we can illustrate the challenges with three representative problems in the context of microbial life: 1) How do we prevent the forward and back contamination of the Earth and other planets? 2) What is a coherent course of action when microbial life is discovered beyond Earth? and 3) What are the ethical and philosophical issues underlying decisions that will have to be made about the intrinsic or instrumental value of microbes as defined in the previous chapter?

Fortunately, and by necessity, the first of these issues has already received considerable attention and offers some guidance on how other issues might be addressed. The problem of biological contamination was raised at the beginning of the Space Age, and has been the subject of a great deal of thought in the course of robotic exploration of the planets in general. It is notable that, like the later astrobiology roadmap, the development of policy for these two cases was a bottom-up process. As early as 1957 Joshua Lederberg, soon to receive the Nobel Prize for his work in genetics, brought the issue of contamination before the scientific community. At his initiative in early 1958, the US National Academy of Sciences urged the International Council of Scientific Unions (ICSU) to develop recommendations to prevent contamination. This it did by sponsoring an ad hoc Committee on Contamination by Extraterrestrial Exploration, which developed standards adopted by ICSU in 1958 (things moved faster in those days!). In 1959 the mandate was transferred to the newly founded Committee on Space Research (COSPAR), a permanent committee of the ICSU, which formulated a policy and implementation requirements as an international standard to prevent forward and back contamination. Most space agencies, including NASA and the European Space Agency, implement these policies through their planetary protection programs. In the case of the United States, NASA seeks the best advice on planetary protection from the Space Studies Board of the National Academy of Sciences and acts accordingly. In 1967 COSPAR planetary protection guidelines were reflected in Article IX of the UN Outer Space Treaty (provisions against "harmful contamination" of celestial bodies and "adverse changes in the environment of the Earth resulting from the introduction of extraterrestrial matter"), still the main source of space law. Since 1999 COSPAR's Panel on Planetary Protection, with international representation, suggests revisions to the COSPAR policy for inclusion by the approval of the COSPAR Bureau and Council.[9]

In short, planetary protection policy began with a prominent scientist who brought the issue to his country's National Academy of Sciences, which then urged international action by the appropriate scientific body. That body (ICSU) did so through its appropriate committee (COSPAR), which established consensus principles regarding biological contamination, and for more than a half-century has kept the policy updated through its international representatives and scientific associates. In the meantime, these principles have also been incorporated in the Outer Space Treaty, one of the five UN treaties and agreements related to outer space, giving them even more force and something approaching a political mandate. Feedback for updates is generated through meetings of the scientific community, including COSPAR-sponsored workshops on general topics such as "Ethical Considerations for Planetary Protection in Space Exploration" (Princeton University, 2010), and "Developing a Responsible Environmental Regime for Celestial Bodies" (George Washington University, 2012), and a more specific focus on the robotic exploration of Mars, Europa, and Enceladus (Bern, 2015), and the human exploration of Mars (Houston, 2016).

Even the planetary protection process, in some ways a model for policy formulation and implementation when it comes to the impact of astrobiology, is far from perfect. As the George Washington University report stated, "complete sterilization of all probes, while maintaining their functionality, turned out to be beyond the limits of the technology available at the time," and it is known that some Earth organisms survived even on the *Viking* landers, the "gold standard" of planetary protection (Figure 9.2). Perfect protection (like perfect safety in human spaceflight) would require that no missions be flown. Compromises therefore have to be made between engineering realities and scientific desires, and biological controls are admittedly limited, especially where planetary environments are thought not to be supportive of life. Still, policy formulation proceeds to the best of our ability. This is a lesson for other impact issues: the perfect should not be the enemy of the good, and the development of strategies and policies should proceed apace.[10]

Due to the historical necessities of the Space Age, we are therefore relatively well prepared to deal with issues of forward and back contamination via microbes. But that is quite different from our second issue, formulating policy for an actual discovery. We can infer from these policies that in the event of actually discovering

Figure 9.2 *Viking 2* lander image showing the spacecraft and part of Utopia Planitia on Mars, looking south. November 2, 1976. The horizon is about two miles distant. The *Viking* landers were subject to planetary protection protocols developed over a long period of time, as more general astrobiology policies, protocols, and strategies may be in the future. Credit: NASA/JPL.

life, a first priority would be to prevent the back contamination of Earth. Beyond that dictum, however, no guidance exists on what to do if microbial life is actually detected, and therefore no coherent course of action. The first step is recognizing the problem, and Margaret Race, a senior scientist at the SETI Institute working with NASA support, has been a pioneer in urging action in this area. As early as 2002 she and bioethicist Richard Randolph called for operating guidelines in the event of such a discovery, arguing that while planetary protection policies address *how* to undertake exploration, they "do not provide clear guidance on what to do *if and when* life is detected." They emphasized that alien microbial discovery has many scenarios: robotic detection of life on sites like Mars, Europa, or Enceladus; astronaut discovery of life on these bodies; and laboratory discovery of extraterrestrial samples on Earth, to name several. Each may require its own protocols. In all of these scenarios, some of the work on planetary protection protocols would come into play in order to avoid contamination. But broader guidelines would need to be developed beyond the contamination issue, and Randolph and Race lay out four ethical principles for such guidelines: 1) cause no harm for planet Earth, its life, and its diverse ecosystems; 2) respect and do not substantively or irreparably alter the extraterrestrial ecosystem; 3) observe "good" science procedures; and 4) ensure the participation of all humankind in the discovery of extraterrestrial life. These ethical principles give rise to eight guidelines, which include informing the United Nations, making all data available to the scientific community, and initiating a moratorium on further space missions prior to international consultation. By comparison with the planetary protection process described earlier, we are today still at step one when it comes to policy for actual discovery. Despite the real possibility of discovering microbial life in the near future, there are no written policies or protocols, with the exception of contamination protocols for a Mars sample return.[11]

Much the same is true of our third issue, astroethics. As shown in the previous chapter, there has been a good deal of discussion of astroethical issues over the past few decades, but no action from a policy point of view. In the context of microbes, it matters whether we adopt an anthropocentric or ratiocentric ethic that confers intrinsic value only on reasoning beings, or a biocentric ethic that values all living things. It matters whether we consider microbes only of scientific value, or whether they are considered to have intrinsic value, in which case microbes have rights too – rights that we do not give them on Earth. The planetary contamination policies seem to confer rights on any microbes we may find on other worlds; the central goal of those policies, after all, is to protect from contamination any planets that might harbor life. That is a kind of biocentric ethic. But it is an unstable one, since by necessity on Earth we stamp out pathogenic microbes while at the same time realizing the microbiome is essential to human health. Thus the status of microbes is one of many ethical dilemmas we will face if and when extraterrestrial microbes are discovered. One has the feeling that, even if a biocentric ethic is adopted in principle, human health will always take priority.

These three issues address only two of the most likely discovery scenarios laid out in Table 2.1: an encounter with alien microbes on Earth, and through robotic explorations even now taking place. Encounters with alien microbes during human exploration of the solar system are also possible, but this will be a much more long-term problem since human missions to Mars or the ocean worlds of the solar system are likely far in the future. This, too is a lesson: in proceeding with policy considerations, it is practical to deal with the most likely scenarios first. The three issues we have identified suffice to show why we should think ahead about the problem, especially since history shows that early actions and reactions may have long-term and deleterious effects. The goals are clear: first to avoid an "Andromeda strain" scenario on the one hand and the contamination or destruction of alien microbes on the other, and second to derive the maximum benefit from the discovery that we can in terms of science and human welfare, employing strong astroethical principles. Whether these goals should come in the form of policy modeled on the Antarctic Treaty of 1961, in particular its Protocol on Environmental Protection added in 1998, or in some other form, is up for discussion.[12]

Issues and Policy for SETI and METI Impact: Who Speaks for Earth?

While the policy issues involved with the discovery of microbes are serious enough, the issues become even more daunting for extraterrestrial intelligence. Once again they depend on the discovery scenario, most urgently in connection with current programs for indirect contact via SETI or METI programs, and most spectacularly in terms of impact if we ever make direct contact with aliens on Earth or in our solar system, even in the form of alien artifacts. Representative of the policy challenges in these cases are the following broad questions: 1) What is the course of action if an intelligent signal is received as a result of a SETI program, whether directed at us or intercepted? 2) Should we initiate messages as part of a METI program, and if so, who speaks for Earth? and 3) What is the course of action in the event of direct contact with alien intelligence?

In addressing these issues, it is important to note that the scientific trajectory of the Search for Extraterrestrial Intelligence (SETI) has been very different from that of astrobiology. Prior to 1992 when the NASA SETI program became operational, it consisted of only very sporadic searches at widely dispersed locations. And after Congress terminated the NASA program in 1993 for political reasons, much the same has been true, with SETI on shoestring budgets. The SETI Institute took over the targeted search portion of the NASA program, but had to rely on private funding, which was useful but not sustained, even at the multimillion-dollar dedicated Allen Telescope Array in California (Figure 9.3). Searches around the world have been even more sporadic. Under these circumstances, a roadmap for SETI science was sponsored by the SETI Institute rather than NASA, which never incorporated SETI into its astrobiology program as a search for intelligent radio signals.[13] The road mapping happened from 1997 to 1999 by a 39-person interdisciplinary team of scientists and engineers, simultaneously but independently of the broader astrobiology roadmap, and resulting in *SETI 2020: A Roadmap for the Search for Extraterrestrial Intelligence*, published in 2002. As in astrobiology, the science goals, which have only partially been achieved, once again percolated up rather than down, based on the input of scientists. And as in astrobiology, the roadmap also briefly addressed societal implications, urging further work on a "Post-detection SETI Plan," and "encouraging active interest among scholars in the societal disciplines."[14]

Figure 9.3 This artistic photo shows the Allen Telescope Array, a dedicated facility for SETI observations. Credit: Seth Shostak.

Unlike the current NASA astrobiology situation, there has been no sustained funding source to carry out the SETI 2020 roadmap. When scientists met at the Green Bank Observatory in 2011 on the fiftieth anniversary of Frank Drake's first SETI project, I well remember Drake's assessment of the prospects for the field as dire, largely due to lack of funding. After an initial infusion of private funds, the SETI Institute's Allen Telescope Array and a few other radio and optical facilities have carried on with shoestring budgets. But not until Russian entrepreneur Yuri Milner pledged $100 million for SETI observations in 2015 was the prospect of a sustained search once again revived. Milner's "Breakthrough Listen" will fund new detectors under the supervision of the University of California at Berkeley; dedicated observing time on the Byrd telescope at Green Bank, West Virginia, and the Parkes radio telescope in Australia; optical searches for laser signals using facilities at the Lick Observatory; and funding for data analysis. "Breakthrough Message" will fund the construction of messages to send aliens, but whether those will ever be used in a real METI program is undecided. These funds have raised hope of a renaissance in SETI, and a new holistic road-mapping exercise is under way that considers SETI in a much broader context, including the evolution of intelligence in known planetary environments, similar to the astrobiology strategy that considers microbes and biosignatures with regard to planetary environments.[15]

That the question of what to do in the event of success in SETI has received considerable attention is again largely because an actual scientific project generates policy concerns, especially when a government program is involved. Unlike planetary protection, the UN Treaty on Outer Space contains no reference to discovery of extraterrestrial intelligence. And although the Committee on the Peaceful Uses of Outer Space (COPUOS) briefly discussed policy for the detection of ETI in 1977, no action was taken. But as early as 1972, US diplomat Michael Michaud began a series of studies that would at least lay out the issues over the next half-century. Individual initiative was crucial; as he noted, during the 1970s "neither the State Department nor the world's other foreign ministries showed any institutional interest in the question."[16]

It was only the NASA SETI program, being developed at an increasing pace in the 1980s, that finally gave impetus to policy considerations higher than the individual level (Table 9.1). Crucial in this process was John Billingham, the head of

Table 9.1 *Levels of Effort in Astropolicy Relevant to Astrobiology and SETI*

Jurisdiction	Organization	Activity/Examples
International	UN General Assembly	
	UN Committee on the Peaceful Uses of Outer Space (COPUOS)	Oversees five UN treaties/agreement
	International Council of Scientific Unions (ICSU) [now International Council For Science]	
	Committee on Space Research (COSPAR)	Planetary Protection Protocols
	International Scientific Organizations	
	International Academy of Astronautics (IAA)	
	SETI Permanent Committee	SETI Post-detection Protocols (SPDP)
	International Astronomical Union (IAU)	SPDP signee
	International Institute of Space Law	SPDP signee
National	National Governments	Executive-Mars Rock Legislative Hearings
	National Scientific Bodies	
	US National Academy of Sciences National Academies of other countries	Reports and Recommends
	Government-sponsored meetings	
	NASA, Royal Society and others	See Table 6.2
Group and Individual Studies	Nongovernment Meetings and Reports	See Table 6.2 Vakoch
	Individual Studies	Harrison Michaud Race Dick

the NASA SETI program, and others who organized sessions at the International Astronautical Congresses in 1986 and 1987 as NASA SETI observations edged closer to becoming operational. Based on this input, Michaud drafted two proposals, one for detection protocols, the other on reply protocols. Only the first one moved forward, and after numerous changes, in 1989, the International Academy of Astronautics (IAA) approved a "Declaration of Principles Concerning Activities Following the Detection of Extraterrestrial Intelligence." This is sometimes referred to as the "first SETI Protocol"; the second protocol, concerning sending our own signals, was never approved.

What are usually referred to simply as the first "SETI post-detection protocols" consist of nine principles, which basically boil down to the following: confirm any putative signal before a public announcement, and then tell everybody.[17] In other words, do not make premature announcements, and do not have a policy of secrecy. While the SETI protocols may seem like a far-reaching and prudent step, and although they have also been endorsed by the International Institute of Space Law, COSPAR, the International Astronomical Union, and the International Union of Radio Science, they have never been acted on by the United Nations. The International Academy of Astronautics did present documents, including the protocols, to the UN Committee on the Peaceful Uses of Outer Space in Vienna in 2000. The Committee's report to the General Assembly noted the receipt of those documents, but no further action was taken at any level. They are therefore not part of any UN treaty and are not legally enforceable. They are most important as an agreement among SETI practitioners, and although Michaud and others have tried to update them to include scenarios such as the discovery of alien artifacts, the IAA has not endorsed such an extension.[18]

While protocols for passive SETI have involved a good deal of astropolitics in the sense of negotiations and compromise, the issue of messaging extraterrestrial intelligence (METI) – also known as active SETI – has been even more controversial. At the most basic level, the policy question is whether sending messages *de novo* to aliens should even be undertaken. The famous *Pioneer* plaques and *Voyager* records of the 1970s caused relatively little controversy, but active SETI is another matter altogether. In contrast to messages aboard spacecraft (and the message of the spacecraft itself!), METI involves intentional and information-rich signals sent to extraterrestrial civilizations, with the ability to reach them at the speed of light. Although stellar distances mean that years, decades, or centuries would elapse before a message could reach even the nearest stars, in astronomical terms, that is the blink of an eye. Even if that were not the case, critics argue, we need to consider the long-term future of Earth.

It is therefore not surprising that METI proposals have done more than raise eyebrows, and deservedly so. Frank Drake's Arecibo message in 1974 caused relatively little stir with one notable exception, British astronomer Sir Martin Ryle, who urged the International Astronomical Union to declare that no further attempts should be made to communicate with other civilizations because of possible hostile consequences. The *New York Times* responded with an editorial titled "Should Mankind Hide?" and concluded mankind should not hide. But in the past several decades, the number of transmissions has ramped up, ranging from multiple Russian broadcasts beginning in 1999 transmitting an abbreviated encyclopedia of human knowledge, to Discovery Channel Canada's 2005 "Calling All Aliens" and NASA's 2008 broadcast of the Beatles' "Across the Universe." Some of these transmissions, widely reported in the press, were too weak to be of real concern, but as transmitters become cheaper and more readily available, the urgency for policy grows.[19]

Foreseeing potential problems and opportunities, in 1995 the IAA SETI Committee – the same one that formulated the SETI Protocols – developed the "Draft Declaration of Principles Concerning Sending Communications with Extraterrestrial Intelligence." The Declaration laid out 10 principles, specifying that any message should be sent on behalf of all humankind rather than from any individual state, that it should reflect the broad interests and well-being of humankind and be available to the public prior to transmission. It stated further that the content should be based on input from a wide variety of people with diverse expertise, and that appropriate international consultations should take place prior to any transmission. The Declaration also encouraged international studies to consider these issues, and urged that these issues should eventually come before the United Nations. These proposals met with less success than the SETI protocols. The SETI Institute's 2002 roadmap recommended against active SETI projects. To this day, the Draft Principles have not achieved consensus even at the IAA and its SETI Committee. The METI declaration remains just that, a draft with no consensus or force.[20]

In the meantime, substantive discussion continues to take place, largely centered around the IAA activities and published in its *Acta Astronautica*, at least partially fulfilling the "international studies" principle of the Declaration. Douglas Vakoch, formerly a long-time researcher at the SETI Institute, has been a leader in pro-METI arguments, arguing that if no one is transmitting, SETI programs are doomed to fail; that having extraterrestrials receive, decode, and interpret messages would facilitate mutual comprehension; and that such a project provides both humans and aliens with a long-term vision for mutual benefit. Others inside and outside the IAA have argued the opposite: Stephen Hawking warned that aliens receiving such a message could retaliate. Jared Diamond wrote in the *New York Times* that extraterrestrials "might proceed to kill, infect, dissect, conquer, displace or enslave us, stuff us as specimens for their museums or pickle our skulls and use us for medical research. My own view is that those astronomers now preparing again to beam radio signals out to hoped-for extraterrestrials are naive, even dangerous." Science fiction writer David Brin has been particularly outspoken, insisting that no action should take place prior to consultation; that a process should take place like that of Asilomar, where brakes were placed on biotechnology in the mid-1970s; and that there is no guarantee of altruism in the universe. To send messages now, he argues, would be like "ignorant children, screaming 'Hello' at the top of their lungs, in the middle of a dark, unknown jungle." These points and more were argued at the Royal Society meeting in 2010, with no consensus. I myself have written that we cannot and should not hide from the universe of which we are a part, and that while governments should not fund METI as a means to propagandize, groups that want to undertake such a project cannot practically be regulated short of legislation on the subject, which does not exist and is unlikely to happen. Given that, it seems to me a responsible group of scholars that wants to undertake a METI project should be free to do so if it seeks consensus inside and outside the group.[21]

Astropolitics in this area reached a peak in the past few years when 28 scientists signed a statement, originating at Berkeley, arguing that METI programs carry "unknown and potentially enormous implications and consequences," and that nothing should be done without international consultation. In 2015 Douglas Vakoch founded METI International, complete with a board of trustees and a large number of distinguished advisors, with the explicit purpose of sending a message to the stars. Both the Berkeley and METI groups agree that consultation prior to any messages being sent is important. The question is at what level the consultation should take place, what the consultation is actually about, and on what principles would a consultation actually be decided. Based on history it is highly unlikely that the United Nations would take up such a consultation, leaving "wise" people, as Brin puts it, to make the decisions.[22]

But who are these wise people? If the Asilomar process is followed, they would be practitioners in the fields of SETI and METI including scientists and social scientists, as well as lawyers, journalists, and government officials. They would break the problem down into its constituent parts and assign risk estimates for specific types of experiments, with guidelines for each. Research would continue, but with strict guidelines about strength of signal, targets, language, and message content, all with the idea of minimizing risk. This sounds good in principle, but there are serious dis-analogies between the biotechnology process in the mid-1970s and METI today. Biotech products were real, but the existence of ET is speculative. Biotech safety mitigation methods could rather easily be devised, less so in METI where the risk is unknown, and mitigation methods more difficult. In the case of METI, action might be delayed for decades rather than for two years, as in the case of biotech (after all, an attempt at consensus on the IAA METI protocols failed in 1995). In the case of biotech, there were many experiments that could be undertaken that were less dangerous but equally interesting, until guidelines could be worked out for specific experiments. In the case of METI, there are of course many possible targets, but not all are equally interesting. Practically speaking, it is difficult to see that SETI or METI practitioners would be expected to forego interesting targets in favor of less promising ones. Finally, if METI practitioners decided not to forego the most interesting targets, there is no enforcement mechanism, short of legislation, that is unlikely to be put in place. By contrast, biotech had government interest in the form of the National Academy of Sciences and the NIH, and the latter might withhold funding if its regulations were not followed. Just the opposite is true for both SETI and METI, which the US government abandoned in 1993 for political reasons.

Paul Berg, one of the four organizers of the Asilomar conference, has argued that today such a process would not work even for biotechnology because of commercial issues, as well as nearly irreconcilable political, religious, and ethical issues in scientific matters. He wrote in 2008, "A conference that sets out to find a consensus among such contentious views would be doomed to acrimony and policy stagnation." The Asilomar process was also criticized for other reasons: DNA co-discoverer James Watson was one among many who complained it delayed science by years; young scientists were worried that their careers would be smothered; still others found the process was good philosophically, but had little bearing on reality. Nevertheless, Berg's parting conclusion was this: "There is a lesson in Asilomar for all of science: the best way to respond to concerns created by emerging knowledge or early-stage technologies is for scientists from publicly-funded institutions to find common cause with the wider public about the best way to regulate – as early as possible. Once scientists from corporations begin to dominate the research enterprise, it will simply be too late." The analog for METI might be that if there is a profit motive, or a reputation motive, or a desire to leave a mark on the universe, it might be too late. While it is possible that these could be motives for SETI and METI, for most practitioners, it seems to me, the motivation is a sincere desire to communicate, to find out if there is intelligent life among the stars. From that point of view, perhaps an Asilomar process could work for METI. But we need to be realistic about the differences between biotechnology and METI, about the vast cultural space that yawns between that era and ours, and about the philosophical ideal and the hard reality when it comes to actually implementing any principles derived from consultation.[23]

As far out as it seems, the METI problem offers many connections to other problems on Earth. Anthropologist Kathryn Denning has pointed a way forward for discussion, suggesting a different framing of the debate, a focus on underlying sociopolitical questions, and analogies to other work that might contribute to the METI debate. She points out that in the end METI and its issues are "the problem of the Commons": "the individual right to action, the accumulated effects of many individual actions upon a society, and appropriate behavior regarding collective resources," in short, a problem of shared resources writ large. That problem, she argued, has no simple technical solution, but it is important to recognize that METI belongs to this class of problems that has a vast literature in economics, the social sciences, and governance, a literature that may be useful for the current debate. And as Mark Lupisella has pointed out, in the end the policy also connects with philosophy and ethics, particularly with cosmocentric thinking as discussed in the previous chapter.[24]

The field of risk assessment also holds lessons. Should we invoke a precautionary principle or is this an irrational aversion to risk-taking? The field is wide open for discussion. In addition, both the SETI and METI issues involve the question "who speaks for Earth?," as Carl Sagan asked in episode 13 of his famous *Cosmos* series. It is an issue sure to test the limits of astropolicy. Yet even as these challenges loom large, message composition continues, begging the question of who should be involved in message composition. In the end METI is part of a larger set of issues having to do with not only the cosmic commons but also with human effects on planet Earth. In his book *Earth in Human Hands: Shaping Our Planet's Future*, astronomer David Grinspoon sees METI in just this way. After detailing the many ways in which humans are affecting Earth on a planetary scale, and debating whether we have entered a new Anthropocene era, he relates that he "began to see the METI debate as emblematic of many of the dilemmas we are facing as part of this planetary transition in which cognition is starting to play a central role in the workings of the planet." He views METI as a global issue, in the same way as planetary protection from harmful organisms, or planetary defense from Near Earth Asteroids, or the rise of the machines of artificial intelligence are. All of these are serious issues, and after wavering on both sides of the arguments, Grinspoon concludes that a "voluntary moratorium" is the best approach until such time as international consultations can take place. As I have intimated, this will not be easy, and such a moratorium might extend indefinitely.[25]

Finally the last of our representative policy issues addresses a very different scenario: direct contact. Interestingly, in the case of biological aliens there may be some overlap with the microbial protocols, because direct contact with biological intelligence could involve problems of contamination. Biological aliens, after all, are likely to have a microbiome no less than Earthlings, and we recall that in *War of the Worlds*, as in European contact with the New World, it was microbes, not intelligence, that destroyed the invaders. And like indirect contact by radio signals or other means, direct contact with biological or postbiological aliens would involve problems of communication and a raft of other issues. In the half-century since Carl Sagan penned his controversial article "Direct Contact among Galactic Civilizations by Relativistic Interstellar Spaceflight," there has been a great deal of controversy about whether aliens would be more likely to communicate by electromagnetic signal or actually travel the universe. The communication paradigm has won out in many quarters, though not on the basis of any principle except dubious alien economics about the cost-effectiveness of interstellar spaceflight. But what if interstellar travel is possible, by either aliens or their robotic surrogates? Or if that is too expensive, what about a universe seeded with active artifacts that could affect Earth as much as direct contact? Serious people are now doing initial planning for humanity's first interstellar probes to the Alpha Centauri system. If we are doing this, another technological civilization may have sent out probes long ago. No one knows what the consequences of such direct encounters would be, but analogs need to be used with extreme caution. Unlike other contact scenarios, the implications of direct contact have not been taken seriously except by a few writers such as Michaud, and in science fiction and the UFO literature. And while much of the UFO literature is not scholarly, and while in my view UFO claims are seriously undermined now that virtually everyone is equipped with a camera on their phone, the impact of direct contact remains a field wide open for study.

As with microbes, many other policy issues arise in connection with other discovery scenarios. But for practical purposes, a good starting point is to develop policy for the most likely scenarios, represented earlier. All these activities represent not policy, but gropings toward policy, a necessary but far from sufficient preparation. In the end the status of policy for issues surrounding the discovery of extraterrestrial intelligence is similar to that of microbial issues: protocols exist for the initial steps of the most likely discovery modes, but beyond the very first steps of "confirm and announce," no protocols exist. That is an unstable situation that raises the risk to humanity. As historian of science Jane Maienschein has emphasized in her role as director of one of the several Biology and Society programs in the United States, in taking on these policy tasks scholars from the humanities, social sciences, and other areas cannot isolate themselves in their discussions and recommendations. Instead they must interact in a meaningful way with both scientists and policymakers. The same holds true in the opposite direction, otherwise the deliberations will remain an academic exercise divorced from political reality. Indeed, preparing for the discovery of life beyond Earth through policymaking bids fair to serve as the leading edge of what Harvard biologist E. O. Wilson has called consilience, the unity of knowledge across the humanities, social sciences, and natural sciences.[26]

Managing Discovery: Ten Stages Involving Scientists, Government, Media, and the Public

Whatever the scenario, the discovery of life beyond Earth will pass through at least 10 stages: confirmation, announcement, assessment of validity and scientific importance, assessment of risk and impact, government involvement, media reporting and opinion, public reaction, public education, continuing scientific research on the subject, and taking action after all due deliberation. These steps will overlap and will unfold on an unknown timetable. We know this not only because it is a logical progression of overlapping events, but also because of history, especially as illustrated in the case of the famous Mars rock ALH84001 in 1996, a process we saw unfold in Chapter 1. That claimed discovery, it is worth remembering, involved only possible fossil life, not living microbes. Nevertheless, as that case and numerous other public policy issues have demonstrated, "managing" is a new and distinct domain from "preparing," and discovery is no exception.

Unlike the agonizingly slow pace of the preparation phase, in the wake of discovery events take on a life of their own. Actions that could be contemplated in the leisure of science meetings now must be formulated in real time as events unfold. And while it is the perpetual hope of policymakers that their policies will lead to optimal or at least beneficial outcomes, that will occur only if policy is well managed and assiduously implemented. Realistically, however, even if policy exists for a specific scenario, the unexpected will undoubtedly force real-time changes. But altering a policy on the fly is clearly better than having no policy at all. As Eisenhower and Churchill both concluded based on their wartime experiences, "Plans are worthless, but planning is everything." This is no less true in the following case, and the following represents my best estimation of the trajectory of events upon discovery of life beyond Earth.

Confirmation

At a scientific level, the first logical step in the wake of any putative discovery, and the consensus first step in SETI's post-detection protocols, will be to confirm the reality of the discovery. This will almost always be an extended affair, first of all because, as I argued in Chapter 2, discovery is always an extended process rather than a eureka moment, and second because reputable scientists do not like egg on their faces as a result of a false announcement. On the other hand, they do not like to be scooped either.

The case of the Mars rock is informative from the point of view of both history and analogy. From the time the rock was recognized as Martian in origin, almost three years of intense study had taken place before NASA scientists dared make the announcement about potential fossil life. As we saw in Chapter 1, for fear of leaks, during that time scientists at Johnson Space Center where the work was being done were circumspect about discussing their results even within NASA. By April 1996 they had enough confidence to submit an article to the prestigious journal *Science*, at which time NASA Headquarters began to play a role. The claim was closely held at the Headquarters level until mid-July, when peer review was complete and publication was scheduled for mid-August. On July 30, the NASA administrator informed the White House, and NASA began receiving news inquiries August 1. Although they had planned for the announcement to coincide with the publication of the *Science* article in mid-August, because word was leaking out at an increasing pace, NASA was forced to move the announcement forward to August 7. As detailed in Chapter 1, the timing of the announcement was also influenced by political events, since the Republican National Convention was scheduled to take place at about the time the announcement was originally scheduled. In these days of social media, it is highly unlikely such a discovery could be kept secret as long as the Mars rock was. And if ever there was a law of human behavior, we can bet that political calculations will be factored in to any decision to announce, or not announce.

Announcement

The second step is the announcement itself. As with the Mars rock, the timing of this step is likely to be overtaken by events. The planning and content of the announcement itself is important, setting the stage for everything that follows. Since the Mars rock research centered on NASA, a government institution, even with the rush to announce, it was well prepared, drawing on NASA's vast experience in public affairs in making such discovery announcements. The August 7, 1996, press conference was hastily called, but the audience included government officials such as the head of the National Science Foundation and the National Academy of Sciences. Speakers included not only the scientists making the claim, but a UCLA professor representing the many who were skeptical of the claim. He argued the evidence was circumstantial, and cited Carl Sagan's dictum that "extraordinary claims require extraordinary evidence." NASA's decision to include a skeptic was important, one that should be duplicated at any future announcements of discovery of life. Uncertainty about the claim is sure to be the hallmark of any such announcement, and perhaps counterintuitively, skepticism adds to the professionalism of the announcement.

The fact that the Mars rock research was centered around a government institution rendered the whole process more susceptible to control. In a nongovernment and therefore less-controlled setting, news of a discovery may well leak even before the confirmation is made. This is exactly what occurred in early summer 1997, when astronomers at the National Radio Astronomy Observatory in Green Bank, West Virginia – the same observatory where Drake carried out Ozma – detected a promising narrow band signal, confirmed as extraterrestrial in origin. It was the most promising such signal ever seen. As SETI astronomer Seth Shostak recalled, despite being very familiar with the SETI protocols, when the *New York Times* was tipped off and called the next morning, he did not lie. He said they were indeed checking out a signal, and to check back three hours later. By that time, they had determined it was a telemetry signal from the Sun-observing satellite SOHO. The rather astonishing lesson, according to Shostak, was that "The SETI protocols, while well intentioned, aren't particularly useful in real life."[27]

A great deal more could be said about the proper way to make important scientific announcements, but suffice it to say both the staging and the content are of utmost importance to all that follows. Early decisions can have a butterfly effect that may seriously undermine the ability to manage subsequent events effectively.

Assessment of Validity and Scientific Importance

The third step, which will undoubtedly overlap with the first two, will be to assess the validity and potential impact of the discovery using one of the scales derived just for this purpose. Flawed as they may be, they are all we have as of now. As discussed in Chapter 6, the London scale is the most comprehensive of these, encompassing both the microbial and intelligence scenarios. The London scale combines factors of scientific importance with credibility, resulting in a scale ranging from 0 to 10, representing insignificance to extraordinary importance, validity, and impact. For example, assessing the Mars rock evidence and credibility in 2010, 14 years after the announcement and much skepticism, the authors of the London scale gave it a preliminary estimate of 3.6 on the index. That was based on values of 2 for the life form (terrestrial with some uncertainty), 2 for the nature of the evidence (fossilized), 4 for the type of discovery (in situ), 4 for the distance of the life form at time of discovery (nearby, on Earth's surface), and 0.3 for the reliability (testable but needs further evidence), yielding $(2 + 2 + 4 + 4) \times 0.3$, or 3.6.[28]

Scientists have also devised a much simpler scale for the case of METI, sending purposeful transmissions from Earth. Known as the San Marino scale, it ranges from 0 for insignificant to 10 for extraordinary. The scale has only two parameters, intensity or strength of the signal sent (ranging from 0 for low to 5 for very high), and character of the transmission (also ranging from 0 for a beacon with no message such as a planetary radar to 4 for a continuous, omnidirectional broadband transmission of a message, and 5 for a reply to an extraterrestrial message). Scientists have also quantified past transmissions based on this scale, assigning

Drake's 1974 Arecibo message as an 8, which characterizes its significance as "far reaching," and a series of three Russian transmissions from the Evpatoria radio telescope in Ukraine as a 7, "high." Near Earth Object radars get a 6 for "noteworthy."[29] This scale does not take into account the probability of being detected. For example, Drake has stated that by the time the Arecibo beacon reaches its target, the central part of the Hercules globular cluster some 25,000 light-years away may well have rotated out of the beam. Omnidirectional signals might be seen everywhere, but for economic and technical reasons the signal strength would likely be much weaker than in the case of a targeted beacon. Nonetheless the London and San Marino scales remain important reference points from which scientists can undertake a more focused discussion.

Assessment of Risk and Impact

Once validity and scientific importance have been assessed, the questions of risk and overall impact will inevitably arise. Managing any discovery – ranging from new microbes to near-Earth asteroids – also involves managing risk, a common practice in many areas of life where a variety of approaches might be beneficially employed for the discovery of alien life. The authors of the London scale peripherally provided four risk categories, ranging from I (no risk) to IV (assumed biohazardous until proven otherwise). But broader studies might also be applicable. The World Economic Forum adopts a scheme that distinguishes preventable, strategic, and external risks. Preventable risks are human mistakes, strategic risks are undertaken voluntarily, and external risks come from the outside. All three of these categories might be applied to alien life. Astrobiology, SETI, and METI might all be considered strategic risks, where it becomes a policy question of risks versus scientific benefit and exploration. External risks could come from a message or direct contact. In all cases, human mistakes will likely be made. Arguably, strategic risks are necessary in life, otherwise humans would never do anything, ranging from going out the front door to flying the space shuttle. The question is always what level of risk is acceptable, and to whom, especially when the risks are global.[30]

A huge literature exists on risk and preparation for extreme events such as climate change, terrorism, and the emergence of dangerous technologies, with an eye toward designing resilience for the aftermath of these events. One group of experts in a collection on *Designing Resilience* defines stages of mitigation, prevention, preparation, response, and recovery. Another collection on *Global Catastrophic Risks* considers potential global risks such as asteroid impacts, nuclear war, global warming, nanotechnology, and artificial intelligence, again with a view toward policy issues and crisis management. There are sure to be lessons in this literature, but little has been done to apply this work to the case of the discovery of life in any form. Again, an urgent task is to determine which global risks are most analogous to those related to SETI and METI, to alter mitigation strategies accordingly, and to apply lessons learned to any given scenario. Given the human penchant to use analogies when any new situation arises, this will likely be the thinking in any case, but preparation before the event occurs tends toward a better outcome.[31]

One obvious question is, who should make the determinations of importance and credibility for the London index, and for the separate categories of risk. For the SETI community, it might be the International Academy of Astronautics SETI Committee, but for non-SETI discoveries, no such body exists, other than the general consensus of the scientific community, which is not an easy thing to pin down and in any case takes time. It might be appropriate to refer some issues to the United Nations, which already assesses pathogens and other bio-risks. But it is often not easy to get the attention of the United Nations until an event occurs. Thus, who should determine the extent of the risks taken remains a wide-open policy question. One thing is certain: it should not be left only to the natural sciences, but to a wide array of scholars and policymakers.

Government Involvement

History and common sense indicate that at the first inkling of an intelligent signal, not to mention an alien spaceship landing on Earth, governments will become actively involved no matter what other protocols are in place. In the SETI arena novels such as Carl Sagan's *Contact* are likely realistic to the extent that they invoke a role for scientific institutions, executive branches of national governments, international agencies, and perhaps the military. But history also indicates the same is true of microbial life, and even fossilized life based on the reaction to the infamous Mars rock in 1996. Questions will immediately be raised as to "who among scientific, governmental, or theological institutions should be involved in making decisions for humankind about the advisability of further contact or interactions with ET life, particularly if it is 'just' microbial." Albert Harrison argues that governments should have a strong central planning effort to deal with the discovery of alien life. The plan would, at a minimum, "establish jurisdictions, assign roles and above all make provision for a coordinated but flexible response." To date no such plan exists.[32]

Role of the Media

An inevitable step – possibly even before confirmation if history is any guide – will be to deal with the media frenzy that will surely occur when extraterrestrial life is discovered. If the Mars rock is any indication, discovery would generate a media frenzy, with information urgently demanded from government and scientific institutions, either in the country making the discovery, or internationally under the proper circumstances. As Morris Jones and others have pointed out, social media will greatly affect the situation in ways the SETI post-detection protocols, written in the 1980s, never anticipated. The new media landscape, including cable news, Facebook, and Twitter, has revolutionized mass communications even since the Mars rock claim. In this online media environment, misinformation, rumors, and hoaxes will abound, while "the presentation and interpretation of a discovery will be hotly debated and contested." Both theories of mass communications and historical cases of announcements of major scientific discoveries or incidents such as nuclear accidents could be helpful by way of preparation.[33]

Public Reaction

The media reaction will be an important factor in the public reaction. Margaret Race has emphasized that the discovery of life will be complicated by "public attitudes, misperceptions, Hollywood-style science fiction, national interests, and ethical/theological considerations." Linda Billings, an expert in communications and culture, argues that the media has conditioned the public to believe in an outcome that may in fact never occur. She believes the cultural environment and the human psyche shape our representations of extraterrestrial life and our potential reactions to its discovery. Since we know nothing about extraterrestrial life, she emphasizes, these expectations may be entirely wrong. This does not exactly instill confidence in our ability to handle a discovery, but it is a wake-up call that events could unfold much differently than we expect.[34]

Public Education

Quite aside from the role of the media, communication, education, and outreach will be crucial to managing expectations and indeed to the impact itself. The World Economic Forum report concluded, "Through basic education and awareness campaigns, the general public can achieve a higher science and space literacy and cognitive resilience that would prepare them and prevent undesired social consequences of such a profound discovery and paradigm shift concerning humankind's position in the universe."[35] Ideally, this would be done prior to discovery, but it will be especially important in the midst of discovery while the media is focusing attention on the subject. If the claim is true, over the long term the public will gradually transform its worldviews – not just individually but collectively, as outlined in Chapter 7.

Further Scientific Research

During all of this activity scientific investigations of the discovery will continue, feeding back into the loop beginning with the first step. Should the discovery turn out to be weak or spurious, research will fizzle and the feedback loop beginning with confirmation and announcement will dampen. If the discovery turns out to be strong, the cycle of steps enumerated here will be repeated again and again, on a schedule and frequency that depend on the timetable of discovery.

Informed Action

Even in the cases when scientific claims turn out to be false positives, actions will often have to be taken: whether to pursue more research, determine how to improve techniques based on mistakes made, or analyze media and public reaction. In the more interesting cases where a discovery turns out to be true, in the midst of and following all of the stages described earlier, informed actions will have to be taken. In the case of the discovery of microbial life: How do we prevent contamination and back contamination? Should sample microbes be returned to Earth and at what biosafety levels, or should they be studied in situ? Which of the ethical principles laid out in Chapter 8 should be applied? Should alien microbes be employed for practical purposes? In the case of the discovery of intelligence: How should we react to the receipt of an intelligent signal? If it contains a message, should we reply and on what principles should a message be constructed? In the absence of a signal, should we initiate one? Whether SETI or METI, who speaks for Earth? In short, all of those issues discussed earlier, and more, will now reach the decision stage, but now reaching beyond the realm of discussion to the realm of action. The most basic decision is whether there should be any action at all, but given the existential stakes and human proclivities, there will almost certainly be action of some sort. The key is that any action should be *informed* action. That is why it is important to discuss, understand, and implement to the best of our abilities each of these stages – before discovery occurs.

The first two sections of this chapter highlight the almost virgin territory open to experts from a variety of fields wishing to undertake studies related to the preparation and management of astrobiological discovery. A beginning has been made, but only a beginning.

Embracing Discovery: From Space Law to Metalaw

Some of the strategies, protocols, and policies we have discussed in this chapter may broadly be considered part of the body of knowledge termed "space law."

Space law is sometimes simply defined as "the international and national laws that govern human activities in outer space," or as the UN Office of Outer Space Affairs puts it in more detail, "a variety of international agreements, treaties, conventions, and United Nations General Assembly resolutions as well as rules and regulations of international organizations." Within its purview, for example, are "the preservation of the space and Earth environment, liability for damages caused by space objects, the settlement of disputes, the rescue of astronauts, the sharing of information about potential dangers in outer space, the use of space-related technologies, and international cooperation." Among its fundamental guiding principles are "the notion of space as the province of all humankind, the freedom of exploration and use of outer space by all states without discrimination, and the principle of non-appropriation of outer space." Planetary protection policies clearly fall within this broader definition, and both COSPAR and the UN Treaty on Outer Space have endorsed some of these policies (Table 9.1). This gives them considerably more force than the SETI Post-detection Protocols, which have not risen above the level of the international scientific organizations in terms of official endorsement. Even with these weak and unenforceable protocols, SETI is way behind the curve when it comes to legal implications of discovering intelligent life.[36]

Alien Microbes and Space Law

The body of space law developed over the past 50 years, including contamination policies and SETI protocols, is a considerable achievement. But it represents only a small portion of the legal issues that will arise in the event of actual discovery of life beyond Earth. The earliest urgent astrobiological issues to demand the attention of space law will likely occur when microbes are discovered. While planetary protection policies are designed to come into force before microbes are discovered, upon discovery the issues change dramatically. The most urgent issue would be a direct exposure on Earth, whether by accidental exposure by sample or astronaut return, or a natural exposure as a result of panspermia or interplanetary transfer of matter. Policies here might be informed by past experience with epidemics, invasive species, and similar analogs.

Less urgent, but no less compelling will be the human or robotic discovery of microbes as a result of the exploration of Mars, comets, or those oceanic satellites of the outer planets. Suppose microbes are discovered on Mars. What should be the procedure, aside from the first directive to "do no harm" to Earth or the indigenous body? Policy and law, informed by astroethical principles, will be needed to deal with these issues, which I touched on in the previous chapter.

We should also recognize that, if we decide that alien microorganisms have instrumental value in addition to intrinsic value – the ethical dilemma discussed in the previous chapter – there may be beneficial uses for humans. This is indeed likely to occur, in the same way the extremophile microbes and viruses such as HIV discovered on Earth in the past few decades have been quickly adapted for use in pharmaceuticals, their initial and continuing purely scientific importance notwithstanding. When that issue comes to the fore, it will involve questions about claims, ownership, patents, and damages. Whether deliberations will be carried out through COSPAR or national institutions remains to be seen, but it is not too early to think about the issues.[37]

Alien Intelligence and Metalaw

While the handling of issues in the wake of the discovery of microbes may be seen as an extension of space law as it now exists, the questions and actions arising with the discovery of extraterrestrial intelligence seem so different in kind as to require a new category of law. As we have seen, the SETI post-detection protocols address only the very short-term issues upon and after discovery. For the longer term we move from space law into the terrain of what Andrew G. Haley in 1956 presciently termed "metalaw," referring to precepts of law that might be universally applicable to all intelligences. Haley, one of the world's first practicing space lawyers, literally wrote the book on *Space Law and Government* in 1963, one of two massive tomes on the subject written in that year. The book dealt largely and originally with the multiple issues then arising with the beginning of the Space Age, ranging from space vehicle regulations and space communications, to liability for property damages in space activities and sovereignty over celestial bodies. But in his last chapter Haley turned to metalaw, a new concept that has received only sporadic attention in the decades since, despite being introduced into popular culture in G. Harry Stine's science fiction novel *A Matter of Metalaw* (1986).

Influenced by the astrobiology work of astronomers Harlow Shapley, Carl Sagan, Frank Drake, and the budding NASA exobiology program, Haley began his chapter on metalaw with the rather shocking statement that "the indefinite projection of anthropocentric law beyond the planet Earth would be the most calamitous act man could perform in his dealings with the cosmos." In his view, metalaw, which he defines as dealing "with all frames of existence – with sapient beings different in kind" from humans, could not simply be an extension of international law. Instead of the usual Golden Rule, "do unto others as you would have them do unto you," Haley proposed the Rule of Metalaw, sometimes called the Interstellar Golden Rule: "We must do unto others as *they* would have done unto them," in other words, we must act as aliens want us to act. At first glance, this seems absurd. In and of itself it seems to dictate that if aliens want us to surrender we should do so, a recipe for the conquest of Earth. Haley was not that naïve. Rather he had in mind that we must not force our legal concepts on other intelligent beings as the Europeans did, for example, on the native American Indians during the Age of Discovery. He explicitly invoked that venerable sixteenth-century culture contact analogy, writing that "Quite apart from all considerations of altruism, we must bear in mind the hapless possibility that the situation might be reversed, and we may turn out to be the savages who are decimated and enslaved."[38]

With that colorful and cautionary (but flawed) analogy in mind, Haley's more robust idea of metalaw was that it would be designed to protect all parties involved. Law, however, can protect people only to the extent there is someone who has the power to enforce it. In other words, a system of law needs to be imposed by force, which entails penalties for noncompliance, otherwise it is not really a law. Like other pioneers in the history of law on Earth, Haley was striving for an ideal. The general principle, he argued, should be that no force of any kind may be used. He proposed the concept of absolute equity between any and all sapient parties, leaving open the possibility of "an indefinite number of frameworks of natural laws." While admitting that it was difficult to formulate principles of metalaw without knowing anything about the other parties, Haley foresaw that "in the not distant future, lawyers, scientists, and sociologists will undertake studies of a substantive statement of metalaw, and as a by-product of these studies our own anthropocentric law undoubtedly will be improved." In short, Haley was grasping for universal principles of law in the same way biologists would soon be striving for a universal biology during the search for life in the universe. Or as a later writer put it, Haley was striving for "legal precepts of theoretically universal application to all intelligences, human and extraterrestrial."[39]

Fifty years later, Haley's vision for metalaw remains a long way from being fulfilled. It was 1970 before Austrian space law pioneer Ernst Fasan would elaborate the idea of metalaw in his pioneering book *Relations with Alien Intelligence: The Scientific Basis of Metalaw*. Space pioneer Wernher von Braun, only a few years from death after a career that both resulted in the V-1 and V-2 bombing of London and propelled humans to the Moon, wrote in his foreword to Fasan's book that the discovery of intelligent life was highly probable even if it remained a "challenging task." Fasan defined metalaw as "the entire sum of legal rules regulating relationships between different races in the universe." Like Haley, he found the "scientific basis" for metalaw in Kant's categorical imperative, an absolute requirement to "act only on such a maxim as you can will that it should become a universal law." Using Kant as the basis Fasan proposed 11 metalegal principles in priority order:[40]

1. No partner of Metalaw may demand an impossibility
2. No rule of Metalaw must be complied with when compliance would result in the practical suicide of the obligated race

3. All intelligent races of the universe have in principle equal rights and values
4. Every partner of Metalaw has the right of self-determination
5. Any act which causes harm to another race must be avoided
6. Every race is entitled to its own living space
7. Every race has the right to defend itself against any harmful act performed by another race
8. The principle of preserving one race has priority over the development of another race
9. In case of damage, the damager must restore the integrity of the damaged party
10. Metalegal agreements and treaties must be kept
11. To help the other race by one's own activities is not a legal but a basic ethical principle

We see immediately that many of the principles of metalaw are designed specifically to prohibit conquest in either direction, of the Earth or by the Earth. Every form of intelligence is allowed to defend itself, and "practical suicide" or genocide are not allowed. Fasan recognized that human interaction with aliens might be complicated, and would depend on the mode of contact (ranging from remote to direct), on the level of intelligence (inferior, equal, and superior), and on the nature of the intelligence ranging from peaceful to aggressive. Nevertheless, all contacts must take place in the framework of metalegal rules: "They alone will provide us and any alien race with the basis for an enriching and rewarding contact between the intelligences in our universe."[41]

It is interesting to note that the metalaw of Haley, Fasan, and their successors boils down to something very similar to Asimov's Three Laws of Robotics. In fact, Fasan, apparently independently of Asimov, found that his 11 metalegal principles could be reduced to three: a prohibition on damaging another race, the right of a race to self-defense, and the right to adequate living space. Some of these are also similar to the Prime Directive of *Star Trek*. In other words, on Earth at least, and to the degree the subject has been discussed at all, there has been a convergence of ideas about how we might treat aliens – a convergence that should feed into our ideas of astroethics as discussed in the previous chapter. Whether aliens would "agree" on these ideas is another question. That hinges in part on the issue of whether universal moral and ethical principles exist that would tend to result in universal legal principles. In the end, as with so many other issues in this volume, metalaw is a search for universality, in this case a kind of "universal law."

Metalaw has been criticized for many reasons, including (in Haley's Interstellar Golden Rule) its seeming attempt to raise alien interests above human interests; its subjective concepts of good and bad; its lack of enforcement; and its adherence to a natural law theory of jurisprudence. The latter holds that just laws are immanent in nature, that they are, like Platonic theories of mathematics, waiting to be discovered rather than invented, and so are independent of the lawmaker. Natural law theory also links universal law directly to morality. Haley and Fasan's principles of metalaw have also been faulted for their potential naïveté from the point of view of the alien. As outward-looking lawyers such as George Robinson and Robert A. Freitas Jr. have invariably noted, one intelligent race's "reason" might not be another's, and in fact not all intelligences might even have developed the concept of law.[42]

All of these are rather basic objections to metalaw. Nevertheless, ideas of metalaw continue to be elaborated sporadically, both in fiction and in reality, where they are tied to specific scenarios. As mentioned earlier, science fiction writer G. Harry Stine (writing under the pseudonym Lee Correy), envisioned one case in his 1986 novel *A Matter of Metalaw*. In his scenario, the Kalocs of Khughar invade the planet Thya, seemingly a clear violation of the canons of metalaw, which Stine lays out in his Preface on Definitions, Rules and Canons. An organization named SIGNET, whose mission is to enforce the canons of metalaw universally accepted by the interstellar community of 1,000 worlds, dispatches an investigation team to the planet after one of its teams has already been destroyed. They find that the Kalocs are a genetically altered race excelling in bioengineering that nevertheless allowed itself to become inbred with a non-survival trait that stimulated pleasure. The details are rather X-rated, but the bottom line is the Kalocs have embraced the idea of conquest by spreading this genetic trait in search of more pleasure. In giving pleasure they believed they were following the canons of metalaw, while the Thyans thought they were following it too by resisting invasion. As in law, metalaw is open to interpretation, especially where different species are involved. The Thyans eventually win because they kill the invaders while the Kalocs continue their single-minded goal of seeking pleasure. Stine cannot resist an analogy: a handful of Spaniards wiped out the Aztecs because the Spaniards were out to destroy the Aztecs, while the Aztecs wanted only a few sacrificial victims. This may be neither good history nor good science fiction, but Stine did succeed in contemplating the implications and ambiguities of metalaw throughout the novel, illustrating what happens when species with very different values and worldviews make contact.

The ideas of metalaw could be applied to many science fiction scenarios, ranging from H. G. Wells's *War of the Worlds* and *Star Wars* to *E.T.* and *Arrival*. As in Stine's novel, the action of the Martians in Wells's novel was a clear violation of Principle 5 in Fasan's canon of metalaw. Invasions are not allowed. But other cases are less clear. Is the Rebel Alliance in *Star Wars* allowed to attack the tyrannical Galactic Empire, when that Empire was clearly in the wrong to start with? Should

E.T. be returned to his home planet, or kept on Earth for further scientific study? Were rogue elements of the military justified in trying to destroy the heptapods in *Arrival*, even though they appeared in the skies of Earth seeking help? Without metalaw no basis exists for answering these questions, aside from moral considerations combined with terrestrial law – which would be stretched beyond its limits to say the least. Most science fiction scenarios are in the situation of the wild, wild West, and in the absence of metalaw we might be too if contact is made with aliens.

The work of devising a system of law dealing with intelligent beings of all kinds remains a task for the future. American lawyer Adam Korbitz has proposed that the process of elaborating such laws should be called "celegistics," combining "cele" (celestial), with "legistics," the drafting of law or legal documents. One way of proceeding would be to employ analogy with current law. To what extent could international law be modified to deal with the specific circumstances of alien contact? To the extent international law has any binding force on Earth, it generally depends on international agreements to enforce certain legal norms, or the willingness of superpowers to enforce their own legal norms unilaterally outside of their borders. Would alien civilizations favor one or the other approach? These may seem like science fiction questions now, but they will quickly become very real once contact is made.

Aside from analogies of this kind, Korbitz points to analogies of another kind as a way forward in celegistics: to borrow legal concepts from fields dealing with "the other," including artificial intelligence, robotics, transhumanism, and animal rights. A considerable literature already exists on legal aspects of these subjects, and once again the question becomes one of the proper use of analogy. Of these possibilities, in my view, the most promising is to study in more detail our relations with artificial intelligence, or what Nick Bostrom has called superintelligence. Not only is this a developing reality on Earth, as I have argued in Chapter 4 and elsewhere, the "postbiological" universe is a likely scenario, given the long timescales intelligence has had to develop. In making such comparisons, of course all the precautions of using analogy set forth in Chapter 3 apply, including possible differences in legal treatment between AI we have created and biological or postbiological aliens. Still, the questions we ask about aliens and those we ask about AI are much the same, including questions of the nature of intelligence, whether aliens and AI will be friendly, and how we should interact with each of them. These kinds of analogical studies could prove immensely illuminating, until such time (likely not until extraterrestrial intelligence is discovered) that metalaw is incorporated into international agreements or law at one of the upper levels of Table 9.1.[43]

The question of timing is an important one. Why, one may well ask, worry about this now? Why not wait until we have a specific scenario that we can focus on and then address problems as they arise? Military planners have an answer for this – they do not wait for a problem to arise to think about it, if only because different scenarios require the deployment of different resources, sometimes quickly. And it is always better to think ahead and have time to consider options in a thoughtful way rather than react in the passion of the moment. Scientists also have an answer. In the United States the Centers for Disease Control, for example, attempts to lay out a strategy for flu season well ahead of the season itself, formulating its vaccine based on recent history, not always with complete success. Together with other organizations such as the World Health Organization, it surveys the world in an attempt to foresee potential health problems, acting the best it can with limited information. Surprising events such as an Ebola outbreak may happen, but the bottom line is that it is always better to be prepared for likely events, with the goal of minimizing risks to humanity. The organizers of the Royal Society of London meeting captured the stark alternatives of timing: "If extraterrestrial life happens to be detected, a coordinated response that takes into account all the related sensitivities should already be in place … Will a suitable process based on expert advice from proper and responsible scientists arise at all, or will interests of power and opportunism more probably set the scene?"[44] Even with a process in place, it is likely to be a combination of both. But with nothing in place, power and opportunism will likely prevail. When the discovery of E.T. does happen, the impact of alien contact is a double-edged sword for jurisprudence: on the one hand the possibility of contact is a strong argument that some form of metalaw must be developed in order to deal with interactions with aliens, whether or not it follows the precepts of Haley and Fasan and their successors in celegistics. On the other hand, in the very process of attempting this, terrestrial jurisprudence will be affected and hopefully improved.

It is important to understand that not all implications of discovering life should be a matter of space law or even public policy. Most would agree that there should be plenty of discussion but no policy about what theologians and particular religions should say and do when alien life is discovered. We cannot dictate policy about how worldviews will change, or how society in general will be affected. That will be a matter of history and cultural evolution. As in other matters of law, metalaw should strive for a balance between freedom and order.

For those areas where policy and law can be addressed, action is called for now, since contact with alien life is, as the World Economic Forum put it, one of the "X factors" for the future of civilization, and one undergirded by scientific principles. It is true that it is occasionally said that METI and SETI are not science, just as it was said in the early days that astrobiology was not science because it had no subject. In my view that is a grave misunderstanding of science, one that could have been applied to almost every scientific endeavor before the desired discoveries were made, including most recently exoplanets and gravitational waves. Science in its grandest vision is exploration, and that is the proper context for the debate as we move forward into a biological universe. I cannot say it any better than Margaret Race already has in the context of one of the many meetings held to grapple with these profound issues: "How we respond in the short or long term to the discovery of extraterrestrial life has significant repercussions for current and future generations on planet Earth. It likewise may have significant impacts on the extraterrestrial life as well. As we plan to communicate about scientific efforts and possible successes in the ongoing search for extraterrestrial life, we must deliberately take steps to consider them in the broadest context – that of responsible exploration for all."[45] Meeting the alien will be an experience we cannot afford to mismanage. In so many ways and more than ever, failure is not an option.

10

Summary and Conclusions

At Home in the Biological (or Postbiological) Universe

Lay down

Your sweet and weary head.

Night is falling.

You have come to journey's end.

"Into the West," Lord of the Rings[1]

Some will consider this book an exercise in fantasy. It is true that we do not know if life exists beyond Earth. But I believe our Copernican and Darwinian presuppositions – that the Earth does not hold a special place in the universe and that life will develop by natural selection wherever conditions are conducive – are the most likely of any set of presuppositions because they invoke what we believe are universal laws and actions. For all that, they are indeed still presuppositions, laid bare for all to see. Yet I have watched as skeptics argued as recently as 30 years ago that no planets exist beyond our solar system. We now know virtually all stars have planets, just as one would expect with universal laws in action. I have watched as skeptics have said, yes, but these are gas planets, not Earth-sized. We now know there are plenty of Earth-sized planets. And I have watched as skeptics have said, yes, they may be Earth-sized, but they orbit close to their parent stars and so are not Earth-like. We now know that a fraction of them are in the habitable zones of their stars, and when this fraction is scaled up to the Galaxy and the universe, the number of Earth-like planets is likely enormous. Diversity in planetary systems is tremendous, and this is exactly what one would expect, at least with the benefit of hindsight.

Indeed, we do not yet know if life actually arose on those Earth-like planets, but biosignatures and other techniques will tell us soon enough. The principle of plenitude, the uniformity of nature's laws, and the principle of mediocrity – that the evolutions of planets, life, and intelligence are typical rather than exceptional – have worked so far, and I believe will be borne out again. Life will be diverse, perhaps beyond our imaginings as even Galileo conjectured 400 years ago, but it does exist in abundance. We will have to deal with it, in the same way we have to deal with life on Earth. The difference is that we may no longer be at the apex of life, as we fancy ourselves on Earth. But we must adjust to that reality, just as we have had to adjust to successive decenterings from anthropocentrism since the Copernican revolution. Indeed, the discovery of life will mark the completion of the Copernican revolution, or at least another step along that revolutionary path. That is *our* worldview adopted in this volume, a view that animates the study of the societal impact of life beyond Earth no less than other X factors in the future of humanity laid out by the World Economic Forum and others.

Given this worldview of what I call in Chapter 7 "the biological universe" – whether microbial, intelligent, or anthropic – it has been our position throughout this book that it is only prudent to contemplate what the implications might be when we do find life out there, even if it is postbiological. I have tried to lay out valid approaches to the problem, urging all due precaution. I have at least recognized our anthropocentrism, our inability to transcend it completely, and yet our need to try in the face of potentially Earth-shaking discoveries. With these tools in hand, I have attempted to lay out various scenarios of how the discovery of life beyond Earth will impact society and perhaps transform our thinking, including our cosmological, theological, and cultural worldviews, as well as our views on ethics and our interactions with alien life in the form of space law and metalaw. In doing so, I have, to the limits of my ability, made use of fields as diverse as philosophy of mind, epistemology, evolutionary psychology and ethics, the philosophy of the natural and social sciences, and law. It is time to gather together my conclusions, both as a source of further contemplation and as a call to action.

Characterizing Discovery and Impact

History, discovery, and analogy converge on at least five solid conclusions. First, the impact of finding life will depend on the discovery scenario. For both microbial and intelligent life, I have characterized these as direct or indirect, terrestrial or extraterrestrial, and specified how these different scenarios might affect humanity. Second, both the discovery and its impact will occur over an extended period of time, again depending on the discovery scenario and the nature of the evidence, and the impact will be proportional to the distance of the encounter. Third, the media will play a significant role in how the discovery is perceived, and public education will help determine to a large extent how the impact plays out. Fourth, the precise impact cannot be predicted, but guidelines are possible, based on history, discovery, and analogy; on societal impact models such as are used in other endeavors; and on a prudent search for universals in fields ranging from biology to social science and law. Finally, we can and must prepare for discovery in order to manage it in a way that is beneficial to humanity, making use of broad research from the humanities and social sciences, as well as from studies of risk and resilience, adapting their results from other areas to the problem at hand and expanding as necessary. The future of humanity may depend on this preparation.

Preparing for Discovery

Preparing for the discovery of life beyond Earth is a subject whose time has come. Chapter 6 examines some of the studies that have been sporadically undertaken by organizations and individuals, but with little notice or impact thus far. More recently a few forward-looking individuals have made specific recommendations. Former US State Department Foreign Service Officer Michael Michaud, culminating his studies on the subject over several decades, offers concrete steps that can be taken now, ranging from practical suggestions like establishing a database of literature on the subject of impact, to composing messages in the event of detection – messages, he hastens to add, that should not be sent without serious consultation. He urges reaching outside narrow disciplines, globalizing the debate to bring in non-Western perspectives, rehearsing the reaction to different discovery scenarios, raising consciousness among policymakers, and actually preparing a plan for international action. In the end, he cautions, we will have to learn to live with ambiguity, recognizing the limitations of prediction, and cultivating patience. But all of this is no excuse for inaction in what he characterizes in Kuhnian terms as a "paradigm break," what could be the greatest of X factors in the history of human civilization.[2]

NASA engineer, biologist, and philosopher Mark Lupisella grounds the subject of preparing for impact in a particular philosophical stance. He argues that it is important to recognize that both biological and cosmic evolution affect our pursuit of values, and presumably the pursuit of values by other beings in the universe, if indeed they pursue value at all – a fundamental question fraught with implications for our interaction with them. As discussed in Chapter 8 in connection with a cosmocentric ethic, he suggests that because all life likely evolves from the same cosmos, the universe can be seen as the ultimate source of creative activity and can inform our ethics. "Extraterrestrial beings may see themselves as a means by which the universe has been bootstrapped into the realm of value, culture, meaning and purpose via beings that bring those properties into the universe, giving rise to a kind of 'bootstrapped cosmocultural evolution,'" he writes. This is also consistent with the philosophy of Clément Vidal, also mentioned in Chapter 8, that the infinite continuation of the evolutionary process is the ultimate good.

Lupisella argues that such factors, including our knowledge of natural selection, cultural diversity, and cosmic evolution, are important in preparing for contact. Based on this philosophical stance, he elaborates 10 practical considerations that follow from this point of view. These include a cosmic humility and open-mindedness as we prepare for the unknown; rational analysis of potential long-term futures of humanity and life on Earth that may inform us of the nature of extraterrestrials; use of these factors in message construction either in reply to a message or in an active SETI (METI) program; the expectation that morals informed by a cosmic worldview might give us a common ground for ethical considerations; realizing the importance of examining the ethical considerations of a postbiological universe; recognizing that self-interest may be a basic feature of all intelligence, and that it may be difficult or impossible for those intelligences to transcend selfishness; and recognizing that because the values of intelligent beings throughout the universe could be extremely diverse, we should study certain nonhuman animals, as well as make use of the fields of evolutionary psychology, cognitive science, and neuroscience to prepare for potential psychologies of nonhuman intelligence. He suggests that analogs from the social and behavioral sciences may be cautiously applied, just as they are in the natural sciences. Finally, Lupisella suggests that intelligence may only be a means to an end, a universe in which not knowledge, but characteristics such as caring, fairness, and diversity are most important. In such a "post-intelligent" universe, he suggests that a "values principle" or "wisdom principle" may be more important than an "intelligence principle," an important counterpoint to those who consider intelligence the ultimate goal of the universe.[3]

The recommendations of Michaud and Lupisella are made largely in the context of extraterrestrial intelligence. I have argued throughout this volume that in order to prepare for discovery, we must include microbial life discovery scenarios and undertake further studies in a variety of areas, including not only the natural sciences but also the social sciences and humanities. This book is only a beginning, a preliminary reconnaissance. The following recommendations, all grounded in the discussions and conclusions in the previous chapters, are my own. But they gain force from the fact that some are also recommendations independently made in other group deliberations, including some of those discussed in Chapter 6 sponsored by NASA, the Library of Congress, and the Royal Society of London.

First, full-scale studies should take place of the promise and problems of analogy in connection with the discovery of life in its many forms, making use of the latest research on analogy in philosophy of science and cognitive science. Facile references to the Aztecs and Incas simply will not do. More sophisticated studies might better determine the strength and weaknesses of the four analogies I have laid out, and that would be most beneficial to studying impact: the microbe analogy, the culture contact analogy, the transmission/translation analogy, and the worldview analogy. Of course, their utility will depend on the discovery scenario. Even better analogies might arise from more full-scale studies, and the results of these deliberations could be of direct benefit in managing the discovery of life beyond Earth. Analogies for some forms of discovery, such as alien artifacts, have not yet been discussed at all. The history of those cases where life was actually thought to have been discovered are the most direct analogies, and should be mined for lessons learned. In a broader sense the entire history of the extraterrestrial life debate can also be searched for lessons learned in the sense of highlighting good or faulty analogical reasoning in areas for which we now know the outcome. The same is true of terrestrial history in general when it comes to such concepts as culture and civilization. Statements to the contrary notwithstanding, we *can* learn lessons from history; indeed, that is one of the chief reasons history is universally taught and its contemplation considered one of the hallmarks of civilization.

Second, while thinking about universal biology is well under way, it is almost totally lacking in the areas of intelligence, culture, civilization, technology, and communication that are so essential to SETI and METI endeavors. This is in part due to the fact that since 1993 NASA no longer supports SETI endeavors and funding for research related to these areas is no longer available from government sources. Thus far the $100 million Breakthrough Listen SETI initiative has gone entirely to observations rather than to any thought about intelligence and cultural universals, research that might actually be essential to any discovery, much less its impact. That should change with the recognition of the utility of the social sciences and humanities to the entire astrobiology endeavor. The SETI Institute, which took over parts of the NASA SETI program almost a quarter century ago, now shows signs of interest in these broader topics that are essential to its work. Research on "nonhuman understanding," on the universality of science and mathematics, and on possible universals in the social sciences will be essential, especially in the event of communication. Whether terrestrial law can be translated into metalaw applicable to all intelligences in the universe, or whether metalaw will require entirely new principles, remains an open question. That is related to the question of astroethics, and whether universal ethical principles can apply to the realm of space.

Third, and related to the second point, research on the role of evolution and possible "universal Darwinism" as they apply to areas outside biology is required in many areas related to astrobiology. There is little doubt evolution is a guiding universal principle in both terrestrial and extraterrestrial biology, but its role has not been sufficiently studied in a variety of other areas applicable to the impact of astrobiology. In addition to overall cosmic evolution, these areas include the role of evolution in culture, ethics, and epistemology, as well as the more global question of the evolutionary continuity between Nature and human nature. These areas, including sociobiology, evolutionary psychology, and evolutionary ethics, have proven very controversial in both the natural sciences and social sciences as applied on Earth, and that controversy will undoubtedly be multiplied when carried to problems beyond Earth. Yet we must try, because the problem at hand is an X factor that may profoundly affect humanity's future.

Fourth, I have argued that research on worldviews is essential to understanding the societal impact of discovering life beyond Earth. Again, there is a good deal of literature on worldviews, much of it related to religion but also to science in the form of Kuhnian paradigms and its variations. But very little of this work has been applied to the discovery of life by way of analogy or dis-analogy. The study of the societal effect of past worldviews such as the Copernican and Darwinian stand ripe for application to our problem. The Ladrièrean models of impact deploying the rise and fall of old worldviews, and the rise of new ones with their accompanying impact on human values, stand ready to be employed or modified for the problem of life beyond Earth. And the rise of the biological universe as a new cosmological worldview, already well under way, as well as the rise of astrotheology, astroculture, and astroethics as new worldviews within this overall cosmic scheme, open new horizons of human thinking that need to be examined before the discovery of life occurs. The increasing awareness of our cosmic connection and our rising cosmic consciousness will demand new ways of thinking beyond our terrestrial history, and perhaps gradually reveal our place in the biological universe.

Fifth, even before the discovery of microbial or intelligent life occurs, and beyond the question of universality, it is prudent that both space law and metalaw be developed to deal with the possibilities that will inevitably arise. The issues include not only planetary protection in the form of forward and back contamination, but also the potential need to interact with intelligent life in a way that is systematic rather than ad hoc. While planetary protection protocols have advanced impressively over the past 50 years, only the most rudimentary thought has been given to interactions with extraterrestrial intelligence, a potentially catastrophic failure of imagination. In the same way that we try to foresee global problems in many areas, including astronomical events like asteroid impacts, we need to address systematically the array of issues that will inevitably arise in the event of contact with alien intelligence. History shows that early interactions are crucial in any culture contacts, and hoping for the best is not a strategy. Nor will neglecting the problem make it go away.

Sixth, media and public education will be crucial at all stages of managing discovery. This is clear not only from history of purported discoveries, but also from the nature of the modern print, electronic, and social media, which now spreads news (and fake news) at the speed of light. A "War of the Worlds" claim these days would spread instantly around the world, with global impact depending very much on how the media portray the supposed event. Historically, and now more than ever, the media have exhibited a trigger finger, latching on to any story that might involve extraterrestrial life, ranging from nanofossils in the Mars rock to anomalous radio signals that could otherwise have been quickly explained, and supposed alien artifacts around a star showing an unusual light curve. As with the *New York Sun* in 1835, most of the media are in the business of making money, and life beyond Earth sells well to almost any audience. In the longer term, education is essential to public understanding of what might be out there – and what might not. The recent trend of the public to ignore facts and of some press to spin them, all too evident in the most recent US election and its aftermath, does not bode well for this endeavor.

Finally, by analogy with what the discipline of astrobiology has achieved beginning 20 years ago, these and other areas of study should be codified in a full Roadmap for the Astrobiological Humanities and Social Sciences. Such a roadmap would provide a core reference point for what has thus far been a scattered and sporadic effort of scholars in many disciplines, with results published in often obscure venues. Like the astrobiology roadmap, now known as the astrobiology strategy document, a Roadmap for the Astrobiological Humanities and Social Sciences could evolve over many years as research in the area progresses. It could be developed by groups such as the newly founded Society for Social and Conceptual Issues in Astrobiology (SOCIA), or under the auspices of a revitalized Focus Group on Astrobiology and Society of the NASA Astrobiology Institute, or by a combination of these and more international groups. Some beginnings have already been made in this direction, including a conference on the subject at the SETI Institute in 2009, and the inauguration of the Baruch S. Blumberg NASA/Library of Congress Chair in Astrobiology, which focuses specifically on the astrobiological humanities. Also by analogy with astrobiology, its two journals, *Astrobiology* and the *International Journal for Astrobiology*, have proven essential to cohering that new discipline. The same would likely be true for the astrobiological humanities and social sciences.

Transforming our Thinking

Finally, I have asked how the discovery of life will transform our thinking, and I have suggested some answers in the previous three chapters. First, the discovery of life will initiate a new cosmological worldview that I have termed the *biological universe*. Whether it will be a microbial or an intelligent biological universe remains to be seen, but both hold out the possibility of an anthropic or biocentric universe in which life is intimately connected to physical reality in its deepest structure and most fundamental properties by way of its physical constants. Why the universe at its core appears to be biofriendly remains one of the greatest mysteries of science. The discovery of life will transform that mystery into a major research program.

Second, as intimated earlier, the discovery of life beyond Earth holds out the possibility of universalizing human knowledge. Philosophical considerations, cognitive science, and evolutionary epistemology indicate that our knowledge even of mathematics and science is most likely not universal. Our knowledge is filtered through the Kantian categories of space and time; our minds are embodied in ways that depend on evolution and thus are contingent on the conditions under which we have evolved. A new discipline of astrocognition – the study of the origin and evolution of cognitive abilities in extraterrestrial environments – must attempt to unravel these connections. Alien minds may have different filters, different perceptions, and perhaps different conclusions about the universe. The discovery of intelligence beyond Earth thus promises to advance the ancient and modern philosophical problem of objective knowledge. Even without the discovery of intelligence, alien microbes will revolutionize biology, leading to a universal biology.

Third, the discovery of life will expand our theological horizons. Indeed astrobiology is already encouraging this expansion, as both scholars and the public contemplate how specific religions and theologies might be affected. The rise of astrotheology has spawned questions about the universality of the Christian doctrines of incarnation and redemption, about the geocentric and anthropocentric nature of Islam, and about the seemingly more universal aspects of other religions such as Hinduism and Buddhism. Would you baptize an extraterrestrial? Is there an extraterrestrial version of Mecca? Is religious enlightenment universally applicable? Do other intelligences have an equivalent to God? As we contemplate these questions based on current human religions, will the discovery of life and our increasing cosmic consciousness spur new theologies, perhaps along the lines of our "cosmotheology," which places humanity near the bottom of the great chain of being? Such theologies would be naturalistic rather than supernatural, would thus have very different ideas of God, would have a moral dimension embracing all life in the universe, and might transform our ideas about human destiny. What forms of astrotheology or astrophilosophy are exhibited in a universe full of extraterrestrials will be one of the great revelations in our explorations of the universe. Contrary to the opinion of many, the discovery of life will likely not destroy religion, and religion might even help humans cope with the new worldviews triggered by the discovery. While major adjustments may be necessary, history demonstrates that the religious impulse is too deeply seated in humans to be jettisoned. This does not mean the transformation will be easy.

Fourth, the discovery of life will move us toward a cosmocentric, rather than an anthropocentric or a biocentric, ethic. Even as we have ambiguous feelings about animal life on Earth, we will be ambiguous about the moral status of alien organisms, ranging from microbes to intelligence. Do they have instrumental or intrinsic value? I have suggested it depends on the theory of ethics employed, ranging from anthropocentric to ratiocentric and the several shades of cosmocentric. I have argued for an expansion of our astroethical horizons toward at least a basic cosmocentric ethic, one that finds value in cosmic consciousness. Astroethics will decenter our ideas about ethics, transforming our parochial ideas about value and moral status into a more universal set of ideas that can be implemented in the event of discovery and will constantly evolve as a result of discovery.

Fifth, the discovery of life will accelerate astroculture, raising our cosmic consciousness to new levels. The Space Age has greatly enhanced our view of the universe and our place in it. The *Apollo* images of Earthrise as seen from the Moon, the Blue Marble Full Earth viewed from a variety of perspectives, the pale blue dot seen from *Voyager*, the awe-inspiring images from the Hubble Space Telescope and other spacecraft – all of these and more reconnect us to the cosmos in intimate ways. We now see our place in 13.8 billion years of cosmic evolution, enhanced and reinforced by the new field of Big History, which places human history in its cosmic context. In the midst of this panoply of cosmic images, our minds have already been captured by aliens, as evidenced in science fiction literature and film, in the UFO debate, and in the scientific discoveries made every day providing circumstantial evidence for the discovery of life. The actual discovery – the reality rather than the idea – will transform our thinking, sparking a global astroculture based on a shared history of cosmic evolution.

Finally, in perhaps the biggest category of all, the unknown unknowns, our worldviews will be changed in ways we cannot yet fathom and that I have therefore not foreseen in this book. If we can survive the slings and arrows of our collective outrageous fortune and misfortune, humanity is on the brink of great discoveries that will perhaps lift us above this mortal coil. We cannot realistically hope for salvation from aliens, but we can hope for perspective – a cosmic perspective already nascent and that will become more robust with the knowledge that we are not alone in the universe, or that we are. Whether the age-old question "Are we alone?" is a plaintive cry of loneliness or an expression of sheer scientific exploration, a definitive statement that "We are not alone" will surely change us. Whether for good or ill remains to be seen, a fate at least partially in our hands.

Envoi

Even if no life is found beyond Earth, the search is already transforming our thinking as we contemplate astrotheology, astroethics, and astroculture. Such studies expand our horizons, force out-of-the-box thinking, and lift us beyond our anthropocentric musings as we are made to generalize some of our most profound concepts about life and intelligence, culture and civilization, technology and communication. Even if that is our only result, this thought experiment will have been well worth it.

Some will remain skeptical of the need for these preparations, recommendations, and adjustments. But resistance is futile; the universe is what it is, not what we want it to be. Cowering on our planet fearful of what is out there is not desirable or sustainable, and not in the best tradition of human exploration. When we come to realize fully that we are part of a biological or a postbiological universe, we must also accept that our goal is to be at home in that universe whatever organisms are out there, whether they be microbes, more complex life, artificial intelligence, or beings beyond our imaginings. After all, both science and Big History emphasize that we share a common heritage in cosmic evolution, tracing back to the Big Bang and the formation and death of stars spewing their life-giving elements. We are indeed made of stardust, and we must embrace astroculture, astrotheology, and astroethics; use astropolicy to lead the way based on history, analogy, and the latest research; and continue to discover new worlds while shaping and embracing whatever future they may hold for humanity.

Notes

1. History

1. Michaud (2011), 308.
2. Huxley (1959).
3. Dick (1996).
4. Michaud (2011) and Dick (2013a), 227–257.
5. The best history of the Moon Hoax is Matthew Goodman's aptly named *The Sun and the Moon* (2008).
6. The full text of the six articles may be found online at www.museumofhoaxes.com/hoax/text/display/the_great_moon_hoax_of_1835_text/. Excerpts with commentary are found in Michael J. Crowe's sourcebook *The Extraterrestrial Life Debate* (2008), 272–296; the quotations here are on pp. 286–287.
7. Crowe (2008), 287, and Goodman (2008), 200.
8. Crowe first suggested the series was written as satire in his article "New Light on the Moon Hoax" (1981), 428–429. He elaborated further in the context of the extraterrestrial life debate in *The Extraterrestrial Life Debate* (1986), 210–215, and reported new evidence in Crowe (2008), pp. 293–295.
9. The German astronomers included Johann Schroeter and Franz von Paulen Gruithuisen; see Crowe (1986), 70–75, 203–205, and Sheehan and Dobbins (2001), chapter 7.
10. Goodman (2008) agrees with Crowe's thesis that the Moon "hoax" was a satire, and tells the story in the context of the birth of tabloid journalism in antebellum New York. On editions of the brochure, see Crowe (2008), 272.
11. Goodman (2008), 12.
12. Crowe (1986), 212–213. The Poe quote is from Poe's article "Richard Adams Locke," cited in Harrison, ed. (1902), pp. 126–137, at 134.
13. Goodman (2008), 178. Goodman describes further reactions on pp. 178ff.
14. John Herschel to Caroline Herschel, January 10, 1837; Lady Herschel to Caroline Herschel, received in London September 26, 1836, and in Hannover

October 1, 1836, both quoted in Sheehan and Dobbins (2001), 92. In an unpublished letter to the editor of the Athenaeum dated August 21, 1836, Herschel spoke of the "incoherent ravings" of the *Sun* series; Ruskin (2002).

15. Crowe (2008), 215.
16. Sheehan and Dobbins (2001), 92.
17. Goodman (2008), 10; on Loomis and Olmstead, see Goodman (2008), 181–182.
18. The most comprehensive history of the event and reaction to it is Gosling (2009). See also Gallop (2011) and Sourcebooks staff (2001).
19. *New York Times*, October 31, 1938, p. 1.
20. *New York Times*, October 31, 1938, p. 1. Cantril (1940).
21. The Lowell canals controversy is the subject of a large literature, most notably in Crowe (1986), 480–546, and Dick (1996), 62–104, and references therein.
22. For the quotations, see Hoyt (1976), 87–88, and Sheehan (1996), 109. Crowe has pointed out that theological concerns may have played a role in the thought of one of the prime debunkers of canals, Greenwich astronomer E. W. Maunder: Crowe (2001).
23. Wells (1908), which particularly cites Lowell (1907). Wells's Lowellian Mars does not necessarily mean Lowell's ideas precipitated the novel as more immediate reasons were at hand; see Dick (1996), 233.
24. Bainbridge (1987), 554.
25. Bartholomew (1998), with numerous references.
26. Pooley and Socolow (2013b), summarized in "The Myth of the War of the Worlds Panic," *Slate*, October 28, 2013. For the backstory to the Cantril volume, see Pooley and Socolow (2013a). The Campbell quote is from Campbell (2010), 26–27.
27. Bartholomew and Radford (2012).
28. All these episodes are described in detail in Gosling (2009).
29. Bell Burnell (1979). A more extensive first-person account is Bell Burnell (1983). Hewish's detailed account is given in his Nobel lecture, cited in note 32. Another detailed account, presented as part of the American Philosophical Society symposium on "Discovery in Astronomy," is Longair (2011).
30. Penny (2013).
31. Penny (2013), 537–538.
32. Hewish, "Pulsars and High Density Physics," Nobel Lecture, p. 178, online at www.nobelprize.org/nobel_prizes/physics/laureates/1974/hewish-lecture .html.
33. Dick (1996), 440; Penny (2013), 538.
34. The CTA-102 episode is analyzed in detail, with citations of original sources, by French radio astronomer Jean Heidmann (1990). The original paper was submitted by Kardashev on December 12, 1963, under the title "The

Communication of Information by Civilizations on Other Worlds." The Shklovskii quote is from his delightful autobiography (1991), p. 253. The best treatment of the WOW! signal is Gray (2012), although recently claims have been made that it is now reasonably explained as due to the release of hydrogen from passing comets; see Harry Pettit, "Solved: Mysterious 'Wow! Signal' in 1977 Was Not a Message from Aliens but Triggered by Gas from Passing COMET," *Daily Mail*, June 6, 2017, at www.dailymail.co.uk/sciencetech/article-4576358/Mysterious-alien-Wow-signal-caused-comets.html. On the *Kepler* star, see www.universetoday.com/122971/seti-institute-undertakes-search-for-alien-signal-from-kepler-star-kic-8462852/.

35. Tarter and Michaud (1990).
36. This and the following paragraphs are drawn from Dick (1998). For a recent personal retrospective from the reporter who broke the story, see Leonard David, "Remembering a Big Scoop about a Small Rock," *Space News*, September 12, 2016, www.spacenewsmag.com/feature/remembering-a-big-scoop-%E2%80%A8about-a-small-rock/.
37. These events are described in detail in Kathy Sawyer's valuable popular book *The Rock from Mars* (2006), and in Dick and Strick (2004), 179–201.
38. Dick and Strick (2004), 188–189. Meyer elaborates on the visit to Gibson and McKay's lab (private communication to author, February 8, 2017): "Looking at the carbonate blobs under the microscope and the evidence they had on hand I made the recommendation that they not present any more information to the community until they had a publication accepted. Furthermore, considering their lack of expertise in biology, I suggested they get a micro-paleontologist to take a look (on a personal note, at the time, they suspected the blobs themselves were microfossils and I was suspicious because of their variability in size). Part of this 'embargo' was driven by Everett's presentation of finding organics in the meteorite, maybe, and he couldn't understand the heightened interest of questioning he got from the press. My conclusion was that any additional exposure would cause a relentless press that could, in the long run, undermine the credibility of the investigators – in other words, the science community would keep seeing stories but no evidence (or worse, piecemeal), and get increasingly frustrated about the lack of material for them to scrutinize. And to their credit, they did go silent until *Science* had accepted their publication. And as feared, we did experience a break in the embargo when they went to the White House and we had to push up the press conference because the story was going viral."
39. The *Science* paper is McKay et al. (1996).
40. Dick and Strick (2004), 190.
41. These events are described in more detail in Sawyer (2006), 114–133.
42. The note to editors is quoted in Jeff Foust, "ALH84001 + 10," online at www.thespacereview.com/article/678/1; "Statement from Daniel S. Goldin, NASA

Administrator," NASA HQ Press Release 96-159, August 6, 1996, online at www.nasa.gov/audience/formedia/archives/MP_Archive_96.html.

43. Donald L. Savage (NASA HQ), James Hartsfield (JSC) and David Salisbury (Stanford), " Meteorite Yields Evidence of Primitive Life on Early Mars," NASA Press Release 96-160, August 7, 1996, online at www.nasa.gov/home/hqnews/1996/96-160.txt.
44. "President Clinton's Statement Regarding Mars Meteorite Discovery," The White House, Office of the Press Secretary, August 7, 1996, 1:15 P.M. EDT, online at www2.jpl.nasa.gov/snc/clinton.html.
45. CNN, "Ancient Meteorite May Point to Life on Mars: Biggest Discovery in the History of Science," August 7, 1996, posted at 1:15 P.M.; John Noble Wilford, "Clues in Meteorite Seem to Show Signs of Life on Mars Long Ago: Startling Find of Organic Molecules from Space," *New York Times*, August 7, 1996, p. A1; *Los Angeles Times* editorial, "Message from Afar: Maybe It's a Lively Universe; Finding Fuels Speculation, along with Mars fever," August 8, 1996, p. 8.
46. Bill Broadway, "Sure, Why Not? Say the Theologians," *Washington Post*, August 10, 1996, p. B6; "Theologians Find Awe in Possibility of Life on Mars," *Los Angeles Times*, August 17, 1996, p. 4. A week earlier, the *Times* had run a front-page article by Marty Martin, "Theology: Life on Mars, What's a Believer to Believe," August 11, 1996. Martin, a professor of religion at the University of Chicago, argued that religions would adapt, at least those that "praise knowledge," and that such religious have prepared the faithful to be ready for any surprise, including fossil life on Mars.
47. US Congress (1996).
48. Logsdon et al. (1997).
49. Surprisingly, the vice president's meeting is not well documented. The account here and at the beginning of this chapter is based on Sawyer (2006), 187–195, and my personal recollections.
50. Smith (2004), 453–454. See also Sawyer (2006), 187ff.
51. Jeff Foust, ALH84001 + 10, *The Space Review*, August 10, 2006, online at www.thespacereview.com/article/678/1.
52. Smith (2004), 454–455.
53. *USA Today*, August 13, 1996.
54. David A. Relman, "Is Unicellular Life Worth Living? Mars Knows," *New York Times*, September 16, 1996, p. A14. In reply to an op-ed in the *New York Times* by Stephen Jay Gould, they argued that recent revelations in bacterial genome structure indicate that the nature of any alien microbes was not a fully predicable outcome of ordinary physics and chemistry.
55. Marozzi (2008), 95. The entire book is a meditation on the lessons of current history through the eyes of Herodotus; see esp. pp. 72–77.
56. Campbell (2010), 43.

2. Discovery

1. Proust (1923).
2. Morrison (1973), 336–337.
3. The details of the discovery and the reaction to it are given in Dick (2003b), 425–429.
4. This controversy, and the whole problem of classification in astronomy, is described in Dick (2013c), 9–30.
5. Kuhn (1962a), 55; Dick (2013c); Caneva (2005); Galison (1987), 76, 126–127.
6. Cho (2012a, 2012b), 115; Vastag and Achenbach (2012); Miller (2012), 12–15. It is notable that the uncertainties of discovery also extend to theory. No fewer than six theorists could claim to have a role in predicting the Higgs mechanism; Cho (2012c), 1286–1289.
7. Aldersey-Williams (2011), 190–197, on helium, and page 152 on hydrogen. For a more detailed history of helium's discovery, see Kragh (2009). For the extended discovery of technetium, see Scerri (2013), 116–143, and for Paul Merrill's role, see Berkowitz (2012). On the Periodic Table, see Gordin (2004) and Scerri (2007).
8. Galileo Galilei to Grand Duke Cosimo II, translated in Van Helden (1974), 105. For more details on the Saturn episode, see Dick (2013c), 39–42, and reference to Van Helden therein.
9. Details of these discoveries, and those of exoplanets, are found in Dick (1996) and its abridgement and update, *Life on Other Worlds* (1998).
10. Latham et al. (1989), 38–40.
11. Mayor and Queloz (1995); Marcy et al. (2007), 926–935.
12. The most reliable and continuously updated list of all extrasolar planets is available at http://exoplanets.org, maintained by Jason Wright at Penn State University; see Jason Wright et al., "The Exoplanet Orbit Database," PASP, 123 (2011), 412–422, preprint online at http://xxx.lanl.gov/abs/1012.5676.
13. See Dick (2013c), 222–232.
14. For more details, see Dick (2013c).
15. www.afi.com/10top10/category.aspx?cat=7
16. Some 10,000 species comprise the human microbiome, including more than 1,000 in the mouth alone, while others inhabit the gut and every nook and cranny of the body. The National Institutes of Health Human Microbiome Project has identified many of these organisms at the *genera* level. Richard Conniff, "Microbes: The Trillions of Creatures Governing Your Health," *Smithsonian Magazine* (May 2013); Marcia Stone, "NIH Builds Substantial Human Microbiome Project," Microbe, 4 (10) (2009), 451–456.
17. Crosby (2003); McNeill (1976).
18. Vidal (2015).

19. Meltzer (2011), 113–246.
20. On the history of panspermia ideas, see Dick (1996), 367–377. On microbes from space, see Hoyle and Wickramasinghe (1979). On the detection of life in Martian meteorites, see Peter Cumpson, "There Could Be Martian Life on Earth," www.realclearscience.com/articles/2016/10/14/there_could_be_martian_life_on_earth_109779.html.
21. See Dick and Strick (2004). The current NASA Astrobiology Strategy is found at https://nai.nasa.gov/media/medialibrary/2015/10/NASA_Astrobiology_Strategy_2015_151008.pdf. The literature on astrobiology is voluminous, including multiple textbooks and two journals. Popular accounts include Darling (2001), Kauffman (2011), and Billings (2013).
22. R. Navarro-Gonzalez et al., "Reanalysis of the Viking Results Suggests Perchlorate and Organics at Mid-latitudes on Mars," *Journal of Geophysical Research*, 115 (2010), E12010; Gilbert V. Levin and Patricia Ann Straat, "The Case for Extant Life on Mars and its Possible Detection by the Viking Labeled Release Experiment," *Astrobiology*, 16 (2016), 798–810; DiGregorio (1997).
23. See, for example, Lisa Kaltenegger et al., "Deciphering Spectral Fingerprints of Habitable Exoplanets," *Astrobiology*, 10 (2010), 89–102; Sara Seager, W. Bains, and J. J. Petkowski, "Toward a List of Molecules as Potential Biosignature Gases for the Search for Life on Exoplanets and Applications to Terrestrial Biochemistry," *Astrobiology*, 16 (2016), 465–485.
24. P. C. W. Davies, "Searching for a Shadow Biosphere on Earth as a Test of the 'Cosmic Imperative'," *Phil. Trans A*, 369, 624–632.
25. Recent work in the field of alien artifacts includes P. C. W. Davies, "Footprints of Alien Technology," *Acta Astronautica*, 73 (2012), 250–257; Paul Davies and R. V. Wagner, "Searching for Alien Artifacts on the Moon," *Acta Astronautica*, 89 (2013), 261–265, and Jason Wright, "Prior Indigenous Technological Species," https://arxiv.org/abs/1704.07263, and *International Journal of Astrobiology*, published online June 2017. An early discussion is R. A. Freitas Jr. and F. Valdes, "The Search for Extraterrestrial Artifacts (SETA)," *Journal of the British Interplanetary Society*, 36 (1983), 501–506, at www.rfreitas.com/Astro/SETAJBISNov1983.htm.
26. On the history of SETI, see Dick (1996), 399–472. On SETI science, see Davies (2010), Billings (2013), and Vakoch and Dowd (2015). For an exciting new vision of a more holistic SETI effort, see Cabrol (2016).
27. On the Drake message, see https://en.wikipedia.org/wiki/Arecibo_message; on the Russian messages and the short history of METI, see Zaitsev (2011); on the goals of METI International, see http://meti.org, including my arguments in favor of messaging at http://meti.org/blog/should-we-message-et.
28. Freeman Dyson, "Search for Artificial Stellar Sources of Infrared Radiation," *Science*, 131 (1960), 1667–1668; Richard A. Carrigan Jr., "Starry Messages:

Searching for Signatures of Interstellar Archaeology," online at https://arxiv.org/abs/1001.5455; on Tabby's star, see www.skyandtelescope.com/astronomy-news/tabbys-star-weird-star-gets-weirder/, Jason Wright, www.seti.org/weeky-lecture/frontiers-artifact-seti-waste-heat-alien-megastructures-tabbys-star, and https://en.wikipedia.org/wiki/KIC_8462852.

29. On December 26, 1996, while *Contact* was still filming, Coppola filed suit claiming Sagan had used numerous ideas in *Contact* that Sagan had originally written under a March 1975 contract to Coppola for a television documentary titled *First Contact*. Coppola went on to produce *Apocalypse Now* (1979) and many other projects, and the *First Contact* documentary was never produced. Following depositions as expert witnesses by Jill Tarter, Andrew Fraknoi, and me in Century City, in February 1998, a Los Angeles Superior Court judge issued a summary judgment in favor of the Sagan estate. Coppola appealed, and in April 2000 a California Court of Appeal also dismissed the suit, saying the suit was brought too late. In any case I had shown in my deposition that all the ideas Coppola claimed were his intellectual property had been put forth by Sagan or others prior to 1975. The episode is mentioned in William Poundstone's *Carl Sagan* (1999), 375. According to Poundstone, Coppola had written Sagan in November 1995, stating he expected to receive a share of Sagan's profits from *Contact*, for which Sagan had received a highly publicized $2 million advance. Sagan wrote a letter stating Coppola's claim was without merit, which went unanswered during Sagan's remaining year of life. See also http://variety.com/1998/film/news/coppola-s-contact-claim-is-dismissed-1117467799/ and http://variety.com/2000/film/news/coppola-loses-contact-1117780544/.
30. Vidal (2015).
31. Dick (2013c), Appendix 2, pp. 347–369.
32. Fry (2015).
33. Gonzalez and Richards (2004).
34. See Dick (2013b). On the popular culture impact, see Achenbach (1999).
35. Jones (2013).

3. Analogy

1. Hofstadter (2001), 499.
2. Denning (2014), 100.
3. McNeill (1986), 138.
4. MacQuarrie (2007), 32. See also Hemming (2003).
5. Gentner and Jeziorski (1993), 448–449.
6. Vidal (2014), 79; Holyoak and Thagard (1995), especially chapter 8. See Gentner and Jeziorski (1993), 450, for six principles of analogical reasoning.
7. Hesse (1966), 1–9.

8. For example, Holyoak and Thagard (1995), chapter 8; North (1989), 285–310. According to Janet Browne, in her two-volume biography, *Charles Darwin: Voyaging,* and *Charles Darwin: The Power of Place* (1995 and 2002), Darwin regarded artificial selection as the best clue to understanding nature. See also Ruse (1998), 35.
9. Bartha (2010), vii–viii.
10. Hofstadter (2001); Hofstadter and Sander (2013).
11. Gaddis (2002), 2; Murphy (2007); "Is Vladimir Putin Acting Like Hitler?" Letters to the Editor, *Washington Post*, March 10, 2014.
12. Mazlish (1965), 12; Hughes (1965), p. 53, note 1. The circumstances of this volume are discussed in Coopersmith (2008). The Launius study is *Historical Analogs for the Stimulation of Space Commerce* (2014).
13. The original frontier thesis was enunciated in Turner (1994). Noted historian of the American West Patricia Nelson Limerick has argued especially vigorously (and not entirely successfully in my view) that the American frontier, with its history of exploitation and conquest, should not be used as an analog for space exploration; see Limerick (1992). My response is that just because exploitation and conquest were indisputably part of the American frontier does not imply that the space frontier needs to follow that course.
14. Bell (1980), pp. 2ff.; Sullivan (2013b); Ross (2013); Denning (2013); Helmreich (2009), 255.
15. For others, see Pyle (2012), pp. 271ff.
16. Biddanda, Nold, and Dick (2012), 13.
17. Harrison (2007), pp. 95ff.; Launius (2013); Capra (1975); Stenger (2011), 122.
18. Traphagan (2014), 132. Denning makes her point on page 100 of this volume, and general archaeological wariness of analogy is argued in Wylie (2002), chapter 9, "The Reaction against Analogy," pp. 136–160.
19. Gest (2004a, 2004b, 2007).
20. Margulis and Sagan (1986).
21. Gest (2004b), 270–272. Leeuwenhoek's observations of animalcules (protozoa and bacteria) in water are collected, translated, and edited in Dobell (1958); see especially chapter 1 for the first observations. Despite his slighting of Hooke and his heroic treatment, Dobell's account of Leeuwenhoek remains valuable.
22. Sapp (2009), 3–7. On the wave of public interest, see Ford (1985), 19.
23. While microscopy "was not a thriving enterprise," it was by no means totally moribund during the nineteenth century. See Schickore (2007), identifying the years around 1830 as the beginning of modern microscopy. Schickore also warns us against "technological determinism," and her thesis is that progress in microscopy was a complex amalgam of technological, conceptual, and institutional factors, as well as methodological and epistemological factors.

24. As microbiologist Thomas D. Brock remarked in his *Milestones of Microbiology*, the history of microbiology is to a large extent the history of bacteriology. And as Brock stated in the 1999 edition of his volume, William Bulloch's *The History of Microbiology* (1938) remains the classic history of microbiology. Brock's volume reprints, edits, and comments on pioneering works in the history of microbiology. It is notable that the term "microbe" dates only to 1876, and the term "bacteria" to Cohn in 1872; Sapp (2009), 46–47. On the many roles of the microbe in the modern world, see Gest (2003).
25. Dick and Strick (2004), 106–107; Sapp (2009), 162–176.
26. Gronstal (2013), 219.
27. Crosby (2003). Almost four decades later Crosby briefly extended his theme to extraterrestrial microbes in "Micro-organisms and Extraterrestrial Travel," in *Humans in Outer Space* (2009). Michael Meltzer documents NASA's planetary protection program in *When Biospheres Collide* (2011).
28. US Congress (1961), pp. 215–216, and notes 225–226; Billingham, Heyns, and Milne (1999).
29. Lowrie (2013).
30. Finney and Bentley (2014).
31. The literature on Neanderthals is large, but a good starting point is Tattersall, *The Last Neanderthal* (1995). On their interbreeding, see Gibbons (2014) and Vernot and Akey (2014).
32. Tattersall (1995), 153.
33. Robert Ascher and Marcia Ascher, "Interstellar Communication and Human Evolution," *Nature* 193, no. 4819 (1962): 940–941, reprinted in Cameron (1963), 306–308, esp. p. 307.
34. Wason (2011).
35. Wason (2011), 44.
36. Wason (2013), 113–129. Perhaps also relevant to our problem are studies of our living ancestors such as chimpanzees, and "intelligence" among dolphins and other species. I address the problem of intelligence in Chapter 4.
37. Martinez (2001).
38. The literature here is large. Other cases are found, for example, in the *Cambridge History of the Native Peoples of the Americas* (1996), and Fitzhugh (1985).
39. Elliott (1970); Parry (1981); Fritze (2002). Among the many histories of this event, see Thomas (2003). The Smithsonian exhibit is online at www.pyramidstudios.com/work/work03e.htm.
40. Axtell (1992), pp. 31ff; Traphagan (2015b), 73–86; Traphagan (2014).
41. Elliott (2006). There are also many myths associated with the encounter; see Restall (2003).
42. Diaz del Castillo (1956), Schwartz (2000). On Diaz del Castillo, see "Diaz del Castillo, Bernal," in *Encyclopedia of Historians and Historical Writing* (London: Fitzroy Dearborn Publishers, 1999), 307–308.

43. Elliott (1992); Blumberg (2011).
44. Kuznicki (2011), 211–212.
45. The definitive history in English of Zheng He and his voyages is by scholar of Chinese history Edward L. Dreyer, *Zheng He* (2007). Also useful, though less definitive, is Levanthes, *When China Ruled the Seas* (1994). The voyages of Zheng He have received increasing attention during the 600th anniversary of the voyages and because of the controversial thesis in Menzies (2002). Menzies's thesis that the Chinese discovered America seven decades before Columbus is not borne out by Chinese records and remains unproven.
46. Dreyer (2007), 3.
47. Dreyer (2007), xii.
48. Both quotations are from Gernet (1999), 401–402.
49. Dreyer (2007), 3–4.
50. Wilson (1999), 121–126, at 124.
51. Finney (1990), 117–121, citing Worseley (1957). For more modern theories of cargo cults, see Lindstrom (1993).
52. Anthropologist Kathryn Denning makes this point; see Denning (2014), 100–101.
53. Hoerder (2002). The fact of numerous other culture contacts does not, of course, mitigate the reality of the disastrous population decline in the century following first contact in the Americas, amounting to something like 90 percent of the 100 million indigenous population, mostly due to European pathogens; Hoerder (2002) 189–190. On the effect of European pathogens, see McNeill (1976). On the recently discovered vast network of trade and other cultural interactions during the first era of globalization, and its parallels with today, see Cline (2014).
54. Billingham et al. (1999); Dick (1995); Toynbee (1957), 241–260.
55. The two quotes are from Lindberg (1992), 203, and Grant (1971), 13–19, at 18. See also Lindberg (1978), Peters (1968), and Rubenstein (2003).
56. Finney and Bentley (2014).
57. Finney and Bentley (2014), 68; Coe (1992).
58. An early attempt of this type was Freudenthal (1960). More recently Carl DeVito has led the way in this method; see DeVito (2014). Embodied mind theorists, however, are skeptical that mathematics is universal in an extraterrestrial context; see Lakoff and Nunez (2000).
59. See Saint-Gelais (2014) and Denning (2014) in the same volume. See also Friedrich (1957), Chadwick (1958), and Robinson (2002). For an account of a famous case of decipherment, see Hinsley and Stripp (1993). I have more to say about this in Chapter 5.
60. Silver (2012), 7.
61. Wilson (1999), 2.
62. On worldviews, see Vidal (2014), 3–57; Vidal (2007); White (1999).

63. Kuhn (1957); Blumenberg (1987); Stimson (1972); Westman (2011). In a contrarian view, Dennis Danielson (2001) has argued that pre-Copernicans, and Copernicus himself, did not view the central Earth as a privileged position in the cosmos.
64. For more detail on this analogy, see Dick (1995). The stages in worldview development I delineated there are motivation, presentation based on observation, elaboration, opposition, exploration of implications outside the field, general acceptance, and final confirmation (or rejection). The literature on the reception of Darwinism is voluminous, but a good *entrée* is Bowler (2009).
65. Berenzden, Seeley, and Hart (1976); Smith (1982).
66. North (1989), 306.

4. Can We Transcend Anthropocentrism?

1. Suvin (1970), 221.
2. Vonnegut Jr. (1994), 146.
3. Wittgenstein (1953), 223.
4. Abbott (1992).
5. Joyce (1994). In 2005 Joyce's lab became the first to produce a self-replicating system, consisting of RNA. The best history of the origins of life controversy is Fry (2000). Popular accounts of current science include Messler and Cleaves II (2016), Lane (2015), and Wills and Bada (2000).
6. Conrad and Nealson (2001).
7. Irwin and Schulze-Makuch (2001); Schulze-Makuch (2015).
8. Bains (2004), 138.
9. Pace (2001).
10. Brandstetter (2012), 952.
11. Bains (2004), 162
12. Ward and Benner (2007).
13. Cleland and Chyba (2007); Pirie (1937). Philosopher Kelly Smith (2016) argues against definitional pessimism. For a sampling of what is now a large literature, see the collection of essays on "What Is Life?" in *Astrobiology*, 10 (2010), 1001–1030.
14. Bacteria and Archaea can range from 200 nanometers to 1 millimeter. Knoll (2004); Grosberg and Strathmann (2007); National Academy of Sciences, National Research Council, Space Studies Board (1999). The latter study was commissioned in the wake of the controversy over the Mars rock. Multicellular prokaryotes in the form of cyanobacteria may have existed 3 billion years ago.
15. www.complex-life.org/
16. Dorminey (2014), citing the work of Victoria Meadows at the University of Washington.

17. Ward and Brownlee (2000).
18. *Nature*, 467, p. 929, reported in "Why Complex Life probably Evolved only Once," *New Scientist*, October 12, 2010.
19. Dick (2013b), 166.
20. Bruce Dorminey, "Complex Life Possible within Cosmos' First Billion Years," December 10, 2013, www.forbes.com/sites/brucedorminey/2013/12/10/complex-life-possible-within-cosmos-first-billion-years/.
21. Conway Morris (1998 and 2003); Gould (1989); Simpson (1964). For a succinct view of Conway Morris's argument, see his article "Predicting What Extra-Terrestrials Will Be Like: And Preparing for the Worst," *Philosophical Transactions of the Royal Society A*, 369 (2011), 555–571. "The worst," for Conway Morris, is, tongue in cheek, that they are either like us or do not exist at all.
22. Bylinksy (1981).
23. Cohen and Stewart (2002); Dickinson (1994); Pickover (1998); Jenkins and Jenkins (1998); Schmidt (1995).
24. For an excellent historical and scientific tour of the Drake Equation, see Vakoch and Dowd (2015).
25. Sagan (1977), 60–79. Sagan's book was updated in Skoyles and Sagan (2002), but books on intelligence in an extraterrestrial context are few and far between.
26. Hawkins and Blakeslee (2004), 97–105.
27. Mithen (1996).
28. Bogonovich (2011), 113–114, and Rospars (2013), 197, both citing R. J. Sternberg, "The Search for Criteria: Why Study the Evolution of Intelligence," in Sternberg and Kaufman (2002), 1–7.
29. For example, as early as 1988 in Reiss (1988), 253, elaborated in Reiss (2011). More recently Lori Marino and Kathryn Denning organized a workshop on "Intelligence and Astrobiology: Deep Integration," at the Astrobiology Science Conference in Atlanta in 2012, part of a series of workshops "intended to integrate the study of intelligence on Earth within an astrobiological framework." The workshops were never completed, but the SETI Institute has recently taken up the challenge. On brain encephalization, see Marino (2002).
30. Bogonovich (2011), 113–122; Marino (2015).
31. Godfrey-Smith (2016); Chorost, *How to Talk to Aliens* (forthcoming).
32. David Gold, "The Diversity and Evolution of Invertebrate Nervous Systems," http://intelligence.seti.org/pages/invertebrates; Maggie Wray, "Social Insects and Collective Intelligence," http://intelligence.seti.org/pages/social_insects; Part I of Chorost (forthcoming) discusses "Bugworld," swarm cognition in connection with insects.
33. Herzing (2014).
34. Sagan (1977), 230.

35. Mayr (1985).
36. Simpson (1964); Dobzhansky (1972); Gould (1989); Conway Morris (1998 and 2003); Chela-Flores (2007).
37. Eiseley (1957), 162–163.
38. On the limitations of the human brain, see Fox (2011). Fox argues that human intelligence may have reached its evolutionary limits based on the laws of physics, including thermodynamic considerations and "the very nature of neurons and the statistically noisy chemical exchanges by which they communicate." This leads to the possibility that the next steps in increasing human intelligence may be hive intelligence or artificial intelligence.
39. On postbiologicals, see Dick (2003a). Also see Martin Rees, "Why Alien Life Will Be Robotic," *Nautilus* (October 22, 2015), http://nautil.us/issue/29/scaling/why-alien-life-will-be-robotic; Davies (1995b), 51–55; Shostak (1998), 103–109; Susan Schneider, "It May Not Feel Like Anything to Be an Alien," *Nautilus* (December 2016), http://cosmos.nautil.us/feature/72/it-may-not-feel-like-anything-to-be-an-alien.
40. For arguments about the human postbiological future, see Moravec (1988 and 1999), Kurzweil (1999 and 2005), and Vinge (1993). On the impact of such a postbiological Earth, see Glenn McDonald, "Stephen Hawking Wants to Prevent AI from Killing Us All," *Seeker*, October 21, 2016, online at www.seeker.com/stephen-hawking-artificial-intelligence-leverhulme-2057637120.html; www.nytimes.com/2016/11/02/technology/new-research-center-to-explore-ethics-of-artificial-intelligence.html?_r=0; www.whitehouse.gov/sites/default/files/whitehouse_files/microsites/ostp/NSTC/preparing_for_the_future_of_ai.pdf.
41. Searle (1984), especially chapter 2, "Can Computers Think?"
42. No critical history of ideas of the origin and evolution of intelligence exists analogous to Fry's detailed and nuanced analysis of issues in the origins of life debate. This constitutes a fundamental lacuna in the field, all the more so because SETI practitioners typically fail to consider the nature of intelligence in their work.
43. Dennett (1996); Ruse (2006), 52–71.
44. Chalmers (2003), 5. On the role of zombies in philosophical discussions, see Tye (2007), 27–29, and *Stanford Encyclopedia of Philosophy*, https://plato.stanford.edu/entries/zombies/.
45. Watts (2006), 323. See also Steven Shaviro's (2016) discussion of the novel in "Thinking Like an Alien," chapter 6 of *Discognition*, a remarkable book about other minds.
46. Crick (1994).
47. Bostrom (2014), p. 26 on the definition, pp. 63–74 on the types, pp. 127–139 on the goals. On brain emulation, see Hanson (2016). Hanson is an economist who looks in detail at the possible societal implications of "ems." Brain

emulation is different from AI because it is "more like porting software from one machine to another machine" rather than writing an entirely new software system (p. 51). For even more exotic forms of postbiology, see Vidal (2014), 214–217, and Freitas Jr. (1979–2010, 2008).

48. Schneider (2014 and 2015).
49. Hawkins and Blakeslee (2004), 209–210.
50. Benner and Sismour (2005). On the problematic idea of progress, see Laudan (1977); Nisbet (1980).
51. For extraterrestrials in science fiction, see Wayne Barlowe's delightful book *Barlowe's Guide to Extraterrestrials* (1979), complete with a description of their physical characteristics, habitat, and culture. See the "Hive Minds," *The Encyclopedia of Science Fiction*, online at www.sf-encyclopedia.com/entry/hive_minds. An example of thinking even further out of the box about intelligence – too far for some – is Vidal (2016), previously elaborated in Vidal (2014), 227–265. If true, this would be an example of what I termed "prediscovery" in Chapter 2. Nagel's 1974 essay "What Is It Like to Be a Bat?" has been republished numerous times, and is online at http://organizations.utep.edu/portals/1475/nagel_bat.pdf.
52. Emilie Reas, "Small Animals Live in a Slow-Motion World," *Scientific American Mind*, July 1, 2014, www.scientificamerican.com/article/small-animals-live-in-a-slow-motion-world/. On subjective time, see Davies (1995a), pp. 272ff. See also David Eagleman, "Brain Time," at www.edge.org/conversation/brain-time and "Time Perception," at https://en.wikipedia.org/wiki/Time_perception.
53. Traphagan (2015b); Denning (2009).
54. Maruyama and Harkins (1975). I discuss this volume more in Chapter 5 in the context of universal human sciences.
55. Morrison, Billingham, and Wolfe (1977), 49–52, at 49.
56. Kroeber and Kluckhohn (1952), 656; Geertz (1973); Kuper (1999); Wilson (1998); Traphagan (2015b), 58–65.
57. An overview of evolutionary models of culture is Laland and Brown (2002). See also Denning (2009). In this article Denning also emphasizes that the terms "social evolution," "cultural evolution," and "sociocultural evolution" may be used interchangeably for most purposes.
58. John Traphagan (2015a) warns against equating culture, civilization, and moral development. Nauman Naqvi, "Civilization," *International Encyclopedia of the Social Sciences* (Thomson-Gale, 2008), vol. 1, 557–559.
59. Traphagan (2015b), 44–52, discusses the concept of civilization in the context of SETI, and criticizes SETI's use of civilization as a global concept.
60. Huntington (1996), 40–44.
61. Jack F. Matlock Jr., "Can Civilizations Clash?," *Proceedings of the American Philosophical Society*, 143 (Sept. 1999), 428–439.
62. Matlock Jr. (1999), 439.

63. Kardashev (1964); Vidal (2014), 223–226; Michio Kaku, "The Physics of Extraterrestrial Civilizations," http://mkaku.org/home/?s=civilizations&x=19&y=11; Basalla (2006), 188–189.
64. Ashkenazi (1998). Vidal (2014) discusses another typology for civilizations – the ability to control small entities, as in biotechnology, nanotechnology, and information technology. See Barrow (1998).
65. Dick and Lupisella (2009); Traphagan (2015a, 141, and 2015b).
66. Tarter (2007), 513; Dyson (1966); Ćirković (2006).
67. Traphagan (2011), pp. 469ff.; Dusek (2006); Kelly (2010).
68. Basalla (2006), 187–189, at 188; Harwit (2013).
69. In the United States in the 1970s a conscious decision was made to emphasize the search (SETI) rather than the much more controversial communications aspect (CETI). As the NASA-sponsored Morrison workshops stated in 1977: "The acronym SETI (Search for Extraterrestrial Intelligence) was adopted by the Workshop and by the Ames Research Center to differentiate our own efforts from those of the Soviet Union and to emphasize the search aspects of the proposal." "The Soviet CETI Report," in Morrison, Billingham, and Wolfe (1977), 211–227, at 211. Especially with the advent of METI, the communications aspect has been emphasized, as in Vakoch (2011d).
70. Gary W. Levvis, "Why We Would Not Understand a Talking Lion," http://digitalcommons.calpoly.edu/cgi/viewcontent.cgi?article=1811&context=bts; Griffin (2001); on dolphin communication, see the references in note 25; Conway Morris (2003), 252.
71. Dick (2012). Figure 4.6 first appeared in Billingham (2000a). See also Dick (2000a).
72. Chiang (2002). The movie differs in many details from the story, but the communication attempts and nonlinear orthography and time remain central to both.
73. Dunér (2011), 123; Clark and Chalmers (1998).
74. McConnell (2001), 197–212; Freudenthal (1960); DeVito (2014).
75. Lemarchand and Lomberg (2011), 371–372.
76. Among his many articles on the subject, see Vakoch (2010, 2011a, 2015). See Vakoch (2011d), part III, for much more on these issues.
77. Minsky (1985).
78. Pitt (1982), 58 and 64.

5. Is Human Knowledge Universal?

1. Ruse (1998).
2. Lakoff and Nunez (2000), 342.
3. Blumberg (2011).

4. A good popular introduction to the art, architecture, and inscriptions in the Library of Congress is Cole (2008). The quotation is on page 9.
5. Among the many histories of philosophy discussing these general developments, see Bertrand Russell's classic *History of Western Philosophy* (1999).
6. As well as treatments in general histories of philosophy, see Lowe (1995).
7. Noonan (1999).
8. Chalton (2008), 100; Gardner (1999), 56–57.
9. Kuehn (2002), 242.
10. Kuehn (2002), 243.
11. Kant, *Critique of Pure Reason*, Introduction to second edition (1787), www.egs.edu/library/immanuel-kant/articles/the-critique-of-pure-reason/preface-to-the-second-edition-1787/; Chalton (2008), 100–101; Papineau (2004), 26.
12. Philosopher Kelly Smith points out that Kant would have said we must *believe* that reason is universal, even if it isn't, because otherwise there would be no justification for believing anything, including science. Private communication, February 9, 2017. This still does not mean, however, that aliens see the world in the same way or that alien science would be the same as ours.
13. Kuehn (2002), 104–105; Dick (1982), 165–171.
14. I am indebted to philosopher Derek Malone-France for this point. He also writes, "There is no reason to believe – on a consistent application of purely Kantian principles – that disparate categorical filters would mutually triangulate on absolute truth in this way. It would be just as likely, given the logic of Kant's employment of the noumenal concept, that all variant perspectival filters were *equally unrepresentative* of the noumenal reality." Private communication, March 3, 2017. On Kant's cosmology, see Dick (1982), 165–175, and Crowe (1986), 47–55, at 52 and 54.
15. Gardner (1999), 63.
16. Derek Malone-France, private communication, June 17, 2016. Another way of putting it is there are metaphysical conditions prior to the transcendental. See Malone-France (2007).
17. If you still do not understand Kant after this discussion, you are not alone. There are probably as many interpretations of Kant as there are philosophers. On relativism, see Laudan (1990); for a succinct account, see https://plato.stanford.edu/entries/relativism/.
18. For a succinct overview of the research programs of cognitive science, see Frankish and Ramsey (2012), part III. More details are found in Sobel (2013), Bermudez (2010), and Friedenberg and Silverman (2012).
19. E. Thelen, quoted in Clark (2012). For an overview of research in embodied cognition, see the entry in the *Stanford Encyclopedia of Philosophy*, http://plato.stanford.edu/entries/embodied-cognition/.
20. Lakoff (2003a).

21. Griffin (2001); Sara J. Shettleworth, "Animal Cognition," in Frankish and Ramsey (2012), 292–311. Griffin was a pioneer in the field known as cognitive ethology. On astrocognition, see Dunér (2011); Dunér, "Extraterrestrial Life and the Human Mind," in Dunér, Parthemore, Persson, and G. Holmberg (2013), 1–25, at 11, also, in the same volume, see Mathias Osvath, "Astrocognition: A Cognitive Zoology Approach to Potential Universal Principles of Intelligence," pp. 49–65.
22. Ruse (1998), 29–66. Those sympathetic to evolutionary epistemology include Konrad Lorenz, Donald Campbell, Karl Popper, and Stephen Toulmin. Donald Campbell first coined the term "evolutionary epistemology" in 1974. See Ruse (1998). For succinct overviews, see *Stanford Encyclopedia of Philosophy*, http://plato.stanford.edu/entries/epistemology-evolutionary/ and the *Internet Encyclopedia of Philosophy*, www.iep.utm.edu/evo-epis/.
23. Ruse (1998), 148–206, at 149 and 175. Derek Malone-France notes that "this perspective represents an updated version of traditional American pragmatism. Peirce, James, et al., were making precisely this point in their epistemology – the environment will reward correctly calibrated cognition and right belief, and it will punish the contrary." Private communication, March 3, 2017. Ruse borrows the term "epigenetic rules" from Charles Lumsden and E. O. Wilson's book on cultural evolution, *Genes, Mind and Culture* (1981). For a critique, see Laland and Brown (2002). One need not accept Lumsden and Wilson's ideas of cultural evolution to accept the idea of innate dispositions.
24. Ruse (1998), 179–184.
25. Ruse (2003), 130–131, and Ruse (1998), pp. 192ff.
26. Malone-France comments that if there is really some coherent underlying structure to reality, then evolutionary epistemology "suggests that the underlying *sameness* behind any and all environmental differences would *also manifest itself* in the context of the evolutionary cognitive development of any and all organisms. Meaning that no successful intelligent life form would pragmatically evolve through Darwinian selection in a way that did not reflect this underlying metaphysical and/or fundamental physical structure. Whatever the diversity of routes of development, it would still be the case, then, that any successful form of life/understanding might share with us at least some basic structures of cognition – I'd nominate temporality, causation, and numerality (as opposed to our specific mathematics), among others." Private communication, March 3, 2017. Such comments demonstrate how philosophers can be essential to the SETI debate when it comes to communication.
27. Rescher (1985), 84.
28. Livio (2009). For more on Platonists versus formalists, see Brown (1999).
29. Wigner (1960). The logic of evolutionary pragmatism mentioned earlier is one answer to Wigner's question.

30. Tegmark (2014); I. Grattan-Guinness, "Solving Wigner's Mystery: The Reasonable (Though Perhaps Limited) Effectiveness of Mathematics in the Natural Sciences." *The Mathematical Intelligencer* 30 (2008), 7–17, online at www.sfu.ca/~rpyke/cafe/reasonable.pdf.
31. Lakoff and Nunez (2000), summarized in Lakoff (2003b), 75–108, at 90–96; DeVito (2011 and 2013).
32. Livio (2009), 234–239.
33. Rescher (1985), 87.
34. Dawkins (1989), 191–192.
35. David Des Marais et al., "The NASA Astrobiology Roadmap," *Astrobiology*, 8, 715–730, online at https://nai.nasa.gov/roadmap/; http://astrobiology.illinois.edu/ for the Institute for Universal Biology.
36. Cleland and Chyba (2007); Mariscal (2015). The latter is based on his PhD dissertation, "Universal Biology," Duke University, Durham, NC.
37. Rescher (1985), 91.
38. Lloyd and Sivin (2002).
39. Fry (2012), 671.
40. Rescher (1985), 89.
41. On the history of the human sciences, see Smith (1997).
42. Gary Gutting, "Michel Foucault," *The Stanford Encyclopedia of Philosophy* (Summer 2013 Edition), Edward N. Zalta (ed.), http://plato.stanford.edu/entries/foucault/.
43. *Stanford Encyclopedia of Philosophy*, http://plato.stanford.edu/entries/foucault/; Foucault (1973), xxii.
44. Malone-France points out that there is a big difference between "universal" and "more universal" in the sense that we can never claim universality in an infinite universe. Strictly speaking, even the physical sciences have not been proven universal, only to work in those parts of the universe we have observed. From the point of view of the philosopher, "No matter where we may travel, or what we may become, in the future, our understandings will always remain finite and, therefore, fallible." He also points out that "this is precisely the conclusion reached by early theological existentialists like Kierkegaard, who contrasted the finitude of human understanding to the infinitude of the divine object. Here, we have simply substituted the concept of an infinite (or at least practically infinite) universe in place of the concept of the divine, as the object of in-traversable epistemic contrast." Private communication, March 4, 2017.
45. Smith (1997), pp. 804ff.; William A. Gorton, "The Philosophy of Social Science," *Internet Encyclopedia of Philosophy*, www.iep.utm.edu/soc-sci/#H2.
46. Rosenberg (2016); on the "laws" of biology, see pp. 253ff. For an overview of these evolutionary perspectives on cultural evolution, see Laland and Brown (2002) and Rosenberg (2016), 237–256.

47. Berry et al. (1992), p. 4, quoted in Harrison (1997), 104–105. On the idea of cultural universals, see Harrison (1997), 151–156.
48. Asimov's definition of psychohistory is given very early in Foundation, "The Psychohistorians," part I, section 4. On Krugman and psychohistory, see www.theguardian.com/books/2012/dec/04/paul-krugman-asimov-economics. See also Gunn (1982), 39–47.
49. Rosenberg (2016), pp. 19ff. and 55ff.
50. Morrison (1973), 333–349, at 336–337.
51. Toffler (1975), vii–xi, especially viii.
52. Toffler (1975), ix.
53. Tax (1975), 200–203.
54. Finney (1990); Harrison et al. (2000); Denning (2009); Battaglia (2009).
55. Vakoch (2014a). For a description of the Symposia, see Vakoch (2009b). A more detailed overview of social science involvement is given in Dick (2013a).
56. Dick (2006), reprinted in Dick, "The Role of Anthropology in SETI," in Vakoch (2013), pp. 49–63.
57. Pass (2016).

6. How Can We Envision Impact?

1. Silver (2012), 448–449.
2. Ladrière (1977), 18.
3. Silver (2012).
4. The literature on societal impact of science and technology is large, but on spaceflight, see Dick and Launius (2007) and Dick (2015a), both available on the NASA History website, along with other relevant publications such as *Cosmos and Culture*. On planetary protection, see Race (2007). On European work regarding the societal impact of spaceflight, see especially Codignola and Schrogl (2009) and Landfester et al. (2011).
5. One attempt at systematically laying out the parameters beyond the usual economic factors is Benoit Goran and Christian Dore, "Measuring the Impacts of Science: Beyond the Economic Dimension," Urbanisation INRS, Culture et Société (Helsinki: Helsinki Institute for Science and Technology Studies, 2005) www.csiic.ca/PDF/Godin_Dore_Impacts.pdf; and Lutz Bornmann, "Measuring the Societal Impact of Research," at www.ncbi.nlm.nih.gov/pmc/articles/PMC3410397/.
6. Ladrière (1977). This volume was prepared following the symposium "Science, Ethics, Aesthetics," organized by the Philosophy Division of UNESCO in Paris, July 9–12, 1974.
7. Stoeger (1996), 151–154.

8. Lewis (1964), 222; Kuhn (1957), 193; Blumenberg (1987); Stimson (1972); Westman (2011).
9. Dick (1982), 86–88.
10. Dick (1995), 521–532; Berenzden, Seeley, and Hart (1976).
11. On the short-term controversy, see Vorzimmer (1970); on the longer-term controversy, see Bowler (1989); Hull (1973); Greene (1981).
12. Stoeger (1996), 153.
13. On the problem of worldviews, see also Vidal (2014). On stages of worldview development, see Dick (1995), 527. On the political connection, see Kissinger (2014), quotations on p. 66. The context of the quotation is the balance achieved after the Congress of Vienna in 1814–1815 in the wake of the French Revolutionary Wars and the Napoleonic Wars.
14. Dick (1989 and 1996).
15. Dominik and Zarneki (2011).
16. There is a voluminous literature on Wells and his novel *War of the Worlds*. On *The Day the Earth Stood Still*, see Ruse (2015), 175–188.
17. Silver (2012), 240–250; Devlin (2002), 288–292. Bayes's Theorem calculates the posterior probability of an event by the equation xy/(xy + z (1–x)), where x is the prior probability, y is the probability given the evidence in any particular case, and z is the probability that the evidence is correct. So in the cancer case where x is 001, y is 0.79, and z is 0.21, the answer is 4.6 percent.
18. One attempt to employ Bayesian statistics in the context of the Drake Equation is Thomas L. Wilson, "Bayes' Theorem and the Real SETI Equation," *QJRAS*, 25 (1984), 435–448. See also Paul Shuch, www.setileague.org/askdr/bayesian.htm.
19. Billingham et al. (1999), 11–14. The results of this meeting were also reported at meetings of the International Astronautical Congress as published in Billingham (1998) and Doyle (1998).
20. Billingham et al. (1999), 33–60, 122.
21. Ibid., 61–82, 122–124.
22. Ibid., 83–94, 124–125.
23. Ibid., 95–120, 125–127.
24. The conclusions of the meeting and much other relevant material were published in Tough (2000). The core of the report is Harrison and Dick (2000). Another result of the Astrobiology Institute call for impact studies was Dick (2000b), a paper given at the 1999 bioastronomy meeting where the seminar was held.
25. Harrison and Connell (2001).
26. Harrison and Connell (2001), 5–10, at 10.
27. All papers from the 2010 Royal Society meeting in London are available online at http://rsta.royalsocietypublishing.org/content/369/1936.toc. A satellite meeting, "Towards a Scientific and Societal Agenda on Extra-

Terrestrial Life," was held in October at the Kavli Royal Society International Centre, Chicheley Hall, Buckinghamshire. It included issues of METI, but was not published.

28. The Library of Congress conference papers are found in Dick (2015b). The congressional hearings are US Congress (2014). Hearings on astrobiology a few months earlier also asked the same question; see US Congress (2013).
29. References to these meetings are given in Dick (2013a), 228.
30. Harrison (1997, 313–314, and 2011a).
31. Miller (1978). A summary of living systems theory is given in G. A. Swanson and James Grier Miller, "Living Systems Theory," Encyclopedia of Life Support Systems, online at www.eolss.net/Sample-Chapters/C02/E6-46–01-03.pdf. See also Vidal (2014), 220–223.
32. Michaud (2007), 338.
33. Binzel (1997); Almar and Tarter (2011); Almar and Shuch (2007). For an online calculator for the Rio scale, devised by the Permanent SETI Committee of the IAA, see http://avsport.org/IAA/riocalc.htm. For a critique of the assumptions behind these scales, see Traphagan (2015b), 73–76, or, more succinctly Traphagan's "SETI and the Meaningless Rio Scale," http://meti.org/blog/seti-and-meaningless-rio-scale, where he argues "the Rio Scale is a good example of what happens when attempts at social science are based on naïve and poorly conceived understandings of human behavior and society."
34. Almar and Race (2011).
35. Race et al. (2012).

7. Astroculture: Transforming Our Worldviews

1. Callicott (1986a), 252.
2. Meynell (1923), 92.
3. Grinspoon (2016).
4. On worldviews, see Vidal (2014), 3–57. Religions seem to be particularly interested in worldviews, especially their own. See Sire (2002), which argues, elegantly but inconclusively in my view, for a theistic Christian worldview.
5. Dick (2013c).
6. Lewis (1964), 222; Zakariya (2017).
7. Dick (1996).
8. B. M. Oliver, "Bio-cosmology: A New NASA Thrust," *Cosmic Search*, 1 (Summer 1979), 16–20; Dick (1989, 13–17; 1996, 541–542; 1997).
9. On the gradual rise of this worldview, see Dick (1982 and 1996), Crowe (1986), and Guthke (1990).

10. Barrow and Tipler (1986); for the Henderson quote, see Henderson (1913), 312. See also Davies (2007).
11. On different versions of the anthropic principle, see Barrow and Tipler (1986). On human destiny, see Dick (2004).
12. Dyson (1988), 50–51. See also Dick (2008).
13. For a succinct history, see Dick (1996), 514–526, and Crowe (1997). For longer versions, see Crowe (1986), and for a select bibliography, see Crowe (2008), pp. 539ff.
14. Raible (1960); Anonymous, "Messages from Space," *America*, 111 (1964), 770; Norman Lamm, "The Religious Implications of Extraterrestrial Life," in *Tradition: A Journal of Orthodox Jewish Thought*, vols. 7 and 8 (1965–1966), 5–56. For more theological reaction to these events, see Dick (1996), 514–526.
15. Christopher Corbally, "Religious Implications from the Possibility of Ancient Martian Life," presented at AAAS meeting in Seattle, February 13–18, 1997, online at www.cyberspaceorbit.com/relimpl.htm.
16. As early as 1965, Rabbi Norman Lamm used the term "exotheology" in the Jewish context in the article cited earlier. The term was still being used among theologians such as Ted Peters in an article in 1995, where he proposed the discovery of ETI would present no significant challenge, at least to Christianity. Peters (1995). In a series of more recent articles, Peters has sensibly proposed that "astrotheology" is a more appropriate term, since it parallels the change in terminology of the original field of exobiology to the now well-established and much broader field of astrobiology. The term "cosmotheology" was first used by Kant, but in a different way.
17. Dick (2000c).
18. O'Meara (2012), 93–94; Wilkinson (2013), 178–184; Consolmagno and Mueller (2014), 249–286, expanded in Consolmagno, "Would You Baptize an Extraterrestrial?" in Dick (2015b), 233–243. See also Peters, Hewlett, Moritz, and Russell (2017), and the series of articles in the journal *Theology and Science*, vol. 15, no. 2 (2017).
19. Peters (2011), 648. The concept of dignity and the moral character of alien intelligence is an ethical issue dealt with in the next chapter.
20. Weintraub (2014), 169–185; Paul (1992); Crowe (1986). On Buddhism and other worlds, see Irudayadason (2013); Traphagan and Traphagan (2015). It remains one of the mysteries of history why Western civilization has been so fascinated by the subject of other worlds, while, at least to my knowledge, for most of history it has been absent or subdued in Eastern cultures. Even so, ways of thinking in non-Western cultures might help us to think out of the box on the subject of this volume, and such studies – both historical and contemporary – are much to be desired.
21. Dick (1996), 514–526; Bainbridge (1983); Ashkenazi (1992); Vakoch and Lee (2000); Peters (2009 and 2013a); Bertka (2013).

22. Bertka (2013), 338.
23. For more on aliens and religion, see McKee (2007), particularly chapter 6, "Christ, Prometheus and Klaatu: Alien Messiahs," pp. 128–150.
24. Davies (2003).
25. Crysdale (2009), 239; McAdamis (2011), 349. On Puccetti and more, see Dick (1996), 524.
26. Dick (2000d, 2005, 2018).
27. Geppert (2012), pp. 6–9, at 8.
28. Geppert (2012), pp. 27ff.
29. Poole (2010); Sagan (1994), 9.
30. White (1987); Bjornvig (2013); Bimm (2014). On *Apollo* and the environmental movement, see Lambright (2015).
31. Sagan (2000), 190. This new edition contains all the original content of the 1973 first edition, plus new contributions by Freeman Dyson, Ann Druyan, and David Morrison. On the rise of cosmic evolution as an idea, see Dick (2009), and in more detail in Zakariya (2017).
32. Christian (2004). The field of Big History, which provides the broadest possible context for our cosmic connection, continues to grow at an accelerating rate; see the series of anthologies *Our Place in the Universe: An Introduction to Big History*, Barry Rodrigue, Leonid Grinin, and Andrey Korotayev, eds. (2015); also by the same editors see *Education and Understanding: Big History around the World* (2016), and *From the Big Bang to Galactic Civilization: A Big History Anthology* (2017).
33. Achenbach (1999); Battaglia (2009), 1.
34. Abel (2012), pp. 28–29. Many books have now analyzed this revolution, ranging from popular treatments such as Man (2002), to more scholarly works including Eisenstein (1983) and Febvre and Martin (1976). For the crucial role of the printing press in the Reformation, see Pettegree (2015).
35. McLuhan (1962).
36. James Gleick's *The Information* (2011) gives the full sweep of this information change history.
37. James A. Dewar, "The Information Age and the Printing Press: Looking Backward to See Ahead," www.rand.org/pubs/papers/P8014/index2.html#fn0; Richard Stacy, "Gutenberg and the Social Media Revolution: An Investigation of the World Where It Costs Nothing to Distribute Information," http://richardstacy.com/2008/11/20/gutenberg-and-the-social-media-revolution-an-investigation-of-the-world-where-it-costs-nothing-to-distribute-information/.
38. Carr (2011). See also Allen Tough, "Positive Consequences of SETI before Detection," *Acta Astronautica*, 44 (1998), 745–748. A new perspective is one of the consequences, as well as new questions.

8. Astroethics

1. Martin Luther King, paraphrasing a longer statement by nineteenth-century abolitionist and Unitarian minister Theodore Parker. King used the quote many times, including during the Selma march in 1965.
2. Schweitzer (1960), 259–260.
3. Rolston III (2014), 211.
4. Stoeger (2013), 56.
5. Leopold (1949); Devall and Sessions (1985); Stoeger (2013), 57–58.
6. Harris (2010), 1–7. See also de Waal (1996) and Rolston III (2014), pp. 214ff.
7. Stoeger (2013), 61.
8. Vidal (2014), pp. 26ff.: 275, 291; Lupisella (in press); Murphy and Ellis (1996), 15–18, 250.
9. Stoeger (2013), 61–62; Daston (2014).
10. Cleland and Wilson (2013), pp. 35ff. Peters (2013b), pp. 210ff.; Lovin (2015).
11. Callicott (1986a), 245–253; Callicott (1986b), 163.
12. Smith (2009 and 2014).
13. Lupisella (2016), pp. 75–91. See also Marshall (1993).
14. Persson (2012).
15. Persson (2012). See also Jacob Haqq-Misra's "Ecological Compass" in the same issue.
16. Cleland and Wilson (2013), 42–43. On animal rights, see http://ieet.org/index.php/IEET/RNHP.
17. Cleland and Wilson (2013), 46–48. See also "The Moral Status of Animals," *Stanford Encyclopedia of Philosophy*, http://plato.stanford.edu/entries/moral-animal/.
18. Shostak (2014).
19. Robert Wright, "Ethics for Extraterrestrials," *New York Times*, May 4, 2010, opinion pages.
20. Ruse (2003b), 142.
21. Dennett (1996), 493. See his chapters 16 and 17 on ethics. Laland and Brown (2002).
22. Shermer (2004). Wilson (1978), 149–167, at 167, and p. x. for his definition. Shermer's book gives a succinct history of the evolution of the field of evolutionary ethics in its Appendix I.
23. Ruse (2003b), 145–146. For a longer treatment of the subject, see Michael Ruse, "Evolutionary Ethics," and "Darwinian Ethics," in Ruse (1998), pp. 67–102, 207–272.
24. Wilson (2004); Alcock (2001).
25. Vakoch (2014b). The two authors are policy analyst George Michael and astronomer Harold Geller.

26. Barkow (2014), 38.
27. Gupta (2013) and Lupisella (2014), 79–92 and 93–110.
28. Rolston III (2014), 221, citing Kant, 1785, on p. 218; Murphy and Ellis (1996).
29. Brin (2011).
30. Potts and Sloan (2010); Thomas Nagel, "What Is It Like to Be a Bat?" (1974), online at http://organizations.utep.edu/portals/1475/nagel_bat.pdf; Puccetti (1968); T. Hibbs, ed. Aquinas: *On Human Nature* (Indianapolis, IN: Hackett, 1998); Impey, Spitz, and Stoeger (2013), 90.
31. Liu (2015).
32. Korbitz (2014b) argues that if such a principle had been applied through human history, nothing would have gotten done.
33. Lederberg (1960).
34. Sagan (1980), 130, 109.
35. Zubrin and Wagner (1996), 132–135. See also McKay and Zubrin (2002).
36. McKay (1990), 194; Lupisella (1997). McKay's article appears in a section on "Ethics and Planetary Engineering" in MacNiven's book.
37. McKay (2013). McKay's views on terraforming were expressed in a joint paper with colleagues Richard O. Randolph and Margaret S. Race (1997). Randolph (2009) has argued that terraforming, or "planetary ecosynthesis," in order to encourage indigenous organisms on Mars can be strongly justified from a Christian ethical perspective. He also contends that introducing new life from Earth should be only a last choice, since indigenous life has priority from a Christian point of view.
38. Sullivan III (2013a), 172; Rolston III (1986).
39. Eric Berger, "Enthusiasts Warn Planetary Protection May Stop Humans from Going to Mars," *Ars Technica*, June 6, 2017, https://arstechnica.com/science/2017/06/mars-enthusiast-planetary-protection-a-racket-should-be-largely-ignored/.
40. Peters (2013b), 200–221.
41. Peters (2013b), 206–210.
42. Ruse (1985), pp. 66ff.
43. Bostrom (2014); Dick (2003a).
44. Hanley (1997), 72–120, at 120; Kurzweil (2005); Gardner (2007). On Spike Jonze's film *Her*, see Schneider (2014).
45. Lupisella and Logsdon (1997). The idea of a cosmocentric ethic was first raised in Haynes (1990), 177. Haynes saw such an ethic as an attempt to resolve the contradiction between superficial views of "evolutionary progress" versus "ecological harmony," a way to balance human environmental imperialism on the one hand and the destructive elimination of all technology on the other, a perspective in which "morality has roots in the very nature of things," giving rise to "the ultimate moral principle of the universe."

46. J. Baird Callicott, "Intrinsic Value in Nature: A Metaethical Analysis," *The Electronic Journal of Analytic Philosophy*, 3 (1995), http://ejap.louisiana.edu/EJAP/1995.spring/callicott.1995.spring.html.
47. Lupisella (2009b), 192–197; Lupisella (2016).
48. Barrow and Tipler (1986).
49. Lupisella (2011).
50. Haqq-Misra (2012).
51. Hart (2013), 396.
52. Ruse (1985), 71.

9. Astropolicy

1. Harrison and Connell (2001), 7.
2. US Congress (2014). Hearings on astrobiology a few months earlier also asked the same question: US Congress (2013).
3. World Economic Forum (2013).
4. See Michaud (2007), 358–376.
5. The original 1998 astrobiology roadmap, its two successors (2003 and 2008), and the 2015 astrobiology strategy are online at https://nai.nasa.gov/roadmap/. On the politics behind the development of astrobiology and its roadmap, see Dick and Strick (2004), 202–220. On the politics of the cancellation of the NASA SETI program, see Garber (2014). On the European astrobiology roadmap, see http://phys.org/news/2016-03-scientific-roadmap-european-astrobiology.html and Horneck et al. (2016).
6. For a preliminary reconnaissance of astrobiology and society issues in the wake of the astrobiology roadmap, see Dick (2000b).
7. The NASA Ames workshop report is Harrison and Connell (2001). For the Astrobiology and Society roadmap, see Race et al. (2012). John Baross, Introduction to *Astrobiology Strategy*, p. xv; Lucas Mix and Connie Bertka, Appendix to the *Astrobiology Strategy*, "Beyond Natural Sciences: Humanities and Social Science Contributions to Astrobiology," online at https://nai.nasa.gov/media/medialibrary/2016/04/NASA_Astrobiology_Strategy_2015_FINAL_041216.pdf, pp. 155–160. On the Blumberg Chair, see www.loc.gov/loc/kluge/fellowships/NASA-astrobiology.html.
8. Dominik and Zarnecki (2011), 503–504; cited hereafter as PTRSA.
9. On the development of planetary protection policies, see Meltzer (2011). The latest COSPAR Planetary Protection Policy can be found in Kminek and Rummel (2015). This article also includes information on the process for policy revisions. COSPAR is a consultative body of the UN Committee on the Peaceful Uses of Outer Space (COPUOS). See also Race (2007), 489.
10. Rummel et al. (2012). The GW report is Pascale Ehrenfreund, Henry Hertzfeld, and Kate Howells, "COSPAR Workshop on Developing a

Responsible Environmental Regime for Celestial Bodies" (March 2013), at www.gwu.edu/~spi/assets/docs/PEX_WorkshopReport_ES_March14_Web%20(1).pdf. The quotation is on page 7. See also documents at www.planetaryprotection.nasa.gov.

11. Race and Randolph (2002); Race (2010). These guidelines are based in part on the SETI protocols I discuss in what follows. See also Race (2015).
12. The UN Outer Space Treaty was in part modeled on the Antarctic Treaty. Even when it comes to treaties, policymakers do not start from scratch, but look for analogies and dis-analogies in already existing law. For a comparison, see Race (2011).
13. NASA did support small SETI efforts where possible, including small research grants. The 2008 Astrobiology Roadmap included as part of its Goal 7 the ability to "identify biosignatures of distant technologies." In order to do so, it recommended that "novel methods should be identified for detecting electromagnetic radiation or other diagnostic artifacts that indicate remote technological civilizations." Similar language was also present in the 2002 Roadmap. The 2015 *Astrobiology Strategy* document states that "While traditional Search for Extraterrestrial Intelligence (SETI) is not part of astrobiology … it is reasonable for astrobiology to maintain strong ties to the SETI community," and that astrobiology can contribute to SETI through the search for "technosignatures," a subset of biosignatures (p. 150). Intriguingly, the Strategy document also contains the following statement: "Therefore, over the coming decades as we develop the ability to discern the nature of exoplanets and examine them for anomalous properties which might be biosignatures, we should also be aware of the possibility of planets with anomalies that are the result of technological activities. Much attention has focused on which qualities of terrestrial life might be universal, and therefore relevant to the search for biosignatures; similarly, it is worth considering which aspects of technological civilization might be universal, how such qualities should be expected to affect the observable aspects of a planet, and how they might be discernible from other biosignatures" (pp. 76–77). Because of SETI's track record as a lightning rod to eliminate government funding, the astrobiology program is very circumspect about full-throated endorsement. I am grateful to John Rummel for calling to my attention the subtleties of this relationship of SETI to astrobiology.
14. Ekers, Cullers, Billingham, and Scheffer (2002).
15. On the history of NASA's SETI program from one of its founders, see Billingham (2013). On the SETI renaissance and road mapping, see Cabrol (2016).
16. Michaud (2007), 359–360.
17. For a firsthand account of the development of the protocols and related matters, see Michaud (2007), 359–363. The original SETI protocols are at http://avsport.org/IAA/protdet.htm and the 1986 and 1987 IAC deliberations

on which they are based in part are found in Tarter and Michaud (1990). Michaud's proposed extension is in Michaud (2005). The brief UN interest in 1977 is discussed in Othman (2011).

18. Dominik and Zarnecki (2011), 504.
19. For a brief history of METI programs from a Russian participant, see Zaitsev (2011).
20. For the Draft Declaration principles, see Annex 2 of http://avsport.org/IAA/position.htm, accessed directly at http://avsport.org/IAA/reply.htm.
21. For the case in favor of METI, see Vakoch (2011b), and Vakoch's edited volume *Communication with Extraterrestrial Intelligence* (2011d). For the case against METI, see Brin (2011), 433; Brin, "Shouting at the Cosmos: How SETI Has Taken a Worrisome Turn into Dangerous Territory," online at www.davidbrin.com/shouldsetitransmit.html. The comments of Hawking are at www.ibtimes.com/why-aliens-shouldnt-contact-us-stephen-hawking-warns-intelligent-extraterrestrial-2420526 and Diamond in the *New York Times Magazine* for December 12, 1999, "To Whom It May Concern," at www.nytimes.com/1999/12/05/magazine/to-whom-it-may-concern.html?scp=1&sq=&st=nyt. My views are detailed at Steven J. Dick, "Should We Message ET?" http://meti.org/blog/should-we-message-et. See also Musso (2012).
22. The Berkeley statement is at http://setiathome.berkeley.edu/meti_statement_0.html. The mission of METI International is at www.meti.org. Full disclosure: I am a member of the board of trustees.
23. Paul Berg, "Meetings That Changed the World: Asilomar 1975: DNA Modification Secured," *Nature*, 455 (September 18, 2008), 290–291, at www.nature.com/nature/journal/v455/n7211/full/455290a.html.
24. Denning (2011c), 252; Lupisella (2011).
25. Grinspoon (2016), 352–404, at 371. For more on the controversy, see Vakoch (2011c); Korbitz (2014b); Harrison (2014); Jones (2011). See also www.universetoday.com/119055/who-speaks-for-earth-the-controversy-over-interstellar-messaging/.
26. Maienschein (2015); Wilson (1998).
27. Seth Shostak, "Aliens on Line 1," *Air and Space*, September 2016, pp. 14–16.
28. Almar and Race (2011); Shostak and Almar, The Rio Scale Applied to Fictional "SETI Detections", IAA-02-IAA.9.1.06, 53rd IAF Congress, Houston, TX, USA, October 2002, http://avsport.org/IAA/abst2002/rio2002.pdf.
29. Almar and Shuch (2007); see also Shuch and Almar (2007a and 2007b).
30. World Economic Forum, http://reports.weforum.org/global-risks-2013/section-one/foreword/#hide/fn-1, as adapted from R. S. Kaplan and A. Mikes, *Managing Risks: A New Framework*, in *Harvard Business Review*, 2012. NASA has faced such decisions on many occasions. See Steven J. Dick and Keith Cowing, eds., *Risk and Exploration: Earth, Sea and the Stars*

(Washington, DC: NASA, 2004), based on a conference held in the wake of the *Columbia* space shuttle accident.

31. Comfort et al. (2010); Bostrom and Ćirković (2008). Astrophysicist Milan Ćirković deals briefly with issues related to the discovery of life in chapter 6 of the latter volume, "Observation Selection Effects and Global Catastrophic Risks," 120–145, at 135–138.
32. Harrison (2011b).
33. Jones (2013); Carol Oliver, "Social Media: Implications for Post-detection Strategies," IAC, Prague, 2010; Sterns and Tennen (2016).
34. Race (2010), 215; Billings (2015). A counterpoint to Billings's view that we know nothing about extraterrestrials is anthropologist Michael Ashkenazi's 400-page tome *What We Know about Extraterrestrial Intelligence* (2017). Ashkenazi's volume was published just as this volume was going to press.
35. http://reports.weforum.org/global-risks-2013/section-five/x-factors/#hide/img-5
36. Kleiman (2013). On international space law, see UN Office of Outer Space Affairs, www.unoosa.org/oosa/en/ourwork/spacelaw/index.html, and Doyle (2002). For US space law, see Gabrynowicz (2010). The actual US space law, known as Title 51, US Code, is found at http://uscode.house.gov/download/annualhistoricalarchives/pdf/2010/2010usc51.pdf. On Title 51, see the *Journal of Space Law*, Special Issue on US Code Title 51: National and Commercial Space Programs, 37 (2011), 1–278.
37. Race (2015), 270.
38. "Metalaw," chapter 12 in Haley (1963), 394–423, especially 394–395. On Haley, see Doyle (2013); Haley cited JPL founder Frank J. Malina and physicist Merle Tuve for helping formulate the "Rule of Metalaw" in Brussels in September 1956, and subsequently elaborated at the Rome meeting of the IAF in that year.
39. Haley (1963), 409–420. The latter quotation is in Korbitz (2014a), 233.
40. Fasan (1970), 214 for the definition, 223 for the Categorical Imperative, pp. 230ff. for the laws, and in priority order on pp. 237ff. Haley quotes Kant's imperative on page 412 of his work. Fasan's book has recently been reprinted as an Appendix to P. M. Sterns and L. I. Tennen (eds.), *Private Law, Public Law, Metalaw and Public Policy in Space* (Switzerland: Springer, 2016), 181–275. Fasan also provides a history of ideas about metalaw up to 1970.
41. Fasan (1970), 246.
42. Korbitz (2014a), 238; Robinson (1969, 2013a, 2013b). Robinson served as an International Relations specialist at NASA, and legal counsel for the Smithsonian Institution. See also Freitas Jr. (1975–1979, 2008).
43. Korbitz (2014a), pp. 245ff., and references therein; Sterns (2000) suggested a treaty should be drafted to develop metalaw.
44. Dominik and Zarnecki (2011), 504.
45. Race (2010), 218.

10. Summary and Conclusions

1. "Into the West," *Lord of the Rings: The Return of the King*, as performed by Annie Lennox in the motion picture, words and music by Fran Walsh, Howard Shore, and Annie Lennox.
2. Michaud (2015).
3. Lupisella (2015), 165–170. See also Lupisella (2009a), 321–356.

Bibliography

Abbott, Edwin A. 1992. *Flatland: A Romance of Many Dimensions*. New York, NY: Dover.

Abel, Richard. 2012. *The Gutenberg Revolution: A History of Print Culture*. New Brunswick, NJ: Transaction Publishers.

Achenbach, Joel. 1999. *Captured by Aliens: The Search for Life and Truth in a Very Large Universe*. New York, NY: Simon and Schuster.

Alcock, John. 2001. *Animal Behavior: An Evolutionary Approach*. Sunderland, MA: Sinauer Associates.

Aldersey-Williams, Hugh. 2011. *Periodic Tales: A Cultural History of the Elements from Arsenic to Zinc*. New York, NY: Harper Collins.

Almar, Ivan. 1995. "The Consequences of a Discovery: Different Scenarios," in Seth Shostak (ed.), *Progress in the Search for Extraterrestrial Life*. San Francisco, CA: Astronomical Society of the Pacific, 499–505.

Almar, Ivan and H. Paul Shuch. 2007. "The San Marino Scale: A New Analytical Tool for Assessing Transmission Risk," *Acta Astronautica* 60, 57–59. Online at http://avsport.org/IAA/acta_60_57-59.pdf.

Almar, Ivan and Margaret S. Race. 2011. "Discovery of Extra-Terrestrial Life: Assessment by Scales of Its Importance and Associated Risks," *Philosophical Transactions of the Royal Society A* 369 (2011), 679–692. Online at http://rsta.royalsocietypublishing.org/content/369/1936/679.

Almar, Ivan and Jill Tarter. 2011. "The Discovery of ETI as a High-Consequence, Low-Probability Event," *Acta Astronautica* 68, 358–361. Online at http://avsport.org/IAA/abst2000/rio2000.pdf.

Arnould, Jacques. 2008. "Does Extraterrestrial Intelligent Life Threaten Religion and Philosophy?" *Theology & Science* 6, 439–450.

Ashkenazi, Michael. 1992. "Not the Sons of Adam: Religious Responses to ETI," *Space Policy* 8, 341–350.

Ashkenazi, Michael. 1998. "On the Applicability of Human Cultural Models to Extraterrestrial Intelligence Civilizations," *Acta Astronautica* 42, 739–743.

Ashkenazi, Michael. 2017. *What We Know about Extraterrestrial Intelligence: Foundations of Xenology*. Cham, Switzerland: Springer.

Axtell, James. 1992. *Beyond 1492: Encounters in Colonial North America*. New York, NY: Oxford University Press.

Bainbridge, William Sims. 1983. "Attitudes toward Interstellar Communication: An Empirical Study," *Journal of the British Interplanetary Society* 36, 298–304.

Bainbridge, William Sims. 1987. "Collective Behavior and Social Movements," in Rodney Stark (ed.), *Sociology*. Belmont, CA: Wadsworth, 544–576.

Bainbridge, William Sims. 2011. "Cultural Beliefs about Extraterrestrials: A Questionnaire Study," in Douglas Vakoch and Albert Harrison (eds.), *Civilizations beyond Earth: Extraterrestrial Life and Society*. New York, NY: Berghahn Books, 118–140.

Bains, William. 2004. "Many Chemistries Could Be Used to Build Living Systems," *Astrobiology* 4, 137–167, at 138.

Barkow, Jerome. 2014. "Eliciting Altruism While Avoiding Xenophobia: A Thought Experiment," in Douglas Vakoch (ed.), *Archaeology, Anthropology and Interstellar Communication*. Washington, DC: National Aeronautics and Space Administration, 37–48.

Barlowe, Wayne. 1979. *Barlowe's Guide to Extraterrestrials: Great Aliens from Science Fiction Literature*. New York, NY: Workman Publishing.

Barrow, John D. 1998. *Impossibility: The Limits of Science and the Science of Limits*. Oxford: Oxford University Press.

Barrow, John D. and Frank J. Tipler. 1986. *The Anthropic Cosmological Principle*. Oxford: Oxford University Press.

Bartha, Paul. 2010. *By Parallel Reasoning: The Construction and Evaluation of Analogical Arguments*. New York, NY: Oxford University Press.

Bartholomew, Robert. 1998. "The Martian Panic Sixty Years Later: What Have We Learned?," *Skeptical Inquirer* (November–December), 40–43.

Bartholomew, Robert and Benjamin Radford. 2012. *The Martians Have Landed: A History of Media-Driven Panics and Hoaxes*. Jefferson, NC: McFarland & Company.

Basalla, George. 1988. *The Evolution of Technology*. Cambridge: Cambridge University Press.

Basalla, George. 2006. *Civilized Life in the Universe: Scientists on Intelligent Extraterrestrials*. Oxford: Oxford University Press.

Battaglia, Debbora. 2009. *E.T. Culture: Anthropology in Outerspaces*. Durham, NC: Duke University Press.

Beck, Lewis. W. 1985. "Extraterrestrial Intelligent Life," in Edward Regis Jr. (ed.), *Extraterrestrials: Science and Alien Intelligence*. Cambridge: Cambridge University Press, 3–18.

Bell, Trudy. 1980. "The Grand Analogy: History of the Idea of Extraterrestrial Life," *Cosmic Search* 2, 1–10. Online at www.bigear.org/CSMO/HTML/CS05/cs05p02.htm.

Bell Burnell, Jocelyn S. 1979. "Little Green Men, White Dwarfs or Pulsars," *Cosmic Search* 1, 16–21. Online at www.bigear.org/vol1no1/burnell.htm. Originally presented as an after-dinner speech with the title of "*Petit Four*" at the Eighth Texas Symposium on Relativistic Astrophysics, published in the *Annals of the New York Academy of Science* 302 (1977), 685–689.

Bell Burnell, Jocelyn S. 1983. "The Discovery of Pulsars," in Ken Kellermann and B. Sheets (eds.), *Serendipitous Discoveries in Radio Astronomy*. Green Bank, WV: National Radio Astronomy Observatory, pp. 160–170.

Benner, Stephen and A. Michael Sismour. 2005. "Synthetic Biology," *Nature Reviews Genetics* 6, 533–543.

Berenzden, Richard, Daniel Seeley, and Richard C. Hart. 1976. *Man Discovers the Galaxies*. New York, NY: Science History Publications.

Berkowitz, Jacob. 2012. *The Stardust Revolution: The New Story of Our Origin in the Stars*. Amherst, NY: Prometheus Books.

Bermudez, Jose Luis. 2010. *Cognitive Science: An Introduction to the Science of the Mind*. Cambridge: Cambridge University Press.

Berry, John W., Ype H. Poortinga, Seger M. Breugelmans, Athanasios Chasiotis, and David L. Sam. 1992. *Cross-Cultural Psychology*. New York: Cambridge University Press.

Bertka, Constance M. (ed.). 2009. *Exploring the Origin, Extent, and Future of Life: Philosophical, Ethical and Theological Perspectives*. Cambridge: Cambridge University Press.

Bertka, Constance M. 2013. "Christianity's Response to the Discovery of Extraterrestrial Intelligent Life: Insights from Science and Religion and the Sociology of Religion," in Douglas Vakoch (ed.), *Astrobiology, History and Society: Life beyond Earth and the Impact of Discovery*. Berlin-Heidelberg: Springer-Verlag, 329–340.

Biddanda, Bopaiah, Stephen C. Nold, and Gregory J. Dick. 2012. "Rock, Water, Microbes: Underwater Sinkholes in Lake Huron are Habitats for Ancient Microbial Life," *Nature Education Knowledge* 3, 13.

Billingham, John. 1998. "Cultural Aspects of the Search for Extraterrestrial Intelligence," *Acta Astronautica* 42, 711–719.

Billingham, John. 2000a. "Summary of Results of the Seminar on the Cultural Impact of Extraterrestrial Contact," in Guillermo A. Lemarchand and Karen J. Meech (eds.), *Bioastronomy '99: A New Era in Bioastronomy*. San Francisco, CA: Astronomical Society of the Pacific, 667–675.

Billingham, John. 2000b. "Who Said What: A Summary and Eleven Conclusions," in Allen Tough (ed.), *When SETI Succeeds: The Impact of High-Information Contact*. Bellevue, WA: Foundation for the Future, 33–39. Online at www.futurefoundation.org/documents/hum_pro_wrk1.pdf.

Billingham, John. 2013. "SETI: The NASA Years," in Douglas Vakoch (ed.), *Astrobiology, History and Society: Life beyond Earth and the Impact of Discovery*. Berlin-Heidelberg: Springer-Verlag, 1–21.

Billingham, John, Roger Heyns, and David Milne (eds.). 1999. *Social Implications of the Detection of an Extraterrestrial Civilization*. Mountain View, CA: SETI Press.

Billings, Lee. 2013. *Five Billion Years of Solitude: The Search for Life among the Stars*. New York, NY: Current.

Billings, Linda. 2015. "The Allure of Alien Life: Public and Media Framings of Extraterrestrial Life," in Steven J. Dick (ed.), *The Impact of Discovering Life beyond Earth*. Cambridge: Cambridge University Press, 308–323.

Bimm, Jordan. 2014. "Rethinking the Overview Effect," *Quest: The History of Spaceflight* 21, 39–47.

Binzel, Richard P. 1997. "A Near-Earth Object Hazard Index," *Annals of the New York Academy of Sciences* 822, 545–551.

Bjornvig, Thor. 2013. "Outer Space Religion and the Overview Effect: A Critical Inquiry into a Classic of the Pro-space Movement," *Astropolitics* 11, 4–24.

Blumberg, Baruch S. 2003. "The NASA Astrobiology Institute: Early History and Organization," *Astrobiology* 3, 463–470.

Blumberg, Baruch S. 2011. "Astrobiology, Space and the Future Age of Discovery," *Philosophical Transactions of the Royal Society A* 369, 508–515. Online at http://rsta.royalsocietypublishing.org/content/369/1936/508.

Blumenberg, Hans. 1987. *The Genesis of the Copernican World*, trans. R. M. Wallace. Cambridge, MA: MIT Press.

Bogonovich, Marc. 2011. "Intelligence's Likelihood and Evolutionary Time Frame," *International Journal of Astrobiology* 10, 113–122.

Bostrom, Nick. 2014. *Superintelligence: Paths, Dangers, Strategies*. Oxford: Oxford University Press.

Bostrom, Nick and Milan M. Ćirković (eds.). 2008. *Global Catastrophic Risks*. Oxford: Oxford University Press.

Bowler, Peter. 1989. *Evolution: The History of an Idea*. Berkeley, CA: University of California Press, chapter 7.

Brandstetter, Thomas. 2012. "Life beyond the Limits of Knowledge: Crystalline Life in the Popular Science of Desiderius Papp (1895–1993)," *Astrobiology* 12, 951–957.

Brin, David. 2011. "A Contrarian Perspective on Altruism," in Paul Shuch (ed.), *Searching for Extraterrestrial Intelligence: SETI Past, Present and Future*. Chichester: Praxis, 429–450.

Brown, James Robert. 1999. *Philosophy of Mathematics*. London and New York, NY: Routledge.

Browne, Janet. 1995. *Charles Darwin: Voyaging*. New York, NY: Alfred A. Knopf.

Browne, Janet. 2002. *Charles Darwin: The Power of Place*. New York, NY: Alfred A. Knopf.

Bulloch, William. 1938. *The History of Microbiology*. Oxford: Oxford University Press.

Bylinksy, Gene. 1981. *Life in Darwin's Universe: Evolution and the Cosmos*. Garden City, NJ: Doubleday and Company.

Cabrol, Nathalie. 2016. "Alien Mindscapes – A Perspective on the Search for Extraterrestrial Intelligence," *Astrobiology* 16, 661–676.

Callicott, J. Baird. 1986a. "Moral Considerability and Extraterrestrial Life," in Eugene C. Hargrove (ed.), *Beyond Spaceship Earth: Environmental Ethics and the Solar System*. San Francisco, CA: Sierra Club Books, 227–259.

Callicott, J. Baird. 1986b. "On the Intrinsic Value of Nonhuman Species," in Bryan Norton (ed.), *The Preservation of Species*. Princeton, NJ: Princeton University Press, 138–172. Reprinted 2014.

Cameron, A. G. W. (ed.). 1963. *Interstellar Communication*. New York, NY: W. A. Benjamin.

Campbell, W. Joseph. 2010. *Getting It Wrong: Ten of the Greatest Misreported Stories in American Journalism*. Berkeley, CA: University of California Press, 26–27.

Caneva, Ken. 2005. "Discovery as a Site for the Collective Construction of Scientific Knowledge," *Historical Studies in the Physical Sciences* 35, 175–291.

Cantril, Hadley. 1940. *The Invasion from Mars: A Study in the Psychology of Panic*. Princeton, NJ: Princeton University Press. Reprinted Transaction Publishers, 2005.

Capra, Fritz. 1975. *The Tao of Physics: An Exploration of the Parallels between Modern Physics and Eastern Mysticism*. London: Wildwood House.

Carr, Nicholas. 2011. *The Shallows: What the Internet Is Doing to Our Brains*. New York, NY: W. W. Norton & Company.

Chadwick, John. 1990. *The Decipherment of Linear B*. Cambridge: Cambridge University Press. Reprinted 1958.

Chaisson, Eric J. 2009. "Cosmic Evolution," in Steven J. Dick and Mark L. Lupisella (eds.), *Cosmos & Culture: Cultural Evolution in a Cosmic Context*. Washington, DC: National Aeronautics and Space Administration, 3–24. Online at http://history.nasa.gov/SP-4802.pdf.

Chalmers, David. 2003. "Consciousness and Its Place in Nature," in Stephen Stich and Ted Warfield (eds.), *Blackwell Guide to the Philosophy of Mind*. London: Blackwell, 102–142. Online at http://consc.net/papers/nature.html.

Chalton, Nicola (ed.). 2008. *Philosophers: Extraordinary People who Altered the Course of History*. New York, NY: Metro Books, 100.

Chela-Flores, Julian. 2007. "Testing the Universality of Biology: A Review," *International Journal of Astrobiology*, 6, 241–248.

Chiang, Ted. 2002. "Story of Your Life," in Ted Chiang (ed.), *Stories of Your Life and Others*. New York, NY: Vintage Books, 91–145.

Cho, Adrian. 2012a. "Higgs Boson Makes Its Debut after Decades-Long Search," *Science* 143, 141–143.

Cho, Adrian. 2012b. "Last Hurrah: Final Tevatron Data Show Hints of Higgs Boson," *Science* 335, 115.

Cho, Adrian. 2012c. "Who Invented the Higgs Boson?" *Science* 337, 1286–1289.

Chorost, Michael. Forthcoming. *How to Talk to Aliens*.

Christian, David. 2004. *Maps of Time: An Introduction to Big History*. Berkeley, CA: University of California Press.

Ćirković, Milan. 2006. "Macroengineering in the Galactic Context: A New Agenda for Astrobiology," in Viorel Badescu, Richard B. Cathcart, and Roelof D. Schuiling (eds.), *Macro-engineering: A Challenge for the Future*. New York, NY: Springer-Verlag, 281–300.

Ćirković, Milan. 2012. *The Astrobiological Landscape: Philosophical Foundations of the Study of Cosmic Life*. Cambridge: Cambridge University Press.

Clark, Andy. 2012. "Embodied, Embedded, and Extended Cognition," in Keith Frankish and William M. Ramsey (eds.), *The Cambridge Handbook of Cognitive Science*. Cambridge: Cambridge University Press, 265–291.

Clark, Andy and D. Chalmers. 1998. "The Extended Mind," *Analysis* 58, 7–19.

Clarke, Arthur C. 1951. *The Exploration of Space*. New York, NY: Harper, 191.

Cleland, Carol E. and Christopher Chyba. 2007. "Does 'Life' Have a Definition?" in Woodruff T. Sullivan III and John A. Baross (eds.), *Planets and Life: The Emerging Science of Astrobiology*. Cambridge: Cambridge University Press, 119–131.

Cleland, Carol E. and Elspeth M. Wilson. 2013. "Lessons from Earth: Toward an Ethics of Astrobiology," in Chris Impey, Anna Spitz, and William Stoeger (eds.), *Encountering Life in the Universe: Ethical Foundations and Social Implications of Astrobiology*. Tucson, AZ: University of Arizona Press, 17–55.

Cleland, Carol E. and Elspeth M. Wilson. 2015. "The Moral Subject of Astrobiology," in Steven J. Dick (ed.), *The Impact of Discovering Life beyond Earth*. Cambridge: Cambridge University Press, 207–221.

Cline, Eric H. 2014. *1177 B.C.: The Year Civilization Collapsed*. Princeton, NJ, and Oxford: Princeton University Press.

Codignola, Luca and Kai-Uwe Schrogl (eds.). 2009. *Humans in Outer Space – Interdisciplinary Odysseys*. New York, NY: Springer.

Coe, Michael D. 1992. *Breaking the Maya Code*. London: Thames and Hudson.

Cohen, Jack and Ian Stewart. 2002. *What Does a Martian Look Like?* New York, NY: Wiley. Published in the United Kingdom as *Evolving the Alien: The Science of Extraterrestrial Life*. London: Ebury Press.

Cole, John Y. 2008. *On These Walls: Inscriptions and Quotations in the Library of Congress*. Washington, DC: Library of Congress.

Comfort, Louise K., Arjen Boin, and Chris D. Demchak (eds.). 2010. *Designing Resilience: Preparing for Extreme Events*. Pittsburgh, PA: University of Pittsburgh Press.

Conrad, P. G. and K. H. Nealson. 2001. "A non-Earth-centric Approach to Life Detection," *Astrobiology* 1, 15–24.

Consolmagno, Guy and Paul Mueller. 2014. *Would You Baptize an Extraterrestrial?* New York, NY: Random House.

Conway Morris, Simon. 1998. *The Crucible of Creation*. Oxford: Oxford University Press.

Conway Morris, Simon. 2003. *Life's Solution: Inevitable Humans in a Lonely Universe*. Cambridge: Cambridge University Press.

Coopersmith, Jonathan. 2008. "Great (Unfulfilled) Expectations: To Boldly Go Where No Social Scientist or Historian Has Gone Before," in Steven J. Dick (ed.), *Remembering the Space Age*. Washington, DC: National Aeronautics and Space Administration, 135–154, SP-2008–4703.

Crick, Francis. 1994. *The Astonishing Hypothesis: The Scientific Search of the Soul*. New York, NY: Scribner.

Crosby, Alfred W. 2003. *The Columbian Exchange: Biological and Cultural Consequences of 1492*. Westport, CT: Praeger Publishers.

Crosby, Alfred W. 2009. "Micro-organisms and Extraterrestrial Travel," in Luca Codignola and Kai-Uwe Schrogl (eds.), *Humans in Outer Space – Interdisciplinary Odysseys*. New York, NY: Springer, 6–13.

Crowe, Michael J. 1981. "New Light on the Moon Hoax," *Sky and Telescope* 62, 428–429.

Crowe, Michael. J. 1986. *The Extraterrestrial Life Debate, 1750–1900: The Idea of a Plurality of Worlds from Kant to Lowell*. Cambridge: Cambridge University Press.

Crowe, Michael. J. 1997. "A History of the Extraterrestrial Life Debate, *Zygon* 32, 147–162.

Crowe, Michael J. 2001. "Astronomy and Religion (1780–1915): Four Case Studies Involving Ideas of Extraterrestrial Life," *Osiris*, 2nd ser., vol. 16, 209–226.

Crowe, Michael. J. (ed.). 2008. *The Extraterrestrial Life Debate, Antiquity to 1900: A Source Book*. Notre Dame, IN: University of Notre Dame.

Crysdale, Cynthia S. W. 2009. "God, Evolution, and Astrobiology," in Constance M. Bertka (ed.), *Exploring the Origin, Extent, and Future of Life: Philosophical, Ethical and Theological Perspectives*. Cambridge: Cambridge University Press, 220–241.

Danielson, Dennis. 2001. "The Great Copernicus Cliché," *American Journal of Physics* 69, 1029–1035.

Darling, David. 2001. *Life Everywhere: The Maverick Science of Astrobiology*. New York, NY: Basic Books.

Daston, Lorraine. 2014. "The Naturalistic Fallacy is Modern," *Isis* 105, 579–587.

Davies, Paul. 1983. *God and the New Physics*. New York, NY: Simon and Schuster.

Davies, Paul. 1995a. *About Time: Einstein's Unfinished Revolution*. New York, NY: Simon and Schuster.

Davies, Paul. 1995b. *Are We Alone? Philosophical Implications of the Discovery of Extraterrestrial Life*. New York, NY: Basic Books.

Davies, Paul. 2003. "ET and God," *The Atlantic Monthly*, September, 112–118. Online at www.theatlantic.com/issues/2003/09/davies.htm.

Davies, Paul. 2007. *Cosmic Jackpot: Why Our Universe Is Just Right for Life*. Boston, MA: Houghton Mifflin.

Davies, Paul. 2010. *The Eerie Silence: Renewing Our Search for Alien Intelligence*. Boston, MA, and New York, NY: Houghton Mifflin Harcourt.

Dawkins, Richard. 1989. *The Selfish Gene*. Oxford: Oxford University Press.

Delano, Kenneth. 1977. *Many Worlds, One God*. New York, MA: Exposition Press.

Dennett, Daniel. 1991. *Consciousness Explained*. Boston, MA: Little, Brown and Company.

Dennett, Daniel. 1996. *Darwin's Dangerous Idea*. New York, NY: Simon and Schuster.

Denning, Kathryn. 2009. "Social Evolution: State of the Field," in Steven J. Dick and Mark L. Lupisella (eds.), *Cosmos & Culture: Cultural Evolution in a Cosmic Context*. Washington, DC: National Aeronautics and Space Administration, 63–124.

Denning, Kathryn. 2011a. "'Is Life What We Make of It?" *Philosophical Transactions of the Royal Society A* 369, 669–678.

Denning, Kathryn. 2011b. "Being Technological," in Paul Shuch (ed.), *Searching for Extraterrestrial Intelligence: SETI Past, Present and Future*. Chichester: Praxis, 477–496.

Denning, Kathryn. 2011c. "Unpacking the Great Transmission Debate," in Douglas Vakoch (ed.), *Communication with Extraterrestrial Intelligence*. Albany, NY: State University of New York Press, 237–252.

Denning, Kathryn. 2011d. "L on Earth," in Douglas Vakoch (ed.), *Communication with Extraterrestrial Intelligence*. Albany, NY: State University of New York Press, 74–83.

Denning, Kathryn. 2013. "Impossible Predictions of the Unprecedented: Analogy, History, and the Work of Prognostication," in Douglas Vakoch (ed.), *Astrobiology, History and Society: Life beyond Earth and the Impact of Discovery*. Berlin-Heidelberg: Springer-Verlag, 301–312.

Denning, Kathryn. 2014. "Learning to Read: Interstellar Message Decipherment from Archaeological and Anthropological Perspectives," in Douglas Vakoch (ed.), *Archaeology, Anthropology and Interstellar Communication*. Washington, DC: National Aeronautics and Space Administration, 95–112.

Des Marais, David J., Joseph A. Nuth III, Louis Allamandola, Alan P. Boss, Jack D. Farmer, Tori M. Hoehler, Bruce M. Jakosky, Victoria S. Meadows, Andrew Pohorille, Bruce Runnegar, and Alfred M. Spormann 2008. "The NASA Astrobiology Roadmap," *Astrobiology* 8, 715–730.

Devall, William and George Sessions. 1985. *Deep Ecology: Living As if Nature Mattered*. Salt Lake City, UT: Gibbs M. Smith, Inc.

DeVito, Carl L. 2011. "On the Universality of Human Mathematics," in Douglas Vakoch (ed.), *Communication with Extraterrestrial Intelligence*. Albany, NY: State University of New York Press, 439–448.

DeVito, Carl L. 2014. *Science, SETI and Mathematics*. New York, NY, and Oxford: Berghahn Books.

Devlin, Keith. 2002. *The Language of Mathematics: Making the Invisible Visible*. New York, NY: W. H. Freeman.

de Waal, Frans. 1996. *Good Natured: The Origins of Right and Wrong in Humans and Other Animals*. Cambridge, MA: Harvard University Press.

Diaz del Castillo, Bernal. 1956. *The Discovery and Conquest of Mexico*. New York, NY: Da Capo Press; with new Introduction by Hugh Thomas, 1996, translation by A. P. Maudslay first published in 1908.

Dick, Steven J. 1982. *Plurality of Worlds: The Origins of the Extraterrestrial Life Debate from Democritus to Kant*. Cambridge: Cambridge University Press.

Dick, Steven J. 1989. "The Concept of Extraterrestrial Intelligence – An Emerging Cosmology," *Planetary Report* 9, 13–17.

Dick, Steven J. 1995. "Consequences of Success in SETI: Lessons from the History of Science," in Seth Shostak (ed.), *Progress in the Search for Extraterrestrial Life*. San Francisco, CA: Astronomical Society of the Pacific, 521–532.

Dick, Steven J. 1996. *The Biological Universe: The Twentieth Century Extraterrestrial Life Debate and the Limits of Science*. Cambridge: Cambridge University Press.

Dick, Steven J. 1997. "The Biophysical Cosmology: The Place of Bioastronomy in the History of Science," in C. B. Cosmovici, Stuart Bowyer, and Dan Werthimer (eds.), *Astronomical and Biochemical Origins and the Search for Life in the Universe*. Bologna: Editrice Compositori, 785–788.

Dick, Steven J. 1998. *Life on Other Worlds*. Cambridge: Cambridge University Press.

Dick, Steven J. 2000a. "Extraterrestrials and Objective Knowledge," in Allen Tough (ed.), *When SETI Succeeds: The Impact of High-Information Contact*. Bellevue, WA: Foundation for the Future, pp. 47–48. Online at www.futurefoundation.org/documents/hum_pro_wrk1.pdf.

Dick, Steven J. 2000b. "Cultural Aspects of Astrobiology: A Preliminary Reconnaissance at the Turn of the Millennium," in Guillermo A. Lemarchand and Karen J. Meech (eds.), *Bioastronomy '99: A New Era in Bioastronomy*. San Francisco, CA: Astronomical Society of the Pacific, 649–659.

Dick, Steven J. (ed.). 2000c. *Many Worlds: The New Universe, Extraterrestrial Life and the Theological Implications*. Philadelphia, PA: Templeton Press.

Dick, Steven J. 2000d. "Cosmotheology: Theological Implications of the New Universe," in Steven J. Dick (ed.), *Many Worlds: The New Universe, Extraterrestrial Life and the Theological Implications*. Philadelphia, PA: Templeton Press, 191–210. Online at www.metanexus.net/essay/many-worlds-cosmotheology.

Dick, Steven J. 2003a. "Cultural Evolution, the Postbiological Universe and SETI," *International Journal of Astrobiology* 2, 65–74. Reprinted as "Bringing Culture to Cosmos: The Postbiological Universe," in Steven J. Dick and Mark L. Lupisella (eds.), *Cosmos & Culture: Cultural Evolution in a Cosmic Context*. Washington, DC: National Aeronautics and Space Administration, 463–488.

Dick, Steven J. 2003b. *Sky and Ocean Joined*. Cambridge: Cambridge University Press, 425–429.

Dick, Steven J. 2004. "The New Universe, Destiny of Life, and the Cultural Implications," in J. Seckbach, Julian Chela-Flores, Tobias Owen, and Francois Raulin (eds.), *Life in the Universe: From the Miller Experiment to the Search for Life on Other Worlds*. Dordrecht: Kluwer Academic Publishers, 319–326.

Dick, Steven J. 2005. "Kosmotheologie – neu betrachtet," in Tobias Daniel Wabbel (ed.), *Leben im All: Positionen aus Naturwissenschft, Philosophie und Theologie*. Dusseldorf: Patmos Verlag, 156–172. Online in English at http://bdigital.ufp.pt/bitstream/10284/778/2/287-301Cons-Ciencias%2002–5.pdf.

Dick, Steven J. 2006. "Anthropology and the Search for Extraterrestrial Intelligence," *Anthropology Today* 22 (2), 3–7.

Dick, Steven J. 2008. "Cosmology and Biology." Proceedings of the 2008 Conference of the Society of Amateur Radio Astronomers. National Radio Astronomy Observatory, Green Bank, West Virginia, Amateur Radio Relay League, 1–16. Online at http://evodevouniverse.com/uploads/f/f3/Dick_2008__Cosmology_and_Biology.pdf.

Dick, Steven J. 2009. "Cosmic Evolution: History, Culture and Human Destiny," in Steven J. Dick and Mark L. Lupisella (eds.), *Cosmos & Culture: Cultural Evolution in a Cosmic Context*. Washington, DC: National Aeronautics and Space Administration, 25–59.

Dick, Steven J. 2012. "Critical Issues in the History, Philosophy, and Sociology of Astrobiology," *Astrobiology* 12, 906–927.

Dick, Steven J. 2013a. "The Societal Impact of Extraterrestrial Life: The Relevance of History and the Social Sciences," in Douglas Vakoch (ed.), *Astrobiology, History and Society: Life beyond Earth and the Impact of Discovery*. Berlin-Heidelberg: Springer-Verlag, 227–257.

Dick, Steven J. 2013b. "The Twentieth Century History of the Extraterrestrial Life Debate: Major Themes and Lessons Learned," in Douglas Vakoch (ed.), *Astrobiology, History and Society: Life beyond Earth and the Impact of Discovery*. Berlin-Heidelberg: Springer-Verlag, 133–173.

Dick, Steven J. 2013c. *Discovery and Classification in Astronomy: Controversy and Consensus*. Cambridge: Cambridge University Press.

Dick, Steven J. 2015a. *Historical Studies in the Societal Impact of Spaceflight*. Washington, DC: National Aeronautics and Space Administration.

Dick, Steven J. (ed.). 2015b. *The Impact of Discovering Life beyond Earth*. Cambridge: Cambridge University Press.

Dick, Steven J. 2015c. "The Drake Equation in Context," in Douglas A. Vakoch and Matthew F. Dowd (eds.), *The Drake Equation: Estimating the Prevalence of Extraterrestrial Life through the Ages*. Cambridge: Cambridge University Press, 1–20.

Dick, Steven J. 2018 "Toward a Constructive Naturalistic Cosmotheology," in Ted Peters, Martinez Hewlett, Joshua Moritz, and Robert John Russell (eds.), *Astrotheology: Science and Theology Meet Extraterrestrial Life*. Eugene, OR: Cascade Books.

Dick, Steven J. and Roger Launius (eds.). 2007. *Societal Impact of Spaceflight*. Washington, DC: National Aeronautics and Space Administration.

Dick, Steven J. and Mark L. Lupisella (eds.). 2009. *Cosmos & Culture: Cultural Evolution in a Cosmic Context*. Washington, DC: National Aeronautics and Space Administration. Online at http://history.nasa.gov/SP-4802.pdf.

Dick, Steven J. and James E. Strick. 2004. *The Living Universe: NASA and the Development of Astrobiology*. New Brunswick, NJ: Rutgers University Press.

Dickinson, Terrence. 1994. *Extraterrestrials: A Field Guide for Earthlings*. Camden East, Ontario: Camden House.

DiGregorio, Barry. 1997. *Mars: The Living Planet*. Berkeley, CA: Frog, Ltd.

Dobell, Clifford. 1958. *Antony van Leeuwenhoek and His "Little Animals": Being Some Account of the Father of Protozoology & Bacteriology and His Multifarious Discoveries in these Disciplines*. New York, NY: Russell and Russell.

Dobzhansky, Theodosius. 1972. "Darwinian Evolution and the Problem of Extraterrestrial Life," *Perspectives in Biology and Medicine* 15, 157–175.

Dominik, Martin and John C. Zarnecki. 2011. "The Detection of Extra-terrestrial Life and the Consequences for Science and Society," *Philosophical Transactions of the Royal Society A*, 369, 499–507. Online at http://rsta.royalsocietypublishing.org/content/369/1936.toc.

Dorminey, Bruce. 2014. "A New Way to Search for Life in Space," *Astronomy* (June), 45–49, citing the work of Victoria Meadows at the University of Washington.

Doyle, Stephen E. 1998. "Social Implications of NASA's High Resolution Microwave Survey," *Acta Astronautica* 42, 721–725.

Doyle, Stephen E. 2002. *Origins of International Space Law*. San Diego, CA: Univelt.

Doyle, Stephen E. 2013. "Andrew G. Haley," in Stephan Hobe (ed.), *Pioneers of Space Law*. Leiden and Boston, MA: Martinus Nijhoff, 70–96.

Drake, Frank. 2013. "Reflections on the Equation," *International Journal of Astrobiology* 12(3), 173–176.

Dreyer, Edward L. 2007. *Zheng He: China and the Oceans in the Early Ming Dynasty, 1405–1433*. New York, NY: Pearson Longman.

Dunér, David. 2011. "Astrocognition: Prolegomena to a Future Cognitive History of Exploration," in Ulrike Landfester, Nina-Louisa Remuss, Kai-Uwe Schrogl, and Jean-Claude Worms (eds.), *Humans in Outer Space – Interdisciplinary Perspectives*. Wien: Springer Verlag, 117–140.

Dunér, David. 2012. *The History and Philosophy of Astrobiology*, special issue of the journal *Astrobiology*, vol. 12, no. 10.

Dunér, David, Joel Parthemore, Erik Persson, and Gustav Holmberg (eds.). 2013. *The History and Philosophy of Astrobiology: Perspectives on Extraterrestrial Life and the Human Mind*. Newcastle-upon-Tyne: Cambridge Scholars Publishing.

Dusek, Val. 2006. *Philosophy of Technology: An Introduction*. Malden, MA, and Oxford: Blackwell.

Dyson, Freeman. 1966. "The Search for Extraterrestrial Technology," in Robert E. Marshak (ed.), *Perspectives in Modern Physics*. New York, NY: Interscience Publishers, 641–655.

Dyson, Freeman. 1988. *Infinite in all Directions*. New York, NY: Harper and Row.

Eiseley, Loren. 1957. *The Immense Journey*. New York, NY: Random House.

Eisenstein, Elizabeth L. 1983. *The Printing Revolution in Early Modern Europe*. Cambridge: Cambridge University Press.

Ekers, Ronald D., D. Kent Cullers, John Billingham, and Louis K. Scheffer (eds.). 2002. *SETI 2020: A Roadmap for the Search for Extraterrestrial Intelligence*. Mountain View, CA: SETI Press.

Elliott, John H. 1970. *The Old World and the New, 1492–1650*. Cambridge: Cambridge University Press. Reprinted 1992.

Elliott, John H. 2006. *Empires of the Atlantic World: Britain and Spain in America, 1492–1830*. New Haven, CT: Yale University Press.

Fasan, Ernst. 1970. *Relations with Alien Intelligence: The Scientific Basis of Metalaw*. Berlin: Berlin Verlag.

Febvre, Lucien and Henri-Jean Martin. 1976. *The Coming of the Book*. London: Verso.

Finney, Ben. 1990. "The Impact of Contact," *Acta Astronautica* 21, 117–127.

Finney, Ben. 2000. "SETI, Consilience and the Unity of Knowledge," in Guillermo A. Lemarchand and Karen J. Meech (eds.), *Bioastronomy '99: A New Era in Bioastronomy*. San Francisco, CA: Astronomical Society of the Pacific, 641–647.

Finney, Ben and Jerry Bentley. 1998. "A Tale of Two Analogues: Learning at a Distance from the Ancient Greeks and Maya and the Problem of Deciphering Extraterrestrial Radio Transmissions," in Douglas Vakoch (ed.), *Archaeology, Anthropology and Interstellar Communication*. Washington, DC: National Aeronautics and Space Administration, 65–78. This article is an expansion of one that appeared in *Acta Astronautica* 42(10–12), 691–696.

Fitzhugh, William H. (ed.). 1985. *Cultures in Contact: The Impact of European Contacts on Native American and Cultural Institutions, AD 1000–1800*. Washington, DC, and London: Smithsonian Institution.

Ford, Brian J. 1985. *Single Lens: The Story of the Simple Microscope*. New York, NY: Harper and Row, 19.

Foucault, Michel. 1973. *The Order of Things: An Archaeology of the Human Sciences*. New York, NY: Vintage Books.

Fox, Douglas. 2011. "The Limits of Intelligence," *Scientific American* (July), 36–43.

Frankish, Keith and William M. Ramsey (eds.), 2012. *The Cambridge Handbook of Cognitive Science*. Cambridge: Cambridge University Press.

Freitas Jr., Robert A. 1975–1979, 2008. *Xenology: An Introduction to the Scientific Study of Extraterrestrial Life, Intelligence, and Civilization*, 1st edn. Sacramento, CA: Xenology Research Institute. Online at www.xenology.info/Xeno.htm.

Freudenthal, Hans. 1960. *Lincos: Design of a Language for Cosmic Discourse*. Amsterdam: North-Holland.

Friedenberg, Jay and Gordon Silverman. 2012. *Cognitive Science: An Introduction to the Study of Mind*. Los Angeles, CA: Sage.

Friedrich, Johannes. 1957. *Extinct Languages*, trans. Frank Gaynor. New York, NY: Philosophical Library.

Fritze, Ronald H. 2002. *New Worlds: The Great Voyages of Discovery, 1400–1600*. Thrupp: Sutton.

Fry, Iris. 2000. *The Emergence of Life on Earth: A Historical and Scientific Overview*. New Brunswick, NJ: Rutgers University Press.

Fry, Iris. 2009. "Philosophical Aspects of the Origin-of-Life Problem: The Emergence of Life and the Nature of Science," in Constance M. Bertka (ed.), *Exploring the Origin, Extent, and Future of Life: Philosophical, Ethical and Theological Perspectives*. Cambridge: Cambridge University Press, 61–79.

Fry, Iris. 2012. "Is Science Metaphysically Neutral?" *Studies in the History and Philosophy of Biological and Biomedical Sciences* 43, 665–673.

Fry, Iris. 2015. "The Philosophy of Astrobiology: The Copernican and Darwinian Philosophical Presuppositions," in Steven J. Dick (ed.), *The Impact of Discovering Life beyond Earth*. Cambridge: Cambridge University Press, 23–37.

Gabrynowicz, Joanne Irene. 2010. "One Half Century and Counting: The Evolution of U. S. National Space Law and Three Long-Term Issues," *Harvard Law & Policy Review* 4, 405–426.

Gaddis, John Lewis. 2002. *The Landscape of History: How Historians Map the Past*. Oxford: Oxford University Press.

Galison, Peter. 1987. *How Experiments End*. Chicago, IL: University of Chicago Press, 76, 126–127.

Gallop, Alan. 2011. *The Martians Are Coming! The True Story of Orson Welles' 1938 Panic Broadcast*. Gloucestershire: Amberly.

Garber, Stephen J. 2014. "A Political History of NASA's SETI Program," in Douglas Vakoch (ed.), *Archaeology, Anthropology and Interstellar Communication*. Washington, DC: National Aeronautics and Space Administration, 23–48.

Gardner, James. 2003. *Biocosm – The New Scientific Theory of Evolution: Intelligent Life is the Architect of the Universe*. Makawao, Maui, HI: Hawaii Inner Ocean Publishing.

Gardner, James. 2007. *The Intelligent Universe: AI, ET and the Emerging Mind of the Cosmos*. Franklin Lakes, NJ: New Page.

Gardner, James. 2009. "The Intelligent Universe," in Steven J. Dick and Mark L. Lupisella (eds.), *Cosmos & Culture: Cultural Evolution in a Cosmic Context*. Washington, DC: National Aeronautics and Space Administration, 361–382.

Gardner, Sebastian. 1999. *Kant and the Critique of Pure Reason*. London and New York, NY: Routledge.

Geertz, Clifford. 1973. *The Interpretation of Cultures* New York, NY: Basic Books.

Gentner, Dedre and Michael Jeziorski. 1993. "The Shift from Metaphor to Analogy in Western Science," in Andrew Ortony (ed.), *Metaphor and Thought*. Cambridge: Cambridge University Press, 447–480. Online at www.psych.northwestern.edu/psych/people/faculty/gentner/newpdfpapers/GentnerJeziorski93.pdf.

Geppert, Alexander C. T. (ed.). 2012. *Imagining Outer Space: European Astroculture in the Twentieth Century*. New York, NY: Palgrave Macmillan.

Gernet, Jacques. 1999. *A History of Chinese Civilization*. Cambridge: Cambridge University Press.

Gest, Howard. 2003. *Microbes: An Invisible Universe*. Washington, DC: American Society for Microbiology Press.

Gest, Howard. 2004a. "The Discovery of Microorganisms by Robert Hooke and Antoni van Leeuwenhoek, Fellows of the Royal Society," *Notes and Records of the Royal Society of London* 58(2), 187–201.

Gest, Howard. 2004b. "The Discovery of Microorganisms Revisited," *ASM News* 70(6), 269–274.

Gest, Howard. 2007. "Fresh Views of 17th-Century Discoveries by Hooke and van Leeuwenhoek," *Microbe* 2(10), 483–488.

Gibbons, Ann. 2014. "Neanderthals and Moderns Made Imperfect Mates," *Science* 31, January, 471–472.

Gleick, James. 2011. *The Information: A History, a Theory, a Flood*. New York, NY: Pantheon Books.

Godfrey-Smith, Peter. 2016. *Other Minds: The Octopus, the Sea, and the Deep Origins of Consciousness*. New York, NY: Farrar, Straus and Giroux.

Gonzalez, Guillermo and Jay Richards. 2004. *The Privileged Planet: How Our Place in the Universe Is Designed for Discovery.* Washington, DC: Regnery Publishing.

Goodman, Matthew. 2008. *The Sun and the Moon: The Remarkable True Account of Hoaxers, Showmen, Dueling Journalists, and Lunar Man-Bats in Nineteenth-Century New York*. New York, NY: Basic Books.

Gordin, Michael D. 2004. *A Well-Ordered Thing: Dmitrii Mendeleev and the Shadow of the Periodic Table*. New York, NY: Basic Books.

Gosling, John. 2009. *Waging the War of the Worlds: A History of the 1938 Radio Broadcast and Resulting Panic*. Jefferson, NC, and London: McFarland & Company.

Gould, Stephen Jay. 1989. *Wonderful Life: The Burgess Shale and the Nature of History*. New York, NY: W. W. Norton & Company.

Grant, Edward. 1971. *Physical Science in the Middle Ages*. New York, NY: Wiley.

Gray, Robert H. 2012. *The Elusive WOW: Searching for Extraterrestrial Intelligence*. Chicago, IL: Palmer Square Press.

Greene, John. 1981. *Science, Ideology and World View*. Berkeley, CA: University of California Press.

Griffin, Donald R. 2001. *Animal Minds: Beyond Cognition to Consciousness*. Chicago, IL: University of Chicago Press.

Grinspoon, David. 2003. *Lonely Planets: The Natural Philosophy of Alien Life*. New York, NY: Harper Collins.

Grinspoon, David. 2016. *Earth in Human Hands: Shaping our Planet's Future*. New York, NY: Grand Central Publishing.

Gronstal, Aaron L. 2013. "Extraterrestrial Life in the Microbial Age," in Douglas Vakoch (ed.), *Astrobiology, History and Society: Life beyond Earth and the Impact of Discovery*. Berlin-Heidelberg: Springer-Verlag, 213–224.

Grosberg, Richard K. and Richard R. Strathmann. 2007. "The Evolution of Multicellularity: A Minor Major Transition?" *Annual Review of Ecology, Evolution and Systematics* 38, 621–654.

Gunn, James. 1982. *Isaac Asimov: The Foundations of Science Fiction*. Oxford: Oxford University Press.

Gupta, Abhik. 2013. "Altruism Toward Non-human: Lessons for Interstellar Communication," in Douglas Vakoch (ed.), *Astrobiology, History and Society: Life beyond Earth and the Impact of Discovery*. Berlin-Heidelberg: Springer-Verlag, 79–92.

Guthke, Karl S. 1990. *The Last Frontier: Imagining Other Worlds from the Copernican Revolution to Modern Science Fiction*. Ithaca, NY: Cornell University Press.

Haley, Andrew G. 1963. *Space Law and Government*. New York, NY: Appleton-Century-Crofts.

Hanley, Richard. 1997. *Is Data Human? The Metaphysics of Star Trek*. New York, NY: Basic Books.

Hanson, Robin. 2016. *The Age of EM: Work, Love, and Life when Robots Rule the Earth*. Oxford: Oxford University Press.

Haqq-Misra, Jacob. 2012. "An Ecological Compass for Planetary Engineering," *Astrobiology* 12, 985–997.

Harris, Sam. 2010. *The Moral Landscape: How Science Can Determine Human Values*. New York, NY: Free Press.

Harrison, Albert A. 1997. *After Contact: The Human Response to Extraterrestrial Life*. New York, NY: Plenum.

Harrison, Albert A. 2007. *Starstruck: Cosmic Visions in Science, Religion, and Folklore*. New York, NY: Berghahn.

Harrison, Albert A. 2011a. "Fear, Pandemonium, Equanimity and Delight: Human Responses to Extra-Terrestrial Life," *Philosophical Transactions of the Royal Society A* 369, 656–668.

Harrison, Albert A. 2011b. "After Contact – Then What?" in Paul Shuch (ed.), *Searching for Extraterrestrial Intelligence: SETI Past, Present and Future*. Chichester: Praxis, 497–514.

Harrison, Albert A. 2014. "Speaking for Earth: Projecting Cultural Values across Deep Space and Time," in Douglas Vakoch (ed.), *Archaeology, Anthropology and Interstellar Communication*. Washington, DC: National Aeronautics and Space Administration, 175–190.

Harrison, Albert A., John Billingham, and Steven J. Dick. 2000. "The Role of Social Science in SETI," in Allen Tough (ed.), *When SETI Succeeds: The Impact of High-Information Contact*. Bellevue, WA: Foundation for the Future, 71–85. Online at www.futurefoundation.org/documents/hum_pro_wrk1.pdf.

Harrison, Albert A. and Kathleen Connell. 2001. *Workshop on the Societal Implications of Astrobiology*. Moffett Field: NASA Ames Research Center. Online at www.astrosociology.org/Library/PDF/NASA-Workshop-Report-Societal-Implications-of-Astrobiology.pdf. See also NASA press release www.nasa.gov/centers/ames/news/releases/1999/99_69AR.html.

Harrison, Albert A. and Steven Dick. 2000. "Contact: Long-Term Implications for Humanity," in Allen Tough (ed.), *When SETI Succeeds: The Impact of High-Information Contact*. Bellevue, WA: Foundation for the Future, 7–31.

Harrison, James A. (ed.). 1902. *Complete Works of Edgar Allan Poe*, vol. XV. New York, NY: Thomas Y. Crowell.

Hart, John. 2013. *Cosmic Commons: Spirit, Science and Space*. Eugene, OR: Cascade Books.

Harwit, Martin. 2013. *In Search of the True Universe: The Tools, Shaping, and Cost of Cosmological Thought*. Cambridge: Cambridge University Press.

Hawkins, Jeff and Sandra Blakeslee. 2004. *On Intelligence*. New York, NY: Henry Holt.

Haynes, Robert. 1990. "Ecce Ecopoiesis: Playing God on Mars," in Don MacNiven (ed.), *Moral Expertise*. London: Routledge, 161–183.

Heidmann, Jean, 1990. "SETI False Alerts as 'Laboratory' Tests for an International Protocol Formulation," in Jill C. Tarter and Michael A. Michaud (eds.), "SETI Post-detection Protocol," special issue of *Acta Astronautica* 21(2), 73–80.

Helmreich, Stefan. 2009. *Alien Ocean: Anthropological Voyages in Microbial Seas*. Berkeley, CA: University of California Press.

Hemming, John. 2003. *The Conquest of the Incas*. Boston, MA: Houghton, Mifflin, Harcourt.

Henderson, Lawrence J. 1913. *The Fitness of the Environment: An Inquiry into the Biological Significance of the Properties of Matter*. Cambridge, MA: Macmillan. Reprinted in 1970 with an introduction by George Wald. Gloucester, MA: Peter Smith, 31.

Herzing, Denise L. 2014. "Profiling Nonhuman Intelligence: An Exercise in Developing Unbiased Tools for Describing Other 'Types' of Intelligence on Earth," *Acta Astronautica* 94, 676–780.

Hesse, Mary B. 1966. *Models and Analogies in Science*. Notre Dame, IN: University of Notre Dame Press, 1–9.

Hinsley, F. Harry and Alan Stripp (eds.). 1993. *Code Breakers: The Inside Story of Bletchley Park*. Oxford: Oxford University Press.

Hoerder, Dirk. 2002. *Cultures in Contact: World Migrations in the Second Millennium*. Durham, NC, and London: Duke University Press.

Hofstadter, Douglas. 2001. "Analogy as the Core of Cognition," in Dedre Gentner, Keith J. Holyoak, and Boicho N. Kokinov (eds.), *The Analogical Mind: Perspectives from Cognitive Science*. Cambridge MA: MIT Press, 499–538. Online at http://prelectur.stanford.edu/lecturers/hofstadter/analogy.html.

Hofstadter, Douglas and Emmanuel Sander. 2013. *Surfaces and Essences: Analogy as the Fuel and Fire of Thinking*. New York, NY: Basic Books.

Holyoak, Keith J. and Paul Thagard. 1995. *Mental Leaps: Analogy in Creative Thought*. Cambridge, MA: MIT Press.

Horneck, Gerda, Nicolas Walter, Frances Westall, John Lee Grenfell, William F. Martin, Felipe Gomez, Stefan Leuko, Natuschka Lee, Silvano Onofri, Kleomenis Tsiganis, Raffaele Saladino, Elke Pilat-Lohinger, Ernesto Palomba, Jesse Harrison, Fernando Rull, Christian Muller, Giovanni Strazzulla, John R. Brucato, Petra Rettberg, and Maria Teresa Capria. 2016. "AstRoMap European Astrobiology Roadmap," *Astrobiology* 16, 201–243.

Hoyle, Fred and Nalin C. Wickramasinghe. 1979. *Diseases from Space*. London: J. M. Dent & Sons Ltd.

Hoyt, William G. 1976. *Lowell and Mars*. Tucson, AZ: University of Arizona Press.

Hughes, Thomas Parke. 1965. "The Technological Frontier: The Railroad," in Bruce Mazlish (ed.), *The Railroad and the Space Program: An Exploration in Historical Analogy*. Cambridge, MA: MIT Press, p. 53, note 1.

Hull, David. 1973. *Darwin and His Critics: The Reception of Darwin's Theory of Evolution by the Scientific Community*. Cambridge, MA: Harvard University Press.

Huntington, Samuel. 1996. *The Clash of Civilizations and the Remaking of the World Order*. New York, NY: Simon and Schuster.

Huxley, Aldous. 1959. "Case of Voluntary Ignorance," in *Collected Essays*. New York, NY: Harper.

Impey, Chris, Anna Spitz, and William Stoeger (eds.). 2013. *Encountering Life in the Universe: Ethical Foundations and Social Implications of Astrobiology*. Tucson, AZ: University of Arizona Press.

Irudayadason, Nishant Alphonse. 2013. "The Wonder Called Cosmic Oneness: Toward Astroethics from Hindu and Buddhist Wisdom and Worldviews," in Chris Impey, Anna Spitz, and William Stoeger (eds.), *Encountering Life in the Universe: Ethical Foundations and Social Implications of Astrobiology*. Tucson, AZ: University of Arizona Press, 94–119.

Irwin, Louis Neal and Dirk Schulze-Makuch. 2001. "Assessing the Plausibility of Life on Other Worlds," *Astrobiology* 1, 143–160.

Jenkins, Susan and Robert Jenkins. 1998. *The Biology of Star Trek*. New York, NY: Harper.

Jones, Morris. 2011. "A Journalistic Perspective on SETI-Related Message Composition," in Douglas Vakoch (ed.), *Communication with Extraterrestrial Intelligence*. Albany, NY: State University of New York Press, 226–235.

Jones, Morris. 2013. "Mainstream Media and Social Media Reactions to the Discovery of Extraterrestrial Life," in Douglas Vakoch (ed.), *Astrobiology, History and Society: Life beyond Earth and the Impact of Discovery*. Berlin-Heidelberg: Springer-Verlag, 313–328.

Joyce, Gerald G. 1994. "Foreword," in David W. Deamer and Gail Fleischaker (eds.), *Origins of Life: The Central Concepts*. Boston, MA: Jones and Bartlett.

Kardashev, N. S. 1964. "Transmission of Information by Extraterrestrial Civilizations," *Soviet Astronomy* 8, 217–221. Online at http://articles.adsabs.harvard.edu/full/1964SvA....8.217K.

Kauffman, Marc. 2011. *First Contact: Scientific Breakthroughs in the Hunt for Life beyond Earth.* New York, NY: Simon and Schuster.

Kelly, Kevin. 2010. *What Technology Wants*. New York, NY: Viking.

Kissinger, Henry. 2014. *World Order*. New York, NY: Penguin Books.

Kleiman, Matthew J. 2013. *The Little Book of Space Law*. Chicago, IL: American Bar Association.

Kminek, Gerhard and John D. Rummel. 2015. "COSPAR's Planetary Protection Policy," *Space Research Today* 193, 7–19. doi: 10.1016/j.srt.2015.06.008. Online at https://cosparhq.cnes.fr/sites/default/files/ppp_article_linked_to_ppp_webpage.pdf.

Knoll, Andrew. 2004. *Life on a Young Planet*. Princeton, NJ: Princeton University Press.

Korbitz, Adam. 2014a. "Altruism, Metalaw, and Celegistics: An Extraterrestrial Perspective on Universal Law-Making," in Douglas Vakoch (ed.), *Archaeology, Anthropology and Interstellar Communication*. Washington, DC: National Aeronautics and Space Administration, 231–247.

Korbitz, Adam. 2014b. "The Precautionary Principle: Egoism, Altruism, and the Active SETI Debate," in Douglas Vakoch (ed.), *Archaeology, Anthropology and Interstellar Communication*. Washington, DC: National Aeronautics and Space Administration, 111–127.

Kragh, Helge. 2009. "The Solar Element: A Reconsideration of Helium's Early History," *Annals of Science* 66, 157–182.

Kroeber, Alfred L. and Clyde Kluckhohn. 1952. *Culture: A Critical Review of Concepts and Definitions*. Cambridge, MA: The Peabody Museum, p. 656.

Kuehn, Manfred. 2002. *Kant: A Biography*. Cambridge: Cambridge University Press.

Kuhn, Thomas S. 1957. *The Copernican Revolution*. Cambridge, MA: Harvard University Press.

Kuhn, Thomas S. 1962a. *The Structure of Scientific Revolutions*. Chicago, IL: University of Chicago Press, 55. Expanded edition, 1970.

Kuhn, Thomas S. 1962b. "The Historical Structure of Scientific Discovery," *Science* 136, 760–764. Reprinted in Kuhn, *The Essential Tension: Selected Studies in Scientific Tradition and Change*. Chicago, IL, and London: University of Chicago Press, 1977, 165–177.

Kuper, A. 1999. *Culture: The Anthropologists' Account*. Cambridge, MA: Harvard University Press.

Kurzweil, R. 1999. *The Age of Spiritual Machines: When Computers Exceed Human Intelligence*. New York, NY: Viking.

Kurzweil, R. 2005. *The Singularity Is Near: When Humans Transcend Biology*. New York: Viking.

Kuznicki, Jason T. 2011. "The Inscrutable Names of God," in Douglas Vakoch (ed.), *Communication with Extraterrestrial Intelligence*. Albany, NY: State University of New York Press, 202–213.

Ladrière, Jean. 1977. *The Challenge Presented to Cultures by Science and Technology*. Paris: UNESCO.

Lakoff, George. 2003a. "How the Body Shapes Thought: Thinking with an All-Too-Human Brain," in Anthony Sanford (ed.), *The Nature and Limits of Human Understanding*. London and New York, NY: T&T Clark, 49–74.

Lakoff, George. 2003b. "How to Live With an Embodied Mind: When Causation, Mathematics, Morality, the Soul, and God Are Essentially Metaphorical Ideas," in Anthony Sanford (ed.), *The Nature and Limits of Human Understanding*. London and New York, NY: T&T Clark, 75–108.

Lakoff, George and Rafael Nunez. 2000. *Where Mathematics Comes From*. New York, NY: Basic Books.

Laland, Kevin N. and Gillian R. Brown. 2002. *Sense and Nonsense: Evolutionary Perspectives on Human Behaviour*. Oxford: Oxford University Press.

Lambright, W. Henry. 2015. "NASA and the Environment: An Evolving Relationship," in Steven J. Dick (ed.), *The Impact of Discovering Life beyond Earth*. Cambridge: Cambridge University Press, 383–426.

Lamm, Norman. 1978. "The Religious Implication of Extraterrestrial Life," in Aryeh Cannell and Cyril Domb (eds.), *Challenge: Torah Views on Science and Its Problems*. New York, NY: Feldheim Publishers, 354–398.

Landfester, Ulrike, Nina-Louisa Remuss, Kai-Uwe Schrogl, and Jean-Claude Worms (eds.). 2011. *Humans in Outer Space – Interdisciplinary Perspectives*. New York, NY: Springer.

Lane, Nick. 2015. *The Vital Question: Energy, Evolution, and the Origins of Complex Life*. New York, NY: W. W. Norton & Company.

Latham, David W., Robert P. Stefanik, Tsevi Mazeh, Michel Mayor, and Gilbert Burki. 1989. "The Unseen Companion of HD114762 – A Probable Brown Dwarf," *Nature*, 339, 38–40.

Laudan, Larry. 1977. *Progress and Its Problems: Towards a Theory of Scientific Growth*. Berkeley, CA: University of California Press.

Laudan, Larry. 1990. *Science and Relativism: Some Key Controversies in the Philosophy of Science*. Chicago, IL: University of Chicago Press.

Launius, Roger. 2013. "Escaping Earth: Human Spaceflight as Religion," *Astropolitics* 11, 45–64.

Launius, Roger. 2014. *Historical Analogs for the Stimulation of Space Commerce*. Washington, DC: National Aeronautics and Space Administration. SP-2014–4554, monographs in aerospace history, no. 54.

Lederberg, Joshua. 1960. "Exobiology: Approaches to Life beyond the Earth," *Science* 132, 398–399.

Lemarchand, Guillermo A. and Jon Lomberg. 2011. "Communicating among Interstellar Intelligent Species: A Search for Universal Cognitive Maps," in Douglas Vakoch (ed.), *Communication with Extraterrestrial Intelligence*. Albany, NY: State University of New York Press, 371–395.

Lemarchand, Guillermo A. and Karen J. Meech (eds.). 2000. *Bioastronomy '99: A New Era in Bioastronomy*. San Francisco, CA: Astronomical Society of the Pacific.

Leopold, Aldo. 1949. *A Sand County Almanac*. New York, NY: Oxford University Press.

Levanthes, Louise. 1994. *When China Ruled the Seas: The Treasure Fleet of the Dragon Throne, 1405–1433*. Oxford: Oxford University Press.

Lewis, C. S. 1964. *The Discarded Image*. Cambridge: Cambridge University Press.

Limerick, Patricia Nelson. 1993. "Imagined Frontiers: Westward Expansion and the Future of the Space Program," in Radford Byerly (ed.), *Space Policy Alternatives*. Boulder, CO: Westview Press, 249–261.

Lindberg, David. 1978. "The Transmission of Greek and Arabic Learning to the West," in David Lindberg (ed.), *Science in the Middle Ages*. Chicago, IL: University of Chicago Press, 52–90.

Lindberg, David. 1992. *The Beginnings of Western Science*. Chicago, IL: University of Chicago Press.

Lindstrom, Lamont. 1993. *Cargo Cult: Strange Stories of Desire from Melanesia and Beyond*. Honolulu, HI: University of Hawaii Press.

Liu, Cixin. 2015. *The Dark Forest*. New York, NY: Tor, 479–484.

Livio, Mario. 2009. *Is God a Mathematician?* New York, NY: Simon and Schuster.

Lloyd, Geoffrey and Nathan Sivin. 2002. *The Way and the Word*. New York, NY, and London: Yale University Press.

Logsdon, John, Bridget R. Ziegelaar, and Anne Marie Burns (eds.). 1997. *Life on Mars: What Are the Implications?* Washington, DC: Space Policy Institute, George Washington University.

Longair, Malcolm. 2011. "The Discovery of Pulsars and the Aftermath," *Proceedings of the APS* 155, 147–157. Online at www.amphilsoc.org/sites/default/files/4-Longair1550204.pdf.

Lovin, Robin. 2015. "Astrobiology and Theology," in Steven J. Dick (ed.), *The Impact of Discovering Life beyond Earth*. Cambridge: Cambridge University Press, 222–232.

Lowe, E. Jonathan. 1995. *Locke on Human Understanding*. London: Routledge.

Lowell, Percival. 1907. *Mars and Its Canals*. New York, NY: Macmillan.

Lowrie, Ian. 2013. "Cultural Resources and Cognitive Frames: Keys to an Anthropological Approach to Prediction," in Douglas Vakoch (ed.), *Astrobiology, History and Society: Life beyond Earth and the Impact of Discovery*. Berlin-Heidelberg: Springer-Verlag, 259–269.

Lumsden, Charles and Edward O. Wilson. 1981. *Genes, Mind and Culture*. Cambridge, MA: Harvard University Press.

Lupisella, Mark L. 1997. "The Rights of Martians," *Space Policy* 13(2), 89–94.

Lupisella, Mark L. 2009a. "Cosmocultural Evolution: The Coevolution of Culture and Cosmos and the Creation of Cosmic Value," in Steven J. Dick and Mark L. Lupisella (eds.), *Cosmos & Culture: Cultural Evolution in a Cosmic Context*. Washington, DC: National Aeronautics and Space Administration, 321–356.

Lupisella, Mark L. 2009b. "The Search for Extraterrestrial Life: Epistemology, Ethics, and Worldviews," in Constance M. Bertka (ed.), *Exploring the Origin, Extent, and Future of Life: Philosophical, Ethical and Theological Perspectives*. Cambridge: Cambridge University Press, 186–204.

Lupisella, Mark L. 2011. "Pragmatism, Cosmocentrism, and Proportional Consultation for Communication with Extraterrestrial Intelligence," in Douglas Vakoch (ed.), *Communication with Extraterrestrial Intelligence*. Albany, NY: State University of New York Press, 319–333.

Lupisella, Mark L. 2014. "Caring Capacity and Cosmocultural Evolution: Potential Mechanisms for Advanced Altruism," in Douglas Vakoch (ed.), *Archaeology, Anthropology and Interstellar Communication*. Washington, DC: National Aeronautics and Space Administration, 93–110.

Lupisella, Mark L. 2015. "Life, Intelligence, and the Pursuit of Value," in Steven J. Dick (ed.), *The Impact of Discovering Life beyond Earth*. Cambridge: Cambridge University Press, 159–174, at 165–170.

Lupisella, Mark L. 2016. "Cosmological Theories of Value: Relationalism and Connectedness as Foundations for Cosmic Creativity," in James S. J. Schwartz and Tony Milligan (eds.), *The Ethics of Space Exploration*. New York, NY: Springer, 75–91.

Lupisella, Mark L. In press. *Cosmological Theories of Value*.

Lupisella, Mark L. and John Logsdon. 1997. "Do We Need a Cosmocentric Ethic?" Paper IAA-97-IAA.9.2.09, IAC, Turin, Italy.

MacQuarrie, Kim. 2007. *Last Days of the Incas*. New York, NY: Simon and Schuster.

Maienschein, Jane. 2015. "Is There Anything New about Astrobiology and Society?" in Steven J. Dick (ed.), *The Impact of Discovering Life beyond Earth*. Cambridge: Cambridge University Press, 248–260.

Malone-France, Derek. 2007. *Deep Empiricism: Kant, Whitehead, and the Necessity of Philosophical Theism*. Lanham, MD, and Boulder, CO: Rowman & Littlefield Publishers.

Man, John. 2002. *Gutenberg: How One Man Remade the World with Words*. New York, NY: Wiley.

Marcy, Geoff, Paul Butler, Eric Williams, Lars Bildsten, James R. Graham, Andrea M. Ghez, and J. Garret Jernigan. 2007. "The Planet around 51 Pegasi," *Astrophysical Journal* 481, 926–935.

Margulis, Lynn and Dorion Sagan. 1986. *Microcosmos: Four Billion Years of Microbial Evolution*. Berkeley, CA: University of California Press.

Marino, Lori, 2002. "Convergence of Complex Cognitive Abilities in Cetaceans and Primates," *Brain, Behavior and Evolution* 59, 21–32.

Marino, Lori. 2015. "The Landscape of Intelligence," in Steven J. Dick (ed.), *The Impact of Discovering Life beyond Earth*. Cambridge: Cambridge University Press, 94–112.

Mariscal, Carlos. 2015. "Universal Biology: Assessing Universality from a Single Example," in Steven J. Dick (ed.), *The Impact of Discovering Life beyond Earth*. Cambridge: Cambridge University Press, 113–126.

Marozzi, Justin. 2008. *The Way of Herodotus: Travels with the Man Who Invented History*. Cambridge, MA: Da Capo Press.

Marshall, Alan. 1993. "Ethics and the Extraterrestrial Environment," *Journal of Applied Philosophy* 10(2), 227–236.

Martinez, A. 2001. "Culture Contact: Archaeological Approaches," in Neil J. Smelser and Paul B. Baltes (eds.), *International Encyclopedia of the Social and Behavioral Sciences*. Amsterdam: Elsevier, 3035–3037.

Maruyama, Magoroh and Arthur Harkins (eds.). 1975. *Cultures beyond the Earth: The Role of Anthropology in Outer Space*. New York, NY: Vintage Books.

Mayor, Michel and Didier Queloz. 1995. "A Jupiter-Mass Companion to a Solar-Type Star," *Nature* 378, 355–359.

Mayr, Ernst. 1985. "The Probability of Extraterrestrial Intelligent Life," in Edward Regis Jr. (ed.), *Extraterrestrials: Science and Alien Intelligence*. Cambridge: Cambridge University Press, 24–30. Reprinted with updated notes in Ernst Mayr, *Toward a New Philosophy of Biology: Observations of an Evolutionist* (Cambridge, MA: Harvard University Press, 1988), 67–74.

Mazlish, Bruce. 1965. *The Railroad and the Space Program: An Exploration in Historical Analogy*. Cambridge, MA: MIT Press.

McAdamis, E. M. 2011. "Astrosociology and the Capacity of Major World Religions to Contextualize the Possibility of Life beyond Earth," *Physics Procedia* 20, 338–352, at 349. Online at www.sciencedirect.com/science/article/pii/S1875389211006006.

McConnell, Brian. 2001. *Beyond Contact: A Guide to SETI and Communicating with Alien Civilizations*. Sebastopol, CA: O'Reilly.

McKay, Christopher P. 1990. "Does Mars Have Rights? An Approach to the Environmental Ethics of Planetary Engineering," in D. MacNiven (ed.), *Moral Expertise: Studies in Practical and Professional Ethics*. London and New York: Routledge, 184–197.

McKay, Christopher P. 2000. "Astrobiology: The Search for Life beyond the Earth," in Steven J. Dick (ed.), *Many Worlds: The New Universe, Extraterrestrial Life and the Theological Implications*. Philadelphia, PA: Templeton Press, 45–58.

McKay, Christopher P. 2009. "Planetary Ecosynthesis on Mars: Restoration Ecology and Environmental Ethics," in Constance M. Bertka (ed.), *Exploring the Origin, Extent, and Future of Life: Philosophical, Ethical and Theological Perspectives*. Cambridge: Cambridge University Press, 245–260.

McKay, Christopher P. 2011. "The Search for Life in our Solar System and the Implications for Science and Society," *Philosophical Transactions of the Royal Society of London A* 594–606.

McKay, Christopher P. 2013. "Astrobiology and Society: The Long View," in Chris Impey, Anna Spitz, and William Stoeger (eds.), *Encountering Life in the Universe: Ethical Foundations and Social Implications of Astrobiology*. Tucson, AZ: University of Arizona Press, 158–166.

McKay, Christopher P. and Robert Zubrin. 2002. "Do Indigenous Martian Bacteria Have Precedence over Human Exploration?" in Robert Zubrin (ed.), *On to Mars: Colonizing a New World*. Burlington, Canada: Apogee, 177–182.

McKay, David S., Everett K. Gibson Jr., Kathie L. Thomas-Keprta, Hojatollah Vali, Christopher S. Romanek, Simon J. Clemett, Xavier D. F. Chillier, Claude R. Maechling, and Richard N. Zare. 1996. "Search for Past Life on Mars: Possible Relic Biogenic Activity in Martian Meteorite ALH84001," *Science*, New Series 273, 5277, 924–930.

McKee, Gabriel. 2007. *The Gospel According to Science Fiction*. Louisville, KY: Westminster John Knox Press.

McLuhan, Marshall. 1962. *The Gutenberg Galaxy*. Toronto: University of Toronto Press.

McMullin, Ernan. 2000. "Life and Intelligence Far from Earth: Formulating Theological Issues," in Steven J. Dick (ed.), *Many Worlds: The New Universe, Extraterrestrial Life and the Theological Implications*. Philadelphia, PA: Templeton Press, 151–175.
McNeill, William H. 1986. "Basic Assumptions of Toynbee's A Study of History," in William H. McNeill (ed.), *Mythistory and Other Essays*. Chicago, IL: University of Chicago Press, 138.
McNeill, William H. 1976. *Plagues and Peoples*. New York, NY: Random House.
Meech, Karen J., J. V. Keane, Michael Mumma, J. L. Siefert, and Dan J. Werthimer (eds.). 2009. *Bioastronomy 2007: Molecules, Microbes and Extraterrestrial Life*. San Francisco, CA: Astronomical Society of the Pacific.
Meltzer, Michael. 2011. *When Biospheres Collide: A History of NASA's Planetary Protection Programs*. Washington, D.C.: National Aeronautics and Space Administration. SP-2011-4234. Online at http://history.nasa.gov/SP-4234.pdf.
Menzies, Gavin. 2002. *1421: The Year China Discovered America*. New York, NY: William Morrow.
Messler, Bill and H. James Cleaves II. 2016. *A Brief History of Creation: Science and the Search for Life*. New York, NY: W. W. Norton & Company.
Meynell, Alice. 1923. "Christ in the Universe," *Poems*. New York, NY, 92.
Michaud, Michael A. G. 2005. "Broadening and Simplifying the first SETI Protocol," *Journal of the British Interplanetary Society* 58, 40–42.
Michaud, Michael A. G. 2007. *Contact with Alien Civilizations: Our Hopes and Fears about Encountering Extraterrestrials*. New York, NY: Copernicus.
Michaud, Michael A. G. 2011. "Seeking Contact: The Relevance of History," in Douglas Vakoch (ed.), *Communication with Extraterrestrial Intelligence*. Albany, NY: State University of New York Press, 307–317, at 308.
Michaud, Michael A. G. 2015. "Searching for Extraterrestrial Intelligence: Preparing for an Expected Paradigm Break," in Steven J. Dick (ed.), *The Impact of Discovering Life beyond Earth*. Cambridge: Cambridge University Press, 286–298.
Miller, James Grier. 1978. *Living Systems Theory*. New York, NY: McGraw-Hill.
Miller, Johanna L. 2012. "The Higgs Particle, or Something Much Like It, Has Been Spotted," *Physics Today* 65, 12–15.
Minsky, Marvin. 1985. "Why Intelligent Aliens Will Be Intelligible," in Edward Regis Jr. (ed.), *Extraterrestrials: Science and Alien Intelligence*. Cambridge: Cambridge University Press, 117–132.
Mithen, Steven. 1996. *The Prehistory of the Mind: The Cognitive Origins of Art, Religion and Science*. London: Thames and Hudson.
Moravec, Hans. 1988. *Mind Children: The Future of Robot and Human Intelligence*. Cambridge, MA: Harvard University Press.
Moravec, Hans. 1999. *Robot: Mere Machine to Transcendent Mind*. Oxford: Oxford University Press.
Morrison, Philip. 1973. "The Consequences of Contact," in Carl Sagan (ed.), *Communication with Extraterrestrial Intelligence*. Cambridge, MA: MIT Press, 333–349, at 336–337.
Morrison, Philip, John Billingham, and John Wolfe. 1977. *The Search for Extraterrestrial Intelligence (SETI)*. Washington, DC: National Aeronautics and Space Administration.

Murphy, Cullen. 2007. *Are We Rome?: The Fall of an Empire and the Fate of America*. Boston, MA: Houghton Mifflin.

Murphy, Nancey and George F. R. Ellis. 1996. *On the Moral Nature of the Universe*. Minneapolis, MN: Fortress Press.

Musso, Paolo. 2012. "The Problem of Active SETI: An Overview," *Acta Astronautica* 78, 43–54.

Naess, Arne. 1973. "The Shallow and the Deep, Long-Range Ecology Movement: A Summary," *Inquiry* 16(1), 95–100.

Nagel, Thomas. 1974. "What Is It Like to Be a Bat?" Online at http://organizations.utep.edu/portals/1475/nagel_bat.pdf.

National Academy of Sciences, National Research Council, Space Studies Board. 1999. *Size Limits of Very Small Microorganisms*. Washington, DC: National Academy Press.

Nisbet, Robert. 1980. *History of the Idea of Progress*. New York, NY: Basic Books.

Noonan, Harold W. 1999. *Hume on Knowledge*. London: Routledge.

North, John D. 1989. "Science and Analogy," in John D. North (ed.), *The Universal Frame: Historical Essays in Astronomy, Natural Philosophy and Scientific Method*. London: Hambledon Press.

Olson, A. Randall and Vladimir V. M. Tobin. 2008. "An Eastern Orthodox Perspective on Microbial Life on Mars," *Theology and Science* 6(4), 421–437.

O'Meara, Thomas F. 2012. *Vast Universe: Extraterrestrial Life and Christian Revelation*. Collegeville, MN: Liturgical Press.

Othman, Mazlan. 2011. "Supra-Earth Affairs," *Philosophical Transactions of the Royal Society* 369, 693–699.

Pace, Norman. 2001. "The Universal Nature of Biochemistry," *Proceedings of the National Academy of Sciences USA* 98, 805–808.

Papineau, David (ed.). 2004. *Western Philosophy*. New York, NY: Metro Books, 26.

Parry, J. H. 1981. *The Age of Reconnaissance: Discovery, Exploration and Settlement, 1450 to 1650*. Berkeley, CA: University of California Press 1st edn. London: Weidenfeld and Nicolson, 1963.

Pass, James. 2016. "An Astrosociological Perspective on the Societal Impact of Spaceflight," in Steven J. Dick (ed.), *Historical Studies in the Societal Impact of Spaceflight*. Washington, DC: National Aeronautics and Space Administration, 535–576.

Paul, Robert. 1992. *Science, Religion and Mormon Cosmology*. Champaign-Urbana, IL: University of Illinois Press.

Penny, Alan John. 2013. "The SETI Episode in the 1967 Discovery of Pulsars," *The European Physical Journal* 38, 535–547, at 537. Preprint online at http://arxiv.org/abs/1302.0641.

Persson, Erik. 2012. "The Moral Status of Extraterrestrial Life," *Astrobiology* 12, 976–984.

Peters, Francis E. 1968. *Aristotle and the Arabs: The Aristotelian Tradition in Islam*. New York, NY: New York University Press.

Peters, Ted. 1994. "Exo-theology: Speculations on Extra-Terrestrial Life," *CTNS Bulletin* 14, 1–9.

Peters, Ted. 1995. "Exo-theology: Speculations on Extraterrestrial Life," in James R. Lewis (ed.), *The Gods Have Landed: New Religions from Other Worlds*. Albany, NY: State University of New York Press,187–206.

Peters, Ted. 2009. "Astrotheology and the ETI Myth," *Theology and Science* 7(1), 3–30.

Peters, Ted. 2011. "The Implications of the Discovery of Extra-Terrestrial Life for Religion," *Philosophical Transactions of the Royal Society A* 369, 644–655.

Peters, Ted. 2013a. "Would Discovery of ETI Provoke a Religious Crisis?" in Douglas Vakoch (ed.), *Astrobiology, History and Society: Life beyond Earth and the Impact of Discovery*. Berlin-Heidelberg: Springer-Verlag, 341–355.

Peters, Ted. 2013b. "Astroethics: Engaging Extraterrestrial Intelligent Life-Forms," in Chris Impey, Anna Spitz, and William Stoeger (eds.), *Encountering Life in the Universe: Ethical Foundations and Social Implications of Astrobiology*. Tucson, AZ: University of Arizona Press, 200–221.

Peters, Ted. 2014. "Astrotheology: A Constructive Proposal," *Zygon* 49, 443–457.

Peters, Ted, Martinez Hewlett, Joshua Moritz, and Robert John Russell (eds.). 2018. *Astrotheology: Science and Theology Meet Extraterrestrial Life*. Eugene, OR: Cascade Books.

Pettegree, Andrew. 2015. *Brand Luther: 1517, Printing, and the Making of the Reformation*. New York, NY: Penguin Press.

Pickover, Clifford. 1998. *The Science of Aliens*. New York, NY: Basic Books.

Pirie, Norman W. 1937. "The Meaninglessness of the Terms 'Life' and 'Living,'" in Joseph Needham and David E. Green (eds.), *Perspectives in Biochemistry*. Cambridge: Cambridge University Press, 11–22.

Pirie, Norman W. 1975. *A Discussion on the Recognition of Alien Life*. London: Royal Society.

Pitt, Joseph C. 1982. "Will a Rubber Ball Still Bounce?," in *Philosophers Look at Science Fiction*. Chicago, IL: Nelson Hall.

Poole, Robert. 2010. *Earthrise: How Man First Saw the Earth*. New Haven, CT: Yale University Press.

Pooley, Jefferson D. and Michael J. Socolow. 2013a. "Checking Up on the Invasion from Mars: Hadley Cantril, Paul Lazarsfeld, and the Making of a Misremembered Classic," *International Journal of Communication* 7. Online at http://ijoc.org/index.php/ijoc/article/view/2117.

Pooley, Jefferson D. and Michael J. Socolow. 2013b. "War of the Words: *The Invasion from Mars* and Its Legacy for Mass Communication Scholarship," in Joy Elizabeth Hayes, Kathleen Battles, and Wendy Hilton-Morrow (eds.), *War of the Worlds to Social Media: Mediated Communication in Times of Crisis*. New York, NY: Peter Lang.

Potts, Rick and Chris Sloan. 2010. *What Does It Mean to Be Human*? Washington, DC: National Geographic.

Poundstone, William. 1999. *Carl Sagan: A Life in the Cosmos*. New York, NY: Henry Holt, 375.

Proust, Marcel. 1923. "La Prisonnière," *Remembrance of Things Past*, volume 5.

Pyle, Rod. 2012. *Destination Mars: New Explorations of the Red Planet*. New York, NY: Prometheus Books.

Puccetti, Roland. 1968. *Persons: A Study of Possible Moral Agents in the Universe*. London: Macmillan.

Race, Margaret S. 2007. "Societal and Ethical Concerns," in Woodruff T. Sullivan III and John A. Baross (eds.), *Planets and Life: The Emerging Science of Astrobiology*. Cambridge: Cambridge University Press, 483–497.

Race, Margaret S. 2008. "Communicating about the Discovery of ETL: Different Searches, Different Issues," *Acta Astronautica* 62, 71–78.

Race, Margaret S. 2009. "The Implications of Discovering Extraterrestrial Life: Different Searches, Different Issues," in Constance M. Bertka (ed.), *Exploring the Origin, Extent, and Future of Life: Philosophical, Ethical and Theological Perspectives*. Cambridge: Cambridge University Press, 205–219.

Race, Margaret S. 2011. "Policies for Scientific Exploration and Environmental Protection: Comparison of the Antarctic and Outer Space Treaties," in Paul Arthur Berkman, Michael A. Lang, David W. H. Walton, and Oran R. Young (eds.), *Science Diplomacy: Antarctica, Science and the Governance of International Spaces*. Washington, DC: Smithsonian Institution Scholarly Press, 143–153.

Race, Margaret S. 2015. "Preparing for the Discovery of Extraterrestrial Life: Are We Ready?," in Steven J. Dick (ed.), *The Impact of Discovering Life beyond Earth*. Cambridge: Cambridge University Press, 262–285.

Race, Margaret S., Kathryn Denning, Constance M. Bertka, Steven J. Dick, Albert A. Harrison, Chris Impey, and R. Mancinelli. 2012. "Astrobiology and Society: Building an Interdisciplinary Research Community," *Astrobiology* 12(10), 958–965.

Race, Margaret S. and Richard O. Randolph. 2002. "The Need for Operating Guidelines and a Decision Making Framework Applicable to the Discovery of Non-intelligent Extraterrestrial Life," *Advances in Space Research* 30(6), 1583–1591.

Raible, Daniel C. 1960. "Rational Life in Outer Space?" *America: National Catholic Weekly Review* 103, 532–535.

Randolph, Richard O. 2009. "God's Preferential Option for Life: A Christian Perspective on Astrobiology," in Constance M. Bertka (ed.), *Exploring the Origin, Extent, and Future of Life: Philosophical, Ethical and Theological Perspectives*. Cambridge: Cambridge University Press, 281–302.

Randolph, Richard O., Margaret Race, and Christopher P. McKay. 1997. "Reconsidering the Theological and Ethical Implications of Extraterrestrial Life," *Center for Theology and the Natural Sciences Bulletin* 17(3), 1–8.

Regis, Edward, Jr. (ed.). 1985. *Extraterrestrials: Science and Alien Intelligence*. Cambridge: Cambridge University Press.

Reiss, Diana. 1988. "Can We Communicate With Other Species on This Planet?" in George Marx (ed.), *Bioastronomy – The Next Steps*. Dordrecht: Kluwer, 253–264.

Reiss, Diana. 2011. *The Dolphin in the Mirror: Exploring Dolphin Minds and Saving Dolphin Lives*. New York, NY: Houghton, Mifflin, Harcourt.

Rescher, Nicholas. 1985. "Extraterrestrial Science," in Edward Regis Jr. (ed.), *Extraterrestrials: Science and Alien Intelligence*. Cambridge: Cambridge University Press, 83–116.

Restall, Matthew. 2003. *Seven Myths of the Spanish Conquest*. Oxford: Oxford University Press.

Robinson, Andrew. 2002. *Lost Languages. The Enigma of the World's Undeciphered Scripts*. New York, NY: Thames & Hudson. Reprinted 2009.

Robinson, George. 1969. "Ecological Foundations of Haley's Metalaw," *Journal of the British Interplanetary Society* 22, 266–274.

Robinson, George. 2013a. "The Biochemical Foundations of Evolving Metalaw: Moving at a Glance to the Biological Basis of Sentient 'Essence,'" *Journal of Space Law* 39, 181–216.

Robinson, George. 2013b. "Metalaw: From Speculation to Humankind Legal Posturing with Extraterrestrial Life," *Journal of Space Philosophy* 2, 49–56.

Rolston III, Holmes. 1986. "The Preservation of Natural Value in the Solar System, in Eugene C. Hargrove (ed.), *Beyond Spaceship Earth: Environmental Ethics and the Solar System*. San Francisco, CA: Sierra Club Books, 140–182.

Rolston III, Holmes. 2014. "Terrestrial and Extraterrestrial Altruism," in Douglas Vakoch (ed.), *Archaeology, Anthropology and Interstellar Communication*. Washington, DC: National Aeronautics and Space Administration, 211–222.

Rosenberg, Alexander. 2016. *Philosophy of Social Science*. Boulder, CO: Westview Press.

Rospars, Jean-Pierre. 2013. "Trends in the Evolution of Life, Brains and Intelligence," *International Journal for Astrobiology* 12, 186–207, at 197.

Ross, Joseph T. 2013. "Hegel, Analogy, and Extraterrestrial Life," in Douglas Vakoch (ed.), *Astrobiology, History and Society: Life beyond Earth and the Impact of Discovery*. Berlin-Heidelberg: Springer-Verlag, 86–102.

Rubenstein, Richard E. 2003. *Aristotle's Children: How Christians, Muslims, and Jews Rediscovered Ancient Wisdom and Illuminated the Dark Ages*. Orlando, FL: Harcourt.

Rummel, John D., Margaret S. Race, and Gerda Horneck. 2012. "Ethical Considerations for Planetary Protection in Space Exploration: A Workshop," *Astrobiology* 12, 1017–1023. Online at www.ncbi.nlm.nih.gov/pmc/articles/PMC3698687/, full report at https://cosparhq.cnes.fr/sites/default/files/ppp_workshop_report_ethical_considerations_princeton_final_v1a_7nov2012.pdf.

Ruse, Michael. 1985. "Is Rape Wrong on Andromeda? An Introduction to Extraterrestrial Evolution, Science and Morality," in Edward Regis Jr. (ed.), *Extraterrestrials: Science and Alien Intelligence*. Cambridge: Cambridge University Press, 43–78.

Ruse, Michael. 1998. *Taking Darwin Seriously*. Amherst, NY: Prometheus Books.

Ruse, Michael. 2003a. "A Darwinian Understanding of Epistemology," in Anthony Sanford (ed.), *The Nature and Limits of Human Understanding*. London and New York, NY: T&T Clark, 111–132.

Ruse, Michael. 2003b. "A Darwinian Understanding of Ethics," in Anthony Sanford (ed.), *The Nature and Limits of Human Understanding*. London and New York, NY: T&T Clark, 133–161.

Ruse, Michael. 2006. *Darwinism and Its Discontents*. Cambridge: Cambridge University Press.

Ruse, Michael. 2015. "'Klaatu Barada Nikto' – or, Do They Really Think Like Us?," in Steven J. Dick (ed.), *The Impact of Discovering Life beyond Earth*. Cambridge: Cambridge University Press, 175–188.

Ruskin, Steven. 2002. "A Newly-Discovered Letter of J. F. W. Herschel Concerning the 'Great Moon Hoax,'" *Journal for the History of Astronomy* 33, 71–74.

Russell, Bertrand. 1999. *History of Western Philosophy*. London: Routledge.

Sagan Carl (ed.). 1973. *Communication with Extraterrestrial Intelligence (CETI)*. Cambridge, MA: MIT Press.

Sagan, Carl. 1977. *The Dragons of Eden: Speculations on the Evolution of Human Intelligence*. New York, NY: Random House.
Sagan, Carl. 1980. *Cosmos*. New York, NY: Random House.
Sagan, Carl. 1987. "The Burden of Skepticism," *Skeptical Inquirer* 12. Online at www.csicop.org/si/show/burden_of_skepticism.
Sagan, Carl. 1994. *Pale Blue Dot: A Vision of the Human Future in Space*. New York, NY: Random House.
Sagan, Carl. 2000. *Carl Sagan's Cosmic Connection: An Extraterrestrial Perspective*. Cambridge: Cambridge University Press.
Saint-Gelais, Richard. 2014. "Beyond Linear B: The Metasemiotic Challenge of Communication with Extraterrestrial Intelligence," in Douglas Vakoch (ed.), *Archaeology, Anthropology and Interstellar Communication*. Washington, DC: National Aeronautics and Space Administration, 79–94.
Sanford, Anthony (ed.), 2003. *The Nature and Limits of Human Understanding*. London and New York, NY: T&T Clark.
Sapp, Jan. 2009. *The New Foundations of Evolution on the Tree of Life*. Oxford: Oxford University Press.
Sawyer, Kathy. 2006. *The Rock from Mars: A Detective Story on Two Planets*. New York, NY: Random House.
Scerri, Eric R. 2007. *The Periodic Table: Its Story and Its Significance*. Oxford: Oxford University Press.
Scerri, Eric R. 2013. *A Tale of Seven Elements*. Oxford and New York, NY: Oxford University Press.
Schickore, Jutta. 2007. *The Microscope and the Eye: A History of Reflections, 1740–1870*. Chicago, IL: University of Chicago Press.
Schmidt, Stanley. 1995. *Aliens and Alien Societies: A Writer's Guide to Creating Extraterrestrial Life Forms*. Cincinnati, OH: Writer's Digest Books.
Schneider, Susan. 2014. "The Philosophy of *Her*," *New York Times*, March 2, 2014, at http://opinionator.blogs.nytimes.com/2014/03/02/the-philosophy-of-her/?_r=0.
Schneider, Susan. 2015. "Alien Minds," in Steven J. Dick (ed.), *The Impact of Discovering Life beyond Earth*. Cambridge: Cambridge University Press, 189–206.
Schulze-Makuch, Dirk. 2015. "The Landscape of Life," in Steven J. Dick (ed.), *The Impact of Discovering Life beyond Earth*. Cambridge: Cambridge University Press, 81–94.
Schwartz, Stuart B. (ed.). 2000. *Victors and Vanquished: Spanish and Nahua Views of the Conquest of Mexico*. New York, NY: Bedford/St. Martin's.
Schweitzer, Albert. 1960. "The Ethic of Reverence for Life," in Charles R. Joy (ed.), *Albert Schweitzer: An Anthology*. Boston, MA: Beacon Press, 259–260.
Searle, John. 1984. *Minds, Brains and Science*. Cambridge, MA: Harvard University Press.
Shaviro, Steven. 2016. *Discognition*. London: Repeater.
Sheehan, William. 1996. *The Planet Mars, A History of Observation and Discovery*. Tucson, AZ: University of Arizona Press.
Sheehan, William and Thomas Dobbins. 2001. *Epic Moon: A History of Lunar Exploration in the Age of the Telescope*. Richmond, VA: Willmann-Bell.
Shermer, Michael. 2004. *The Science of Good and Evil*. New York, NY: Henry Holt.
Shklovskii, Iosif S. 1991. *Five Billion Vodka Bottles to the Moon: Tales of a Soviet Scientist*. New York, NY: W. W. Norton & Company.

Shostak, Seth (ed.). 1995. *Progress in the Search for Extraterrestrial Life*. San Francisco, CA: Astronomical Society of the Pacific.

Shostak, Seth. 1998. *Sharing the Universe: Perspectives on Extraterrestrial Life*. Berkeley, CA: Berkeley Hills.

Shostak, Seth. 2014. "The Second Signal," *Communications of the ACM* 57 (January), 128–129.

Shuch, Paul (ed.). 2011. *Searching for Extraterrestrial Intelligence: SETI Past, Present and Future*. Chichester: Praxis.

Shuch, Paul and Ivan Almar, 2007a. "Quantifying Past Transmissions Using the San Marino Scale," http://avsport.org/IAA/IAC07-A4.2.04.pdf.

Shuch, Paul and Ivan Almar. 2007b. "Shouting in the Jungle: The SETI Transmission Debate," *Journal of the British Interplanetary Society* 60, 142–146. Online at http://avsport.org/IAA/jbis07a.pdf.

Silver, Nate. 2012. *The Signal and the Noise: Why So Many Predictions Fail – but Some Don't*. New York, NY: Penguin Press.

Simpson, George Gaylord. 1964. "The Non-prevalence of Humanoids," *Science* 143, 769–775. Reprinted in *This View of Life: The World of an Evolutionist*. New York, NY: Harcourt, Brace & World, 253–271.

Sire, James W. 2002. *The Universe Next Door*. Downers Grove, IL: Intervarsity Press.

Skoyles, John and Dorion Sagan. 2002. *Up from Dragons: The Evolution of Human Intelligence*. New York, NY: McGraw-Hill.

Smith, David H. 2004. "The Origins of NASA's Astrobiology Program," in Pascal Ehrenfreund, W. M. Irvine, Tobias Owen, Luann Becker, Jen Blank, J. R. Brucato, Luigi Colangeli, Sylvie Derenne, Anne Dutrey, Didier Despois, Antonio Lazcano, and Francois Robert (eds.), *Astrobiology: Future Perspectives*. Netherlands: Kluwer, 444–465.

Smith, Kelly. 2009. "The Trouble with Intrinsic Value: An Ethical Primer for Astrobiology," in Constance M. Bertka (ed.), *Exploring the Origin, Extent, and Future of Life: Philosophical, Ethical and Theological Perspectives*. Cambridge: Cambridge University Press, 261–280.

Smith, Kelly. 2014. "Manifest Complexity: A Foundational Ethic for Astrobiology?," *Space Policy* 30, 209–214.

Smith, Kelly. 2016. "Life Is Hard: Countering Definitional Pessimism Concerning the Definition of Life," *International Journal of Astrobiology* 15(4), 277–289.

Smith, Robert. 1982. *The Expanding Universe: Astronomy's "Great Debate."* Cambridge: Cambridge University Press.

Smith, Roger. 1997. *The Human Sciences*. New York, NY: W. W. Norton & Company.

Sobel, Carolyn P. 2013. *The Cognitive Sciences: An Interdisciplinary Approach*. Los Angeles, CA: Sage.

Sourcebooks Staff. 2001. *The War of the Worlds: Mars' Invasion of Earth, Inciting Panic and Inspiring Terror from H. G. Wells to Orson Welles and Beyond*. Naperville, IL: Sourcebooks.

Stenger, Victor. 2011. *The Fallacy of Fine Tuning: Why the Universe Is Not Designed for Us*. Amherst, NY: Prometheus, 122.

Sternberg, Robert J. and James C. Kaufman (eds.). 2002. *The Evolution of Intelligence*. Mahwah, NJ: Lawrence Erlbaum Associates.

Sterns, Patricia Margaret. 2000. "SETI and Space Law: Jurisprudential and Philosophical Considerations for Humankind in Relation to Extraterrestrial Life," *Acta Astronautica* 46, 759–763.

Sterns, Patricia Margaret and Leslie I. Tennen. 2016. "SETI, Metalaw, and Social Media," in Patricia Margaret Sterns and Leslie I. Tennen (eds.), *Private Law, Public Law, Metalaw and Public Policy in Space*. Switzerland: Springer, 159–179.

Stimson, Dorothy. 1972. *The Gradual Acceptance of the Copernican Universe*. Gloucester: Peter Smith.

Stoeger, William R. 1996. "Astronomy's Integrating Impact on Culture: A Ladrierean Hypothesis," *Leonardo: Journal of the International Society for the Arts, Sciences and Technology* 29(2), 151–154.

Stoeger, William R. 2013. "Astrobiology and Beyond: From Science to Philosophy and Ethics," in Chris Impey, Anna Spitz, and William Stoeger (eds.), *Encountering Life in the Universe: Ethical Foundations and Social Implications of Astrobiology*. Tucson, AZ: University of Arizona Press, 56–79.

Sullivan, Woodruff T., III. 2013a. "Planetocentric Ethics: Principles for Exploring a Solar System that May Contain Extraterrestrial Microbial Life," in Chris Impey, Anna Spitz, and William Stoeger (eds.), *Encountering Life in the Universe: Ethical Foundations and Social Implications of Astrobiology*. Tucson, AZ: University of Arizona Press, 167–177.

Sullivan, Woodruff T., III. 2013b. "Extraterrestrial Life as the Great Analogy, Two Centuries Ago and in Modern Astrobiology," in Douglas Vakoch (ed.), *Astrobiology, History and Society: Life beyond Earth and the Impact of Discovery*. Berlin-Heidelberg: Springer-Verlag, 73–83.

Sullivan, Woodruff T., III and John A. Baross (eds.). 2007. *Planets and Life: The Emerging Science of Astrobiology*. Cambridge: Cambridge University Press.

Suvin, Darko. 1970. "The Open-Ended Parables of Stanislaw Lem and 'Solaris,'" Afterword to Stanislaw Lem (ed.), *Solaris*. New York, NY: Berkley Medallion Books, 212–223.

Tarter, Jill C. 2007. "Searching for Extraterrestrial Intelligence," in Woodruff T. Sullivan III and John A. Baross (eds.), *Planets and Life: The Emerging Science of Astrobiology*. Cambridge: Cambridge University Press, 513–536.

Tarter, Jill C. and Michael A. Michaud. 1990. "SETI Post-Detection Protocol," *Acta Astronautica* 21, 69–154.

Tattersall, Ian. 1995. *The Last Neanderthal: The Rise, Success, and Mysterious Extinction of our Closest Human Relatives*. New York, NY: Westview Press.

Tax, Sol. 1975. Afterword to Magoroh Maruyama and Arthur Harkins (eds.), *Cultures beyond the Earth: The Role of Anthropology in Outer Space*. New York, NY: Vintage Books, 200–203.

Tegmark, Max. 2014. *Our Mathematical Universe: My Quest for the Ultimate Nature of Reality*. New York, NY: Alfred A. Knopf.

Thomas, Hugh. 2003. *Rivers of Gold: The Rise of the Spanish Empire, from Columbus to Magellan*. New York, NY: Random House.

Toffler, Alvin. 1975. Foreword to Magoroh Maruyama and Arthur Harkins (eds.), *Cultures beyond the Earth: The Role of Anthropology in Outer Space*. New York, NY: Vintage Books.

Tough, Allen (ed.). 2000. *When SETI Succeeds: The Impact of High-Information Contact.* Bellevue, WA: Foundation for the Future. Online at www.futurefoundation.org/documents/hum_pro_wrk1.pdf.

Toynbee, Arnold. 1957. *A Study of History*, Abridgement by D. C. Sovervell, vol. 2. London: Oxford University Press.

Traphagan, John W. 2011. "Culture, Meaning and Interstellar Message Construction," in Douglas Vakoch (ed.), *Communication with Extraterrestrial Intelligence*. Albany, NY: State University of New York Press, pp. 469ff.

Traphagan, John W. 2014. "Anthropology at a Distance: SETI and the Production of Knowledge in the Encounter with an Extraterrestrial Other," in Douglas Vakoch (ed.), *Archaeology, Anthropology and Interstellar Communication*. Washington, DC: National Aeronautics and Space Administration, 131–142, at 132.

Traphagan, John W. 2015a. "Equating Culture, Civilization and Moral Development in Imagining Extraterrestrial Intelligence: Anthropocentric Assumptions?," in Steven J. Dick (ed.), *The Impact of Discovering Life beyond Earth*. Cambridge: Cambridge University Press, 127–142.

Traphagan, John W. 2015b. *Extraterrestrial Intelligence and Human Imagination: SETI at the Intersection of Science, Religion, and Culture*. Heidelberg: Springer.

Traphagan, John W. 2016. *Science, Culture and the Search for Life on Other Worlds*. Heidelberg: Springer.

Traphagan, John W. and Julian W. Traphagan. 2015. "SETI in Non-Western Perspective," in Steven J. Dick (ed.), *The Impact of Discovering Life beyond Earth*. Cambridge: Cambridge University Press, 299–307.

Trigger, Bruce G. and Wilcomb E. Washburn (eds.). 1996. *Cambridge History of the Native Peoples of the Americas*. Cambridge: Cambridge University Press.

Turner, Frederick Jackson. 1994. "The Significance of the Frontier in American History," in John M. Faragher (ed.), *Rereading Frederick Jackson Turner: The Significance of the Frontier in American History and Other Essays*. New Haven, CT: Yale University Press.

Tye, Michael. 2007. "Philosophical Problems of Consciousness," in Max Velmans and Susan Schneider (eds.), *The Blackwell Companion To Consciousness*. Oxford: Blackwell Publishing, 22–35.

US Congress. US House of Representatives Committee on Science and Astronautics. 1961. *Proposed Studies on the Implications of Peaceful Space Activities for Human Affairs*, prepared for NASA by the Brookings Institute. Report of the Committee on Science and Astronautics, US House of Representatives, 87th Congress, 1st session, March 24, 1961.

US Congress. 1996. Life on Mars? Hearing before the Subcommittee on Space and Aeronautics of the Committee on Science, US House of Representatives, 104th Congress, 2nd sess. Washington, DC: Government Printing Office. September 12, p. 1. Online at https://archive.org/details/lifeonmarshearin00unit.

US Congress. House Committee on Science, Space, and Technology. 2013. Hearings on "Astrobiology: Search for Biosignatures in our Solar System and Beyond," December 4. Online at www.gpo.gov/fdsys/pkg/CHRG-113hhrg86895/html/CHRG-113hhrg86895.htm.

US Congress. House Committee on Science, Space, and Technology. 2014. Hearings on "Astrobiology and the Search for Life in the Universe," May 21. Online at http://science.house.gov/hearing/full-committee-hearing-astrobiology-and-search-life-universe.

Vakoch, Douglas. 1998. "Constructing Messages to Extraterrestrials: An Exosemiotic Perspective." *Acta Astronautica* 42, 697–704.

Vakoch, Douglas. 1999. "The View from a Distant Star: Challenges of Interstellar Message-Making," *Mercury* 28, 26–32.

Vakoch, Douglas. 2000. "Roman Catholic Views of Extraterrestrial Intelligence: Anticipating the Future by Examining the Past," in Allen Tough (ed.), *When SETI Succeeds: The Impact of High-Information Contact*. Bellevue, WA: Foundation for the Future, 165–174.

Vakoch, Douglas. 2009a. "Encoding Our Origins: Communicating the Evolutionary Epic in Interstellar Messages," in Steven J. Dick and Mark L. Lupisella (eds.), *Cosmos & Culture: Cultural Evolution in a Cosmic Context*. Washington, DC: National Aeronautics and Space Administration, 415–439.

Vakoch, Douglas. 2009b. "Anthropological Contributions to the Search for Extraterrestrial Intelligence," in Karen J. Meech, J. V. Keane, Michael Mumma, J. L. Siefert, and Dan J. Werthimer (eds.), *Bioastronomy 2007: Molecules, Microbes and Extraterrestrial Life*. San Francisco, CA: Astronomical Society of the Pacific, 421–427.

Vakoch, Douglas. 2010. "An Iconic Approach to Communicating Musical Concepts in Interstellar Messages," *Acta Astronautica* 67, 1406–1409.

Vakoch, Douglas. 2011a. "The Art and Science of Interstellar Message Composition: A Report on International Workshops to Encourage Multidisciplinary Discussion," *Acta Astronautica* 68, 451–458.

Vakoch, Douglas. 2011b. "Asymmetry in Active SETI: A Case for Transmissions from Earth," *Acta Astronautica* 68, 476–488.

Vakoch, Douglas A. 2011c. "Responsibility, Capability, and Active SETI: Policy, Law, Ethics, and Communication with Extraterrestrial Intelligence," *Acta Astronautica* 68, 512–519.

Vakoch, Douglas (ed.). 2011d. *Communication with Extraterrestrial Intelligence*. Albany, NY: State University of New York Press.

Vakoch, Douglas (ed.). 2013. *Astrobiology, History and Society: Life beyond Earth and the Impact of Discovery*. Berlin-Heidelberg: Springer-Verlag.

Vakoch, Douglas (ed.). 2014a. *Archaeology, Anthropology and Interstellar Communication*. Washington, DC: National Aeronautics and Space Administration. National Aeronautics and Space Administration. SP 4413.

Vakoch, Douglas. 2014b. *Extraterrestrial Altruism: Evolution and Ethics in the Cosmos*. Heidelberg and New York: Springer.

Vakoch, Douglas. 2015. "Communicating with the Other: Infinity, Geometry, and Universal Math and Science," in Steven J. Dick (ed.), *The Impact of Discovering Life beyond Earth*. Cambridge: Cambridge University Press, 143–154.

Vakoch, Douglas A. and Matthew F. Dowd. 2015. *The Drake Equation: Estimating the Prevalence of Extraterrestrial Life through the Ages*. Cambridge: Cambridge University Press.

Vakoch, Douglas and Harrison, Albert (eds.). 2011. *Civilizations Beyond Earth: Extraterrestrial Life and Society*. New York, NY, and Oxford: Berghahn Books.

Vakoch, Douglas A. and Yuh-Shiow Lee. 2000. "Reactions to Receipt of a Message from Extraterrestrial Intelligence: A Cross-Cultural Empirical Study," *Acta Astronautica* 46, 737–744.

Van Helden, Albert. 1974. "Saturn and His Anses," *JHA* 5, 105–121.

Vastag, Brian and Joel Achenbach. 2012. "Scientists' Search for Higgs Boson Yields New Subatomic Particle," *Washington Post*, July 4.

Vernot, Benjamin and Joshua M. Akey. 2014. "Resurrecting Surviving Neanderthal Lineages from Modern Human Genomes," *Science*. Published online January 29, at www.sciencemag.org/content/early/2014/01/28/science.1245938.

Vidal, Clément. 2007. "An Enduring Philosophical Agenda: Worldview Construction as a Philosophical Method." Online at http://cogprints.org/6048.

Vidal, Clément. 2011. "Metaphilosophical Criteria for Worldview Comparison," *Metaphilosophy*.

Vidal, Clément. 2014. *The Beginning and the End: The Meaning of Life in a Cosmological Perspective*. New York, NY: Springer.

Vidal, Clément. 2015. "A Multidimensional Impact Model for the Discovery of Extraterrestrial Life," in Steven J. Dick (ed.), *The Impact of Discovering Life beyond Earth*. Cambridge: Cambridge University Press, 55–75.

Vidal, Clément. 2016. "Stellivore Extraterrestrials? Binary Stars as Living Systems," *Acta Astronautica* 128, 251–256.

Vinge, Vernor. 1993. "The Coming Technological Singularity: How to Survive in the Post-human Era." Published in *Vision-21*, NASA Conference Publication 10129, pp. 11–22. Online at http://ntrs.nasa.gov/archive/nasa/casi.ntrs.nasa.gov/19940022855.pdf.

Vonnegut, Kurt, Jr., 1994. *Slaughterhouse Five: Or the Children's Crusade*. New York, NY: Delacorte Press, 146.

Vorzimmer, Peter. 1970. *Charles Darwin, the Years of Controversy: The Origin of Species and Its Critics, 1859–1882*. Philadelphia, PA: Temple University Press.

Wabbel, Tobias Daniel (ed.). 2005. *Leben im All: Positionen aus Naturwissenschft, Philosophie und Theologie*. Dusseldorf: Patmos Verlag, 156–172.

Ward, Peter D. and Steven A. Benner. 2007. "Alien Biochemistries," in Woodruff T. Sullivan III and John A. Baross (eds.), *Planets and Life: The Emerging Science of Astrobiology*. Cambridge: Cambridge University Press, 537–544.

Ward, Peter D. and D. Brownlee. 2000. *Rare Earth: Why Complex Life Is Uncommon in the Universe*. New York, NY: Springer-Verlag.

Wason, Paul K. 2011. "Encountering Alternative Intelligences: Cognitive Archaeology and SETI," in Douglas Vakoch and Albert Harrison (eds.), *Civilizations beyond Earth: Extraterrestrial Life and Society*. New York, NY, and Oxford: Berghahn Books, 42–59.

Wason, Paul K. 2013. "Inferring Intelligence: Prehistoric and Extraterrestrial," in Douglas Vakoch (ed.), *Astrobiology, History and Society: Life beyond Earth and the Impact of Discovery*. Berlin-Heidelberg: Springer-Verlag, 113–129.

Watts, Peter. 2006. *Blindsight*. New York, NY: Tor.

Weintraub, David. 2014. *Religions and Extraterrestrial Life: How Will We Deal With It?* Heidelberg and New York, NY: Springer.

Wells, H. G. 1908. "The Things That Live on Mars," *Cosmopolitan Magazine*, 335–343.

Westman, Robert S. 2011. *The Copernican Question: Prognostication, Skepticism, and Celestial Order.* Berkeley, CA: University of California Press.

White, Frank. 1987. *The Overview Effect: Space Exploration and Human Evolution.* Boston, MA: Houghton Mifflin Company.

White, Frank. 1990. *The SETI Factor: How The Search for Extraterrestrial Intelligence is Changing our View of the Universe and Ourselves.* New York, NY: Walker and Company.

Wigner, Eugene. 1960. "The Unreasonable Effectiveness of Mathematics in the Natural Sciences," *Communications in Pure and Applied Mathematics* 13(1).

Wilkinson, David. 2013. *Science, Religion, and the Search for Extraterrestrial Intelligence.* Oxford: Oxford University Press.

Wills, Christopher and Jeffrey Bada. 2000. *The Spark of Life: Darwin and the Primeval Soup.* Cambridge, MA: Perseus.

Wilson, David Sloan. 2004. *Darwin's Cathedral: Evolution, Religion, and the Nature of Society.* Chicago, IL: University of Chicago Press.

Wilson, Edward O. 1978. *On Human Nature.* Cambridge, MA: Harvard University Press.

Wilson, Edward O. 1998. *Consilience: The Unity of Knowledge.* New York: Alfred A. Knopf.

Wilson, Samuel M. 1999. *The Emperor's Giraffe, and Other Stories of Cultures in Contact.* Boulder, CO: Westview Press.

Wittgenstein, Ludwig. 1953. *Philosophical Investigations.* New York, NY: MacMillan, 223.

World Economic Forum. 2013. *Global Risks 2013*, Lee Howard (ed.). Geneva: World Economic Forum. Online at http://reports.weforum.org/global-risks-2013/section-five/x-factors/#hide/img-5.

Worseley, P. M. 1957. *The Trumpet Shall Sound: A Study of Cargo Cults in Melanesia.* London: MacGibbon and Kee.

Wright, Robert. 2010. "Ethics for Extraterrestrials," *New York Times*, May 4, opinion pages.

Wylie, Alison. 2002. *Thinking from Things: Essays in the Philosophy of Archaeology.* Berkeley, CA: University of California Press.

Zaitsev, Alexander L. 2011. "METI: Messaging to ExtraTerrestrial Intelligence," in Paul Shuch (ed.), *Searching for Extraterrestrial Intelligence: SETI Past, Present and Future.* Chichester: Praxis, 399–428.

Zakariya, Nasser. 2017. *A Final Story: Science, Myth and Beginnings.* Chicago, IL: University of Chicago Press.

Zubrin, Robert and Richard Wagner. 1996. *The Case for Mars.* New York, NY: The Free Press.

Index